Symbol	Name	Page
	Vector Notation	
$\mathbf{v}$	Vector	187
$(v_1, v_2, ..., v_n)$	Vector	187
$\mathbf{x} + \mathbf{y}$	Sum	189
a	Scalar	190
$a\mathbf{v}$	Scalar times vector	190
$\mathbf{x} \cdot \mathbf{y}$	Dot product	191
$|\mathbf{v}|$	Norm	192
$\sum_{i=m}^{n} v_i =$ $\Sigma_{i=m}^{n} v_i$	Sigma notation	195
$\prod_{i=m}^{n} v_i =$ $\Pi_{i=m}^{n} v_i$	Pi notation	201
	Matrix Notation	
$C = [c_{ij}]$	Matrix	233
$c_{ij} = C[i, j]$	Element	233
aC	Multiplication by scalar	234
$A + C$	Addition	235
$\mathbf{O}_{mxn}$	Zero matrix	236
$-A$	Additive inverse	237
$A_{mxq} B_{qxn}$	Matrix product	244
$\mathbf{I}_n$	Identity matrix	246
A^{-1}	Multiplicative inverse	261
	Graphs	
v	Vertex or node	272
$e = (u, v)$	Edge	272
$\deg(v)$	Degree	273
$e = (u, v)$	Directed edge	276
(S, I, O, f_N, f_O)	Finite-state machine	311
	Formal Grammars	
$::=$	BNF replacement	209
$<x>$	BNF nonterminal	209
$x \mid y$	BNF "or"	209
xy	BNF concatenation	209
(V, V_T, P, S)	Formal grammar	210
V	Vocabulary	210

Symbol	Name	Page
V_T	Set of terminal symbols	210
P	Set of productions	210
S	Start symbol	210
$\rightarrow$	Replacement	210
a^n	String of n a's	315
ab	Concatenation of a and b	315
$a \vee b$	a or b	315
λ	Empty string	315
a^*	$\{\lambda, a, aa, aaa, ...\}$	315
	Boolean Algebra	
$[B, +, \cdot, ', 0, 1]$	Boolean algebra	327
$x + y$	Addition	327
$x \cdot y$	Multiplication	327
x'	Complement	327
0	Additive identity	327
1	Multiplicative identity	327
	OR gate	341
	AND gate	341
	Inverter	341
	Miscellaneous	
x^n	Base to exponent	6
$\approx$	Approximately equal	15
MHz	Megahertz	106
$a \, R \, b$	Related	119
(S, R)	Partially ordered set	127
$a \mid b$	Divides	128
$[x]$	Equivalence class	137
$\equiv_n$	Congruence modulo n	137
$x \equiv y \pmod{n}$	Congruent modulo n	137
$O(g)$	Big oh	409

Discrete Mathematics
for Computer Science

Discrete Mathematics for Computer Science

ANGELA B. SHIFLET
Lander College

WEST PUBLISHING COMPANY
St. Paul   New York   Los Angeles   San Francisco

Copyediting: Virginia Dunn
Text Design: Lucy Lesiak Design
Illustrations: Perfectplot
Composition: Interactive Composition Corporation
Cover design: Lucy Lesiak Design

Cover Art

Three Trees by Nelson L. Max, Lawrence Livermore National Laboratory (operated by the University of California for the Department of Energy), and Jules Bloomenthal, Xerox Palo Alto Research Center. Graphics courtesy N. L. Max. The scene on the cover was created using computer graphics, applying technology for an appreciation and understanding of natural forms. A different type of tree carries special significance in computer science and mathematics (see Section 8.1). More than 12,000 triangles were used to model the tree trunk with about 1500 twelve-sided polygons for the leaves. (See "Modeling the Mighty Maple" by Jules Bloomenthal, *SIGGRAPH '85 Proceedings* 19(3): 305–311, July 1985.) The shading on the bark was calculated from measured bark thicknesses. The drawing algorithm is capable also of casting shadows in the hazy atmosphere so that light beams are visible in the air beneath the trees. (See the Chapter 1 picture and "Shadows for Bump Mapped Surfaces" by Nelson L. Max, *Advanced Computer Graphics: Proceedings of Computer Graphics Tokyo '86*, T. L. Kunii, ed., 145–156, Tokyo: Springer Verlag, 1986.)

Chapter Opening Art

Chapter 1 art: *Binary Numbers in a Doorway* by Nelson L. Max, Lawrence Livermore National Laboratory, University of California. Graphics courtesy N.L. Max.

Chapter 2 art: *Mandelbrot Set.* IBM/Germany. From *The Beauty of Fractals* by H. O. Peitgen and P. H. Richter, New York: Springer Verlag, 1986. Photo courtesy IBM Thomas J. Watson Research Center.

Chapter 3 art: *Mathematics by the Measure.* Photo by and courtesy of James E. Stoots, Jr., Lawrence Livermore National Laboratory, University of California. Music courtesy of Anna Laura Page.

Chapter 4 art: *Hours of Daylight* by Stanley L. Grotch, Lawrence Livermore National Laboratory, University of California. Photo courtesy S. L. Grotch.

Chapter 5 art: *Graph of* $z = f(x, y) = |x - y||x + y|$ by Stanley L. Grotch, Lawrence Livermore National Laboratory, University of California. Photo courtesy S. L. Grotch.

Chapter 6 art: *Ice Crystal* by Nelson L. Max, Fujitsu Limited, Tokyo, Japan, from the IMAX®/Omnimax® film *We Are Born of Stars*, © 1985, Fujitsu LTD. Reprinted with permission from N. L. Max and IMAX Systems Corporation, Toronto, Ontario.

Chapter 7 art: *Tree.* Photo by and courtesy of James E. Stoots, Jr., Lawrence Livermore National Laboratory, University of California.

Chapter 8 art: *MILNET.* Courtesy of the Data Defense Network, Department of Defense.

Chapter 9 art: *Fluorinert.* Photo by and courtesy of James E. Stoots, Jr., Lawrence Livermore National Laboratory, University of California.

Chapter 10 art: *Cover of Voyager Record.* Photo courtesy the Jet Propulsion Laboratory.

Chapter 11 art: *Foggy Chessmen* by Turner Whitted and David M. Weimer, Bell Laboratories. Courtesy of Bell Laboratories.

Library of Congress Cataloging-in-Publication Data

Shiflet, Angela.
  Discrete mathematics for computer science.

  Includes index.
  1. Electronic data processing—Mathematics.
I. Title.
QA76.9.M35S54 1987   510   86-19112
ISBN 0-314-28513-X

Dedicated to

my husband
George
and my parents,
Isabell and Carroll Buzzett

CONTENTS

Preface

CHAPTER **1** **Numbers and Operations**   1

1.1   Fundamental Concepts: Numbers and Exponents   2
1.2   Error   13
1.3   Operator Precedence   18
1.4   Polish Notation   24
      Chapter Review   28
      References   30

CHAPTER **2** **Sets and Logic**   31

2.1   Properties of Sets   32
2.2   Operations on Sets   38
2.3   Algebra of Sets   46
2.4   Algebra of Propositions   53
2.5   Logic   63
2.6   Proof Techniques   70
      Chapter Review   78
      References   81

CHAPTER **3** **Counting**   83

3.1   Counting Principles   84
3.2   Permutations and Combinations   91
3.3   Loop Computations   99
3.4   Computer Measurements   105
      Chapter Review   113
      References   115

CHAPTER **4**   Relations   117

   4.1   Relation Definition   118
   4.2   Partial Order   125
   4.3   Equivalence Relations   133
         Chapter Review   140
         References   141

CHAPTER **5**   Functions   143

   5.1   Function Definition   144
   5.2   Some Mathematical Functions   152
   5.3   Built-in Functions   164
   5.4   Composition of Functions   170
   5.5   Mod Function   175
         Chapter Review   182
         References   183

CHAPTER **6**   Subscripts   185

   6.1   Vectors or One-Dimensional Arrays   186
   6.2   Sigma and Pi   194
   6.3   Recursion   204
   6.4   Induction   216
         Chapter Review   227
         References   229

CHAPTER **7**   Matrices   231

   7.1   Matrices or Two-Dimensional Arrays   232
   7.2   Multiplication of Matrices   242
   7.3   Systems of Linear Equations   253
   7.4   Matrix Inverses   260
         Chapter Review   267
         References   269

CHAPTER **8**   Graph Theory   271

   8.1   Graphs   272
   8.2   Representations   280
   8.3   Trees   287
   8.4   Paths and Spanning Trees   297
   8.5   Finite-State Machines   310
         Chapter Review   319
         References   322

CHAPTER **9**   **Boolean Algebra**   325

9.1   Boolean Algebra Definition   326
9.2   Properties   334
9.3   Logic Gates   340
9.4   Logic Circuit   348
      Chapter Review   356
      References   357

CHAPTER **10**   **Binary and Hexadecimal Number Systems**   359

10.1   Binary Number System   360
10.2   Hexadecimal Number System   366
10.3   Conversion from Decimal   373
10.4   Two's Complement   381
10.5   Arithmetic   387
       Chapter Review   394
       References   395

CHAPTER **11**   **Analysis of Algorithms**   397

11.1   Exponentials and Logarithms   398
11.2   Complexity   408
       Chapter Review   418
       References   419

APPENDIX **A**   **Partial Table of Ascii Data Representation**   420

APPENDIX **B**   **Largest Prime Less Than a Given Number**   421

Glossary   422

Answers to Selected Exercises   430

Index   445

COMPUTER SCIENCE APPLICATIONS

Application	Page
1 as 0.9999999 on computer printout	3–4, 18
Primes discovered with computer	5
K, memory size	6, 11
Computer speed	7–9, 11–12
Decoder	10–11, 354, 365
Floating point numbers	13
Error in computer arithmetic	15–18
Operator precedence in computer languages	18–23, 63–64, 68
Stacks and Polish notation	24–28
Mandelbrot set	31
Sets in Pascal	32–33, 37, 39, 41, 46
Computer solid modeling with set operations	42–43
Proof techniques improving software	48
Theorem-proving program	50–51
Logical or Boolean expressions in programs	54–56, 61, 70, 329, 338
Number of characters encoded by n bits, ASCII and EBCDIC encoding schemes	85, 88
Number of possible variables in a language	89
Number of user-ids for a computer	92
Traveling salesperson problem	97, 309, 413
Combinational (switching) function or circuit	98, 150–52, 333, 344–55
Number of possible pairs out of order in an array	98
Number of possible colors in a graphics application	98
Looping	99–105
Delays	103–5
Sequential search	104, 408, 411

Application	*Page*
Approximation of zeros of a function with computer	105
Cycle time and frequency of a computer	105–6, 112
Floating-point operations per second	106–7, 112
Data-transfer rate	107–8, 113
Tape density, speed, data-transfer rate, gaps	108–10, 112–13
Disk storage	110–11, 113
Array as a relation	120
File as a relation	120–21
Relational data base	123–24
Hierarchy diagram as a partial ordering	127
Scheduling as a partial ordering	127
Alphabetizing as a linear ordering	130
Sorting on one or two fields as a partial ordering	131
Topological sort of program statements	132–33
Partition of memory	134
UNIX operating system	134, 370–71, 373
Partitioning of a file or data base	134–135
EQUIVALENCE in FORTRAN	136, 139
Computer program to find anagrams involving prime numbers and equivalence relation	137–39
Functions written in computer languages (e.g., SQUARE, error, ABSOLUTE, percent, average)	145–47, 152–53, 160–61
Function in a relational data base	147
Encoding schemes as one-to-one functions	148
Relative file organization, one-to-one function	148
Key in file, one-to-one function	149
Odd parity as a function	150
Voltage to bit function	150
Predicate function	151
Projection function in computer graphics	154
Percent, mean, median used in computer literature	155–63
SPSS example	164
Computer functions MAX, MIN, TRUNC, ROUND	164–75
Truncation and rounding in computer languages	166
Pigeonhole principle, floor, ceiling applied to parallel processing	167
LISP functions CAR, CDR, CONS	168–69, 175
Composition of computer functions	170–71, 175
ORD and CHR, SUCC and PRED as inverses	173
Public–key cryptosystems using modulo arithmetic	176, 181–82
Linear congruential random number function	177–80
Hashing function	178–80
Program to find the day of the week	180
EVEN function	180
Program to isolate digits	180–81

Application

Page

One–dimensional arrays in programming languages

186–88, 193–94

Vector processing in CRAY-2

188–89

Vectors in computer graphics

190–91

Sums and indices in computer languages

194–95

SUM in MACSYMA

197, 203

Sigma notation in computer science literature

203

Recursion in programming languages

204, 206–7, 213–16

Stack, stack overflow

206–7

Backus-Normal form (BNF)

209–10

Formal grammars

210–15, 225, 289–90, 294

Program verification

223–24, 226

Matrix for storing colors for computer graphics

232–3

Two-dimensional arrays in programming languages

233, 236–37, 241

Column- and row-major order

238–39

Dithering

241

Scaling, translation, rotation in computer graphics

246–48, 250–53

Computer applications of systems of equations

259–60

Matrix inverses in APL

262

Cryptography

264–67

Allocation of computer resources

280

Linked list, doubly linked list

280–81, 284

Program on length of shortest path through network

287

Parse tree

289–90, 294

Binary tree to alphabetize

290–92, 295

Solid computer graphics using trees

293

Storing trees in arrays

295–96

Hamiltonian path applied to Gray code

298–99, 309

Euler path applied to de Bruijn sequence

303–5, 309

Minimal spanning tree through a computer network

306

Binary tree for encoding

309

Finite-state machines

310–18

Computer implementation of Pascal sets

329–31, 334

Hamming codes

333–34

Gates, logic circuits

340–55

Half-adder

349–50, 355

Circuit for odd parity

351

Full-adder

353–54

Circuitry for Gray code

354–55

Binary representation of numbers

360–93

Range of unsigned integers expressible in n bits

364–65, 371–72

Memory locations addressable with n bits

364–65

Encoder

365

Shifting to double or halve binary numbers

365–66, 391–92

Hexadecimal number system

366–393

Application   *Page*

Memory dump   368, 372
ASCII   370, 372
Octal number system   370–73
Branching with assembler statements   372, 390–91, 393
Testing equality of real numbers in the computer   379, 381
Microcomputer control of an audio system   380
Representation of signed numbers in the computer   381–393
Range of decimal integers in a computer   386–87, 404, 406
Arithmetic in the computer   387–393
Overflow   389–90
Memory addressing   393
Computer implementation of real number exponents   405
Complexity, analysis of algorithms   408–17
Binary search   411–14, 416
QUIK sort   415–16
Insertion sort   416–17

PREFACE

Discrete Mathematics for Computer Science can be used in a one-quarter to one-year course in discrete mathematics for freshman or sophomore level students. This text covers the mathematical concepts that students will encounter in computer science courses. A good knowledge of high school algebra is prerequisite; a class comparable to the first major college course in programming is corequisite.

In writing this text, I have emphasized the *interrelationship of mathematics and computer science*, which is displayed in the following pedogogical features:

Many examples and exercises found throughout the text contain **applications** of mathematics to computer science along with brief descriptions of the computer science concepts. A list of the applications in the text can be found immediately following the table of contents.

Microcomputer icons in the margins of the text mark the location of **computer-language examples and exercises.** Many examples and exercises comment on program segments from a variety of computer languages, especially Pascal. The program segments are explained so that *there is no need for the student or professor to be proficient in, or even to have heard of, the programming languages.*

To enhance the subject's appeal and provide a feeling of authenticity and significance which is so important in a course at this level, I have integrated **historical notes** throughout the text. Such notes are indicated by punched-card icons in the margins.

ORGANIZATION

Because the level of discrete mathematics courses and the order in which the material is covered differs a great deal from course to course, I have written this book to flexibly accommodate many courses. The following paragraphs provide an overview of the material in each chapter and the options available to professors.

Chapter 1   Section 1.1 primarily contains review material examined from the standpoint of its application to computer science. Exercises involving error (Section 1.2) appear in Section 1.3 and in Chapter 5. Section 1.4 on Polish Notation is optional.

Chapter 2   Sections 2.1 and 2.2 cover set properties and operations. The algebras of sets and propositions are studied in Sections 2.3–2.5. Coverage of these sections may vary from a discussion of the basic properties and the use of truth tables to proofs of the properties and tests of the validity of arguments. Since the student is first exposed to proofs in the book in Section 2.3, several examples and exercises lead the student through set proofs from the definitions (see Exercises 3–5) and from the properties (see Exercises 16–18 and 30). Section 2.6 discusses various proof techniques. For a more application-oriented course, this section can be omitted.

Chapter 3   The Fundamental Counting Principle (Section 3.1) and to a lesser extent permutations and combinations (Section 3.2) are used throughout the book. Sections 3.3 and 3.4 provide additional applications. The former covers computations involving programming loops, while the latter introduces various computer measurements of time, space, and rate.

Chapter 4   The material on relations in Section 4.1 is needed for the next chapter on functions. The study of partial order (Section 4.2) and equivalence relations (Section 4.3) can be omitted or delayed, however, at the professor's discretion. Exercise 2 of Section 4.2 does introduce congruence modulo n, a topic that is expanded in Section 5.5 on the mod function.

Chapter 5   The section on the function definition (Section 5.1) is important. A professor, however, may choose to cover, omit, or integrate into other material some of the special functions discussed in Sections 5.2 and 5.3. The amount of coverage of these two sections will affect which exercises are assigned in the section on composition of functions (Section 5.4). The section on the mod function (Section 5.5) with applications to random number generation could be omitted or covered at a later time.

Chapter 6   Coverage of this chapter on vectors can easily occur earlier or later in a particular discrete mathematics course. Section 6.1 on vectors may be discussed briefly; vector notation and the dot product are necessary for the study of matrix multiplication in Chapter 7. Sigma notation (Section 6.2) is used in the rest of Chapter 6, Chapter 7, the exercises of Sections 8.1–8.3, and Section 11.2. For a class emphasizing applications and not proofs, the section on mathematical induction (Section 6.4) can be omitted.

Chapter 7 Material on vectors, dot product, and sigma notation (Sections 6.1–6.2) is prerequisite to this chapter on matrices. Sections 7.3 and 7.4 on systems of linear equations and matrix inverses are not referred to later in the book. Matrix addition and multiplication (Sections 7.1–7.2) are used in representations of graphs (Section 8.2).

Chapter 8 Section 8.1 discusses some of the fundamental definitions and theorems of graphs and has no prerequisite other than sigma notation. The material on graph representations (Section 8.2) assumes a knowledge of matrix multiplication (Section 7.2) but is not used in later sections.

Chapter 9 This chapter on Boolean algebra ties in with Chapter 2 and so can be studied after that chapter if the students have a sufficient knowledge of functions. Logic gates and circuits (Sections 9.3–9.4) provide an application of Boolean algebra to computer science.

Chapter 10 Some professors prefer examining the binary and hexadecimal number systems early in the course, others prefer to do so late in the course, while others prefer postponing coverage until another course, such as computer organization. Since it is self-contained, Chapter 10 provides that flexibility.

Chapter 11 For the student who is comfortable with exponential and logarithmic functions, Section 11.1 can be used as a review to prepare for the next section on complexity. Material on the ceiling function, sigma notation, arrays, and trees is needed as well for the last section (Section 11.2).

LEARNING FEATURES

- **Examples** An example is usually given before and after any definition or theorem. The format of going from the specific to the general helps students understand abstract mathematical statements. The end of each example is indicated by a closed square in the second color. ■

- **Exercises** Almost 1700 exercises, many containing applications of mathematics to computer science, are presented at the end of the sections. **Answers** to the exercises numbered in the second color can be found following the Glossary.

- **Algorithms** Algorithms are emphasized and provide an organized approach for the students to solve problems.

- **Additional topics** Enrichment topics not included in the main body of the chapters are developed within the exercises and can be assigned at the professor's discretion. For instance, the application of matrices to rotations in computer graphics is introduced in the Exercises 65–69 of Section 7.2.

- **Problem solving and proofs** To aid student understanding of word problems and proofs, a number of exercises show how to approach problem solving and the proving of statements. (See, for example, Exercise 58 in Section 1.1, Exercise 9 in Section 2.6, Exercise 17 in Section 4.3, and Exercises 7–8 in Section 8.1.) Theorems are displayed in boxes for easy reference.

- **Reviews** At the end of each chapter the most important terms, notation, and methods are summarized in outline form to aid students in reviewing important material. A page number and comment are included for each item.

- **Glossary** A glossary of computer science terms is located at the end of the book for convenient reference.

- **References** References appear at the end of each chapter to guide students in further study.

- **Symbols** A list of symbols appears on the inside front cover of the textbook for quick and easy reference.

SUPPLEMENTARY MATERIALS

- **Instructor's Manual** An Instructor's Manual, written by the author, contains solutions to text exercises not found in the appendix, and contains additional test problems with answers.

- **Transparency Masters** Transparency Masters are available from the publisher. These masters present selected algorithms, examples, and figures from the textbook.

ACKNOWLEDGMENTS

There are many people who should be acknowledged for their valuable assistance in the completion of this book.

I wish to thank the administration of Lander College, for granting me a one-semester sabbatical to write. Special thanks go to my colleague, John Hinkel, whose knowledgeable review and suggestions were most helpful. I appreciate the corrections found by the students in the classes that used the book in draft form. Special thanks go to my proofreader, Robin Fox.

The opportunities provided by Lawrence Livermore National Laboratory the past four summers have had an enormous impact upon this project. Ted Einwohner, with whom I worked, taught me much. Bob Cralle was a fount of interesting information and suggestions; he also revealed the world of word processing and drawing on the Macintosh microcomputer. George Michael, with many ideas, also facilitated a number of contacts. With the eye of an artist, Jim Stoots suggested and shot many photographs. Nelson Max created beautiful computer graphics and altered pictures to conform to the needs of the book. Stan Grotch made available his portfolio of computer graphic data displays to me. Barbara Costello helped in locating historical photographs.

The Computer Museum in Boston is an enriching and enjoyable experience. Director Gwen Bell and her staff were most helpful in my research for the book.

Appreciation goes to Nell Dale, University of Texas at Austin, who suggested, "Why not write a book on discrete mathematics?"

The staff of West Publishing Company deserve many thanks. I truly appreciate the work, help, and ideas provided by Peter Marshall, Executive Editor, and Jane Gregg, Associate Production Editor. Virginia Dunn did a fine job of copy editing the manuscript. Gary Rockswold, Mankato State University, provided valuable assistance in checking the answers to all examples, exercises, and test questions. Lucy Lesiak created a beautiful design. My thanks go to Becky Tollerson for her help as well.

I am grateful to the following reviewers who offered many valuable constructive criticisms:

John T. Annulis   *University of Arkansas*
David Berman   *University of New Orleans*
Dorothee Blum   *Virginia Commonwealth University*
Dick J. Clark   *Portland Community College*
Carl Droms   *James Madison University*
Ross Gagliano   *Georgia State University*
Edward Gaughan   *New Mexico State University*
Paul Gormley   *Villanova University*
John Hinkel   *Lander College*
Elmo Moore   *Humboldt State University*
Sharon Ross   *DeKalb Community College*
Robin Schweitzer   *Gordon College*
Fred Shuurmann   *Miami University*
Rick Smith   *University of Florida*
John Watson   *Arkansas Tech*
Nancy Zumoff   *Kennesaw College*

A number of individuals aided the project by responding to a survey: John T. Annulis, University of Arkansas; David Berman,

University of New Orleans; Douglas Compbell, Brigham Young University; Dick J. Clark, Portland Community College; Max Garzon, Memphis State University; Edward Gaughan, New Mexico State University; Donald Goldsmith, Western Michigan University; Jerrold Grossman, Oakland University; William E. Hinds, Midwestern State University; Wendell Johnson, University of Akron; M. S. Jacobson, University of Louisville, Belknap; Nicholas Krier, Colorado State University; Arnold McEntire, Appalachian State University; Elmo Moore, Humboldt State University; Matthew O'Malley, University of Houston, Houston Park; Albert D. Otto, Illinois State University; Harold Reiter, University of Maryland; C. A. Rodger, Auburn University; Robin Schweitzer, Gordon College; Rick Smith, University of Florida; Paul Weiner, St. Mary's College.

My very personal and special appreciation goes to my husband, George, for all his love, time, help, understanding, encouragement, and prayers.

NOTE TO THE STUDENT

Digital computers can express numbers with a finite number of digits only; that is, a computer deals with a discrete number system as opposed to a continuous one. To illustrate the difference, a clock with two hands indicates time in a continuous, unbroken way, while a digital clock shows time in a discrete manner, from one minute to the next. Similarly, a thermometer outside a house would probably have a column of liquid, smoothly rising and falling, to indicate the temperature. But as you pass the time-and-temperature sign in front of a bank, one moment it might register 78 degrees Fahrenheit, the next it might jump to 79 degrees.

Both how numbers are expressed in the computer and how computers themselves are built are based on discrete mathematics. Discrete mathematics actually encompasses a number of areas of mathematics including set theory, logic, combinatorics, matrix algebra, graph theory, and abstract algebra. This book covers some of the mathematical concepts that will be useful to you as you study and apply computer science.

A number of computer science examples are given in the text to illustrate the importance of discrete mathematics to this subject. In such cases a brief description of the computer science concept is given to help you understand the topic. It is not expected that you will already know the particular term or concept, and a glossary of computer science terms is included at the end of the book for your reference. A colored closed square ends each example and proof.

There also are many examples, exercises, and comments that use program segments from a variety of computer languages. These are indicated by microcomputer icons in the margin. Again, these short examples are explained so that there is no need for you to be proficient in or even to have heard of these languages.

A special feature of the book is the interweaving of elements of the history of computer science and mathematics throughout the text. The punched card icons in the margin and colored open squares delimit such historical notes.

As well as reading the book, you are urged to do the exercises. Mathematics and computer science are both disciplines in which the working of problems is essential to the understanding of the material. Be patient with word problems; do not expect to see how to do them immediately. Mark any problem you do not understand, find out from someone how to work it, and then go back and do it again without looking at what you were shown unless you get stuck; continue attempting the problem until you can work it unassisted. Keep up with the exercises as you are going along, for mathematics builds on itself. Then, in studying for a test, work problems again, looking at what you did previously only to check your answers. This subject, as with any, will be more interesting if you study it as you go along.

The answers to exercises marked by colored problem numbers can be found in the Answers section following the Appendices. Exert a reasonable amount of effort towards solving a problem before you look up the answer, and rework any problem you miss.

Discrete Mathematics
for Computer Science

1 NUMBERS AND OPERATIONS

Binary Numbers in a Doorway. With calculations performed on a CRAY-1 supercomputer, computer graphics are employed to display this picture of the two numbers that are the most significant in the study of computer science.

Fundamental Concepts: Numbers and Exponents

This section covers some fundamental concepts of numbers and exponents that will be useful throughout the book. Since this is primarily review material examined from the standpoint of its application to computer science, a detailed study of this section may not be necessary. But for future reference students should review the summary of terminology and notation at the end of the section.

One fundamental structure of mathematics is a **set** or a collection of things. In Chapter 2 we examine various sets and properties of sets, but in this section we consider only several specific sets of numbers useful in mathematics and/or computer science.

Almost any computer language allows users to work with integers. The **set of integers** $\mathbb{I}$ is

$$\mathbb{I} = \{\ldots, -3, -2, -1, 0, 1, 2, 3, \ldots\}$$

where the ellipses indicate that the sequence continues in the same fashion forever and the braces indicate a set.

In mathematics, an appropriately placed decimal makes no difference: the integer $9 = 9. = 9.0$. But in a computer language, such as Pascal, if a variable is declared to be of integer data type, it cannot be assigned either of the last two representations since decimal points are not allowed. The **data type** of a variable is a description of the kind of values that variable can have. If X is declared to be an integer variable in Pascal, the assignment statement X := 9 is legal, but X := 9.0 is not. Moreover, because of the way numbers are stored in the computer, most computer languages can represent only a finite number of integers. In Pascal, the built-in identifier MAXINT has as its value the largest integer that can be represented in the computer on which the language is running; the smallest integer is usually one less than the negative of MAXINT. $\square$ For example, if the largest value is 32,767, the set of integers would be

$$\mathbb{I}_c = \{-32{,}768, \ldots, -3, -2, -1, 0, 1, 2, 3, \ldots, 32{,}767\}.$$

Unlike $\mathbb{I}$, which is infinite, $\mathbb{I}_c$ is a finite set that does *not* allow equivalent numbers with decimal points, such as 9.0. As is explained in Chapter 10, an unusual range such as this ($-32{,}768$ to $32{,}767$) exists because of the way numbers are stored in the computer.

Another set of integers is $\mathbb{N}$, the **set of natural numbers**

$$\mathbb{N} = \{0, 1, 2, 3, \ldots\}.$$

LANGUAGE EXAMPLE

Since zero is neither positive nor negative, $\mathbb{N}$ is not the same as the **set of positive integers**

$\mathbb{I}^+ = \{1, 2, 3, \ldots\}.$

Later, we examine the importance of two other counting sets, the **binary set**

$\mathbb{B} = \{0, 1\}$

and the **hexadecimal set**

$\mathbb{H} = \{0, 1, 2, 3, 4, 5, 6, 7, 8, 9, A, B, C, D, E, F\}$

where A, B, C, D, E, and F are actual numbers. As with $\mathbb{I}_c$, we do not consider decimal expansions such as 1.0 in $\mathbb{B}$ and $\mathbb{H}$.

In mathematics, the **set of rational numbers** has two definitions:

1. $\mathbb{Q} = \{p/q \mid p \text{ and } q \text{ are integers}; q \neq 0\}$, the set of all fractions p/q such that p and q are integers and the denominator is not 0. In **set-builder notation** the vertical bar stands for "such that", and the statement after the bar gives the condition that the term before the bar must satisfy to be included in the set. Another definition for the set of rationals is obtained by considering the result of the division of q into p from the first definition:
2. $\mathbb{Q} = \{x \mid x \text{ has a repeating or terminating decimal expansion}\}$.

EXAMPLE 1

The following are examples of rational numbers

$1/2 = 0.5$

$-4/7 = -0.571428\overline{571428} = 4/-7 = -8/14$

The bar over the sequence 571428 indicates this pattern continues, repeating forever.

$6 = 6/1 = 30/5 = 6.0.$ ■

EXAMPLE 2

The number one (1), with various representations, is a rational number.

$1 = 1.0 = 72/72 = 0.9\overline{9} = 0.99\overline{9}.$

The fact that $0.9\overline{9}$ is exactly equal to 1 seems impossible, but can easily be proven by the following argument.

Solution

Let $x = 0.99\overline{9}$. We will show x also equals 1. Multiplying x by 10 we have

$$10x = 9.99\overline{9}$$

Thus, $10x - x = 9.99\overline{9} - 0.99\overline{9} = 9$

$$9x = 9$$

$$x = 1.$$

Sometimes on a computer printout, a number you know to be 1.0 (or another integer i, such as $i = 4.0$) is printed as 0.9999999 (or $i - 1$ followed by .9999999, such as 3.9999999). This result is in part a function of the fact that only a finite number of the 9's can be stored in the computer. ■

Notice that 1 and every other integer is a rational number, but there are rational numbers, such as $1/2$ and $-4/7$, that are not integers.

LANGUAGE EXAMPLE

Variables that are declared to be of REAL data type in Pascal, that are described with a decimal in the computer language COBOL, or that have a FIXED DECIMAL attribute in PL/I, can actually contain only rational numbers with terminating decimal expansions. □ There are only a certain number of digits that can be stored, and there are limits as to how big and small these numbers can be. We use the notation $\mathbb{R}_c$ to stand for the set of these decimal numbers in the computer:

$$\mathbb{R}_c = \{x \mid x \text{ is expressed with a terminating decimal expansion}$$
$$\text{and } x \text{ can be stored in the computer under discussion}\}.$$

We see in Chapter 10 that some terminating decimal numbers, such as 0.1, cannot be represented exactly in the computer. For now, however, assume that any terminating decimal number is in $\mathbb{R}_c$.

The **set of real numbers** $\mathbb{R}$ in mathematics is more extensive:

$$\mathbb{R} = \{x \mid x \text{ has a decimal expansion}\}.$$

The numbers 5 and $7/2$ are members of $\mathbb{R}$ because they have decimal expansions 5.0 and 3.5, respectively. Since 5 and $7/2$ are written without a decimal, they are not included in $\mathbb{R}_c$, but $\mathbb{R}_c$ does contain 5.0 and 3.5. The set $\mathbb{R}$ includes not only all rational numbers but numbers that are not rational, called **irrational.** Two examples of such numbers are $\pi = 3.141592\ldots$ and $\sqrt{2} = 1.414213\ldots$, where the ellipses indicate nonterminating, nonrepeating decimal expansions. The **set of irrational numbers** $\mathbb{IR}$ is

$$\mathbb{IR} = \{x \mid x \text{ is a real number but is not rational}\}.$$

The existence of irrational numbers was discovered by followers of Pythagoras in Greece about 500 B.C. They were horrified that there were numbers that could not be expressed as the ratio of whole numbers, so they tried to keep the discovery a secret, and, according to legend, drowned a follower who dared to reveal the news. $\sqrt{2}$ was the first irrational number found. □

Actually, the square root of any positive integer that is not a **perfect square,** the square of an integer, is irrational. Since $25 = 5^2$ is a perfect square, its square root is rational, $\sqrt{25} = 5$. But 24, not being a perfect square, has a square root that is irrational, $\sqrt{24} = 4.89897948.\ldots$ Irrational real numbers and rational numbers with no terminating decimal expansion cannot be expressed exactly in the computer except with symbols. Any infinite expansion must be either rounded off or cut off, which is called **truncated.**

Another set of numbers useful in computer science is the **set of prime numbers** $\mathbb{P}$.

DEFINITION

A **prime number** p is an integer greater than 1 whose only positive integer divisors are 1 and p.

That is, 1 and p are the only positive integers that divide into p leaving a zero remainder. Thus, 5 is a prime since 1 and 5 divide 5; and division of 5 by 2, 3, or 4 gives a remainder of 1, 2, or 1, respectively. 6 is not a prime since 2 and 3, as well as 1 and 6, divide 6. The first few primes are 2, 3, 5, 7, 11, 13, 17, 19, 23.

The set of primes $\mathbb{P}$ is infinite, but, since the time of ancient Greece, much computation has been done by hand to determine larger and larger elements of $\mathbb{P}$. In 1876, $2^{127} - 1$, a 39-digit number, was shown to be prime. It was not until 1952 with the use of the EPSAL computer that the next prime was found. Discovery of this 79-digit number, $180(2^{127} - 1)^2 + 1$, was only the beginning of an ongoing investigation into large primes using the computer. □

You may be asked in one of your classes to write a computer program to determine if a number is prime. Or you might need a prime number to write a hashing function, one that decides placement of items in a file. Hashing functions are discussed in Chapter 5.

Mention of the two large prime numbers cited above leads us into a discussion of exponential notation, a handy way of representing some very large and very small numbers. Exponential notation has many applications in computer science from a form of printing real numbers to a calculation of how many bits can be stored in the memory of a computer. A **bit** or **binary digit** is a 0 or 1. Everything is stored in the computer in strings or sequences of bits. We are used to the decimal system of ten basic digits, 0, 1, 2, 3, 4, 5, 6, 7, 8, 9, where powers of 10 have great significance. But since only two bits are used in the computer, powers of 2 are also important. Moreover, as is seen in Chapter 10, an abbreviation for binary notation, called **hexadecimal representation,** involves one of $16 = 2^4$ digits: 0, 1, 2, 3, 4, 5, 6, 7, 8, 9, A, B, C, D, E, F.

Before looking at some specific examples, let's review some of the basic properties of exponents in Table 1.1. In x^n, x is called the **base** and n the **exponent.**

One application of exponents is the memory size of a computer. In the computer a character, such as A, 3, or ?, is encoded with a string or sequence of bits, called a **byte.** In most, though not all, computers there are eight bits in a byte. The memory size of a computer is often expressed as the number of bytes it contains. Because this number is quite large, memory is calculated in terms of chunks of $2^{10} = 1024$ bytes. Since 1024 is close to 1000, the metric

TABLE 1.1
Properties of Exponents

Property (Let n and m be integers)	Example
1. $x^n = \underbrace{x \cdot x \cdots x}_{n \text{ factors}}$	$2^3 = 2 \times 2 \times 2 = 8$
2. $x^n \cdot x^m = x^{n+m}$	$2^2 \times 2^3 = (2 \times 2)(2 \times 2 \times 2) = 2^5 = 32$ or $2^2 \times 2^3 = 4 \times 8 = 32$
3. $(x^n)^m = x^{nm}$	$(2^2)^3 = 2^6 = 64$ or $(2^2)(2^2)(2^2) = (2 \times 2)(2 \times 2)(2 \times 2) = 64$
4. $x^{-n} = 1/x^n$ for $x \neq 0$	$2^{-3} = 1/2^3 = 1/8$
5. $x^n/x^m = x^{n-m}$ for $x \neq 0$	$2^5/2^2 = 2^3 = 8$ or $(2 \times 2 \times 2 \times 2 \times 2)/(2 \times 2) = 2 \times 2 \times 2 = 8$
6. $x^0 = 1$ for $x \neq 0$, 0^0 undefined	$2^0 = 1$
7. $(x \cdot y)^n = x^n \cdot y^n$	$(3 \times 5)^2 = 3^2 \times 5^2 = 225$ or $(3 \times 5)(3 \times 5) = 3 \times 3 \times 5 \times 5 = 225$
8. $(x/y)^n = x^n/y^n$ where $y \neq 0$ and for $n < 0$, $x \neq 0$	$(3/5)^2 = 3^2/5^2 = 9/25$ or $(3/5) \times (3/5) = (3 \times 3)/(5 \times 5) = 9/25$
9. $x^{1/n} = \sqrt[n]{x}$, where $x \geq 0$ for even n	$8^{1/3} = \sqrt[3]{8} = 2$ since $2 \times 2 \times 2 = 8$
10. $x^{n/m} = \sqrt[m]{x^n} = (\sqrt[m]{x})^n$, $x \geq 0$ for even m	$8^{2/3} = \sqrt[3]{8^2} = \sqrt[3]{64} = 4$ $8^{2/3} = (\sqrt[3]{8})^2 = 2^2 = 4$

notation **K** for 1000 or kilo is used. A 64K microcomputer contains $64 \times 1024 = 65{,}536$ bytes or in powers of 2: $2^6 \times 2^{10} = 2^{16}$ bytes. You can estimate the number of bytes as approximately $64 \times 1000 = 64{,}000$ bytes.

EXAMPLE 3

Suppose a computer's memory contains 262,144 bytes. How many K does it have?

Solution

Without calculation we could estimate that our answer should be near 262K since 262,144 is about 262 thousand and 1K is about one thousand. It is always a good idea to check to see if your answer seems reasonable.

Now to compute the exact answer, recall that there are 1024 bytes in every K or $1 = 1024$ bytes/K. The placement of units can help you to decide whether to multiply or divide. Since multiplication by 1 does not effect the number, and since K needs to be in the numerator of the final answer, we have

$$262{,}144 \text{ bytes} = 262{,}144 \text{ bytes} \times \frac{K}{1024 \text{ bytes}} = 256K. \quad \blacksquare$$

The metric notation K, standing alone, means 2^{10} bytes, but as a prefix to a unit, such as seconds, K or kilo indicates 10^3. Based on powers of 10, the metric system is used to express, among other things, timing in the computer. Some of the prefixes in the metric system are given in Table 1.2 and on the inside cover of this book.

The necessity of considering such large and small numbers comes from the size and speed of modern digital computers as well as scientific applications. For example, memory size will be expressed in terms of megabytes for a **mainframe** or large computer. And, as indicated for the machine in Figure 1.1, the time it takes to execute an instruction on a **supercomputer,** the fastest type, is measured in nanoseconds. To get a concept of just how short a nanosecond is, if an average person could take one step each nanosecond, that person could circle the earth about 20 times a second! (See Exercise 60 for the computation.)

TABLE 1.2
The Metric System

Unit	Abbreviation	Meaning	
giga	g	$10^9 = 1{,}000{,}000{,}000$	billion
mega	M	$10^6 = 1{,}000{,}000$	million
kilo	K	$10^3 = 1000$	thousand
milli	m	$10^{-3} = 1/1000$	thousandth
micro	μ	$10^{-6} = 1/1{,}000{,}000$	millionth
nano	n	$10^{-9} = 1/1{,}000{,}000{,}000$	billionth
pico	p	$10^{-12} = 1/1{,}000{,}000{,}000{,}000$	trillionth

CDC 7600 supercomputer.
Compliments of the
Computer Center Museum,
Computing History Group,
Lawrence Livermore
National Laboratory, and
the U.S. Department of
Energy. The CDC 7600 can
execute about 18 million
instructions per second or
on the average one
instruction every 55 nsec. It
costs about nine million
dollars

EXAMPLE 4

A 32-megabyte computer has how many bytes of main storage?

Solution

$$32 \text{ Mbytes} = 32 \text{ Mbytes} \times \frac{10^6 \text{ bytes}}{\text{Mbyte}} = 32{,}000{,}000 \text{ bytes.}$$

Note: Sometimes 1 megabyte is meant to represent
$1024 \times 1024 = 2^{10} \times 2^{10} = 2^{20}$ bytes. By 1 megabyte, however, we
will always mean 10^6 bytes. ■

EXAMPLE 5

A disk pack stores data for a mainframe much as a diskette is a
secondary storage medium for a microcomputer. Suppose a disk
pack has a capacity of 7×10^9 bytes. How many kilobytes can it
contain?

Solution

$$7 \times 10^9 \text{ bytes} = 7 \times 10^9 \text{ bytes} \times \frac{1 \text{ Kbyte}}{1000 \text{ bytes}}$$

$$= 7 \times 10^6 \text{ Kbytes.} ■$$

EXAMPLE 6

Suppose a **subroutine,** a section of a program that may be used
several times, takes 0.09847 seconds (sec) to execute. How many
microseconds (μsec) will it take?

Solution

$$0.09847 \text{ sec} = 0.09847 \text{ sec} \times \frac{10^6 \ \mu\text{sec}}{1 \text{ sec}} = 98{,}470 \ \mu\text{sec.}$$

Since $1\ \mu\sec = 10^{-6}$ sec, there are $10^6\ \mu$sec in 1 sec. The smaller the unit, such as μsec, the larger the associated number, here 10^6. ■

TERMINOLOGY AND NOTATION

Set	Collection of things
Set of integers	$\mathbb{I} = \{\ldots, -3, -2, -1, 0, 1, 2, 3, \ldots\}$
Set of computer integers	$\mathbb{I}_c = \{x \mid x \text{ is an integer expressed without a decimal and } x \text{ can be stored in the computer under discussion}\}$
Set of natural numbers	$\mathbb{N} = \{0, 1, 2, 3, \ldots\}$
Set of positive integers	$\mathbb{I}^+ = \{1, 2, 3, \ldots\}$
Set of binary numbers	$\mathbb{B} = \{0, 1\}$
Set of hexadecimal numbers	$\mathbb{H} = \{0, 1, 2, 3, 4, 5, 6, 7, 8, 9, A, B, C, D, E, F\}$
Set of rational numbers	$\mathbb{Q} = \{p/q \mid p \text{ and } q \text{ are integers}, q \neq 0\} = \{x \mid x \text{ has a repeating or terminating decimal expansion}\}$
Set of computer real numbers	$\mathbb{R}_c = \{x \mid x \text{ is expressed with a terminating decimal expansion and } x \text{ can be stored in the computer under discussion}\}$
Set of real numbers	$\mathbb{R} = \{x \mid x \text{ has a decimal expansion}\}$
Set of irrational numbers	$\mathbb{IR} = \{x \mid x \text{ is a real number but is not rational}\}$
Set of prime numbers	$\mathbb{P} = \{p \mid p > 1 \text{ and } p \text{ is an integer whose only positive integer divisors are 1 and } p\}$
Bit	0 or 1
Byte	String or sequence of bits to encode a character, usually 8 bits
Base	x in x^n
Exponent	n in x^n
Properties of exponents	See Table 1.1
Metric system	See Table 1.2

EXERCISES 1.1

1. List the even primes.

2. List the primes that are divisible by 3, i.e., the primes that when divided by 3, leave a remainder of 0.

Give all the sets, $\mathbb{I}$, $\mathbb{I}_c$, $\mathbb{N}$, $\mathbb{I}^+$, $\mathbb{Q}$, $\mathbb{IR}$, $\mathbb{R}$, $\mathbb{R}_c$, $\mathbb{P}$, $\mathbb{B}$, *or* $\mathbb{H}$, *to which the numbers in Exercises 3–23 belong.*

3. 7   **4.** 7.0   **5.** 0.75   **6.** $\sqrt{7}$ **7.** 7/5

8. 7.4×10^9 **9.** -7 **10.** $-7\frac{1}{10}$ **11.** $-7.1\overline{1}$

12. 7π **13.** 3.1875   **14.** 22/7   **15.** $3.14\overline{14}$

16. 1/3   **17.** $3.3\overline{3}$ **18.** 3.9375   **19.** 0

20. 0.0   **21.** $\sqrt{400}$ **22.** $\sqrt{35}$ **23.** 12.569447 . . .

24. There are exactly 100 prime numbers whose digits are strictly increasing; 2, 59, and 1,235,789 are examples.
  a. How many such primes are there below 100?
  b. Find three such primes above 100.
  Note: To check whether a number x is prime, you only need to determine if the primes less than or equal to $\sqrt{x}$ divide x. This statement is true because if a prime p divides x, there is an integer q such that $\sqrt{x} \times \sqrt{x} = x = pq$. If p is greater than $\sqrt{x}$, q must be less than $\sqrt{x}$. Thus, to see if 101 is a prime, we need only to see if 2, 3, 5, or 7 divides 101.

25. a. Double 2^5, expressing the answer with and without an exponent.
  Note: Notice that increasing the exponent of 2 by 1 doubles the number!
  b. Halve 2^5, expressing the answer with and without an exponent.

26. Repeat Exercise 25 using 2^9.

27. a. Express as a power of 2 the number of bits in a byte (assume 8 bits to a byte).
  b. Express as a power of 2 the number of bits in 4 bytes.

28. Suppose you take 10^4 pennies to the bank to get them exchanged.
  a. If you want dimes, how many will you get?
  b. Dollars?   **c.** Ten-dollar bills?
  d. Hundred-dollar bills?   **e.** Thousand-dollar bills?

29. Suppose you have 8 one-hundred dollar bills. If you exchange them for pennies, how many will you have, expressed as a number times a power of 10?

30. a. What is the smallest number that is a power of 2 and is greater than or equal to 19?
  b. 48?
  c. 113?

31. An **n-to-2^n decoder,** which is used in the construction of a computer, is a circuit with n input data lines and 2^n output lines. Suppose a decoder has 8 lines coming out of it. The number of input lines is the exponent of 2 when 8 is expressed as a power of 2.

 a. How many incoming lines are there?
 b. If there are 7 incoming lines, how many output lines are there?

32. How many bytes are in a 64K memory? Express your answer as a power of 2.

33. Repeat Exercise 32 using 256K.

34. How many bytes and gigabytes are in 56 megabytes?

Express as a power of 2 the number of K in Exercises 35–37.

35. 2^{23} bytes   **36.** 16,384 bytes   **37.** 131,072 bytes

For Exercises 38–41 express the times in picoseconds, nanoseconds, and microseconds.

38. 0.00045 sec   **39.** 892 psec

40. $3 \times 10^5 \ \mu$sec   **41.** 14.75 nsec

42. Suppose there are a total of 1000 bits in one inch of magnetic tape. If this tape is auxiliary storage to a computer which has a 10-bit byte, how many bytes per inch are on the tape?

43. a. Express 16^3 as a power of 2.   **b.** Of 8.

44. a. Express 8^{20} as a power of 2.   **b.** Of 16.

Express the numbers in Exercises 45–56 as powers of 10 and then without the exponents.

45. $10^3 \times 10^2$ **46.** $(10^3)^2$ **47.** $10^3/10^2$ **48.** $10^2/10^3$

49. $10^2/10^2$ **50.** $(1/10^2)^{-3}$ **51.** $1000^{5/3}$ **52.** 1000^5

53. $10,000^{3/2}$ **54.** $10,000^{-3/2}$ **55.** $2^7 \times 5^7$ **56.** $30^8/3^8$

57. Compare the add times for the following early computers. For each pair, how many times faster is the second computer in performing addition than the first?
 a. PILOT ACE, 0.54 msec;   Whirlwind, 0.05 msec
 b. Harvard Mark I, 0.3 sec;   ENIAC, 0.2 msec
 c. Zuse Z3, 2 sec;   EDSAC, 1.4 msec

58. A particular **algorithm** (method for doing a problem) when coded into a language takes about 5×10^4 **computer clock cycles** (a measurement of time for execution) to run. Two programming teams have a contest to see which can best improve the speed of this algorithm. Team A reduces the number of cycles by a factor of 10, while Team B reduces the number by 7500 cycles. Which team is the winner? *Note:* When working a word problem, you may need to read the problem several times. Then pick out the key elements and write them down, such as follows:

Algorithm 5×10^4 cycles
Team A   reduce by factor of 10
Team B   reduce by 7500
Winner?

You always have a place to start with a word problem because you can write down what you are given and what you want to find. Then ask yourself questions: *What do we want to find?* We need to figure out the winner. *What does "winner" mean?* It means the team that can improve the speed of this algorithm the most. *What does this statement mean?* It means which team produces the fastest algorithm. *How can we determine the fastest algorithm?* Here is where we begin the calculations of the speed in cycles of the two algorithms. Then, the one that takes the shortest amount of time is the winner. *What does "reduce by a factor" mean?* Divide. *What does "reduce by" mean?* Subtract.

59. In "Mathematics and Computer Science: Coping with Finiteness" (*Science* 194 (4271): 1235–1242, December 17, 1976, © Copyright 1976 by the AAAS), Donald E. Knuth, using the tables below, compared the changes in transportation and computation speeds.

a. How many times faster is plane flight than walking?

b. How many times faster is the supersonic jet than the snail?

c. A medium-speed computer, vintage 1976, can add how many times faster than we can?

d. A fast computer from 1976 is how many times faster than a calculator in addition?

Note: Knuth asked us to "consider how much a mere factor of 10 in speed provided by the automobile, has changed our lives. . . . Computers have increased our calculation speeds by . . . more than the ratio of the fastest airplane velocity to a snail's pace."

Typical Transportation Speeds

Snail	0.006	mile/hour
Man walking	4	mile/hour
U.S. automobile	55	mile/hour
Jet plane	600	mile/hour
Supersonic jet	1200	mile/hour

Computation Speeds (10-digit numbers)

Man (pencil and paper)	0.2	/sec
Man (abacus)	1	/sec
Mechanical calculator	4	/sec
Medium-speed computer	200,000	/sec
Fast computer	200,000,000	/sec

60. The earth has a diameter d of about 7918 miles.

a. Find an approximate value, expressed in miles, for its circumference, which is πd.

b. There are 1760 yards in a mile. Find the approximate circumference in yards.

c. Suppose the length of a person's step is 1 yard. If a person could take one step each nanosecond, how long would it take to circle the earth?

d. How many times could this hypothetical person circle the earth in 1 sec?

e. If someone could take a step every picosecond, how many times could this person circle the earth in a second?

1.2

Error

Many computer languages allow real numbers to be printed in exponential form as a decimal fraction times a power of 10. For instance, output of 9.843600 E 02 from a program written in the computer language FORTRAN means $9.8436 \times 10^2 = 984.36 = 0.98436 \times 10^3$. Usually real numbers are stored in the computer in two parts, a **fractional part** or **mantissa**, like 98436, and an exponent, like 3. Stored in this manner, these real numbers are called **floating point numbers.** The numbers are stored in binary, and the base is typically 2 or 16 instead of 10; but study of exponential notation in the decimal system will help us to understand other bases.

A **normalized** number in exponential notation will have the decimal point immediately preceding the first nonzero digit, as in 0.594368×10^3 or 0.594368×10^{-3}. Recall that

$$0.594368 \times 10^3 = 594.368$$

and

$$0.594368 \times 10^{-3} = 0.000594368.$$

The following is an **algorithm** or method for converting numbers like 64.8 and 0.00648 to normalized exponential notation.

**Algorithm to Convert a Number
to Normalized Exponential Notation**

1. Place the decimal before the first nonzero digit to obtain the fraction (0.648).

2. The exponent of 10 is positive or negative the number of positions the decimal was moved (here 2).

3. Calculate the sign of the exponent of 10 by comparing the new fraction with the old number.
   a. If you need to move the decimal to the right to get back the original number (as in $0.648 \rightarrow 64.8$), the exponent is positive
   $$(64.8 = 0.648 \times 10^2).$$

   b. If you need to move the decimal to the left to return to the first number (as in $0.648 \rightarrow 0.00648$), the exponent is negative
   $$(0.00648 = 0.648 \times 10^{-2}).$$

EXAMPLE 1

Solution

Express 43882 and 0.000051 in normalized exponential notation.

$$43882 = 43882. = 0.43882 \times 10^?.$$

Since the decimal was moved 5 places, the exponent is either $+5$ or -5. Looking at the answer, we want to move the decimal to the right to get the original number of 43882. Thus

$$43882 = 0.43882 \times 10^5.$$

Or, because 43882 is larger than 0.43882, we need a positive exponent for 10.

Now, converting 0.000051 we have

$$0.000051 = 0.51 \times 10^?$$

where ? is $+4$ or -4. To move the decimal back to the left, the exponent must be -4. ∎

When a number is expressed in normalized exponential notation, all the digits of the fractional part are what we call significant. Thus, for numbers in $\mathbb{I}_c$ (integers written without a decimal), all the digits except leading and trailing zeros are **significant digits**; for other numbers all digits are significant except leading zeros. For example, there are four significant digits in $003,704,000 = 0.3704 \times 10^7$, 3, 7, 0, and 4. The **most significant digit** is the leftmost one, 3. The most significant digit of $0.09200 = 0.9200 \times 10^{-1}$ is 9 since the leading zero is not significant. All other digits, however, are significant in this number.

Precision is the number of significant digits. So 003,704,000 and .09200 each have a precision of 4. **Magnitude** is an indication of the relative size of the number and is the power of 10 when the number is expressed in normalized exponential notation. Therefore, 0.3704×10^7 has a magnitude of 10^7. In BASIC language on the Apple microcomputer, precision of real numbers is about 9 decimal digits, while the magnitude ranges from about 10^{-39} to 10^{38}.

EXAMPLE 2

Solution

Calculate $(0.356 \times 10^8)(0.228 \times 10^{-3})$.

Since the order of multiplication does not matter, group the powers of 10 together, adding exponents, while multiplying the fractional parts:

$$0.356 \times 0.228 \times 10^8 \times 10^{-3} = 0.081168 \times 10^5.$$

Now, normalizing we obtain

$$0.81168 \times 10^{-1} \times 10^5 = 0.81168 \times 10^4. \quad \blacksquare$$

But suppose our computer only had a precision of 3, allowing only 3 digits in the fraction. Usually, the computer will truncate, or chop off, the extra digits of lesser significance so the result in Example 2 would be 0.811×10^4. Thus, an error has been introduced. Absolute error is the difference between the correct answer and the result:

$$\text{absolute error} = 0.81168 \times 10^4 - 0.811 \times 10^4$$
$$= 0.00068 \times 10^4 = 6.8.$$

Relative error is the ratio of the absolute error and the correct answer:

$$0.00068 \times 10^4/(0.81168 \times 10^4) \approx 0.0008378.$$

To express the relative error as a percentage, we multiply by 100% to obtain 0.08378%.

DEFINITION

If n_a is the correct answer and n_r is the result obtained, then

$$\textbf{absolute error} = n_a - n_r$$
$$\textbf{relative error} = (n_a - n_r)/n_a$$
$$= (n_a - n_r)/n_a \times 100\%.$$

If the result had been rounded instead of truncated, the error would have been less. To **round** to three significant digits, look at the digit in the fourth position. If the digit is 5 or more, increase the third digit by one and truncate to three places. 0.81168×10^4 rounded to three significant digits is 0.812×10^4. The absolute error is

$$0.81168 \times 10^4 - 0.812 \times 10^4 = -0.00032 \times 10^4 = 3.2$$

and the relative error is

$$-0.00032 \times 10^4/(0.81168 \times 10^4) \approx -0.0003942 = -0.03942\%.$$

Errors can also arise in addition or subtraction. Consider $0.684 \times 10^3 + 0.950 \times 10^{-2}$. Unlike in multiplication, the decimal points must be aligned for addition, so we have

$$0.684 \times 10^3 = \quad 684.0000 = \quad 0.6840000 \times 10^3$$
$$0.950 \times 10^{-2} = + \quad \underline{0.0095} = + \underline{0.0000095 \times 10^3}$$
$$684.0095 = \quad 0.6840095 \times 10^3.$$

**LANGUAGE
EXAMPLE**

If our computer only allows for three significant digits, the normalized result is 0.684×10^3, and the effect of the 0.950×10^{-2} is lost.

When such errors occur in **loops** (repetitive sections of a program) they create even greater problems that can cause the result to vary significantly from the correct answer. For example, let X be **initialized** to be zero, that is, assigned a beginning value of 0.0. Since the symbol for assignment in Pascal is ":=", we have X := 0.0. Suppose the number 0.9389 is to be accumulated into X, or added to X, 5 times. This repetitive process, which will result in the calculation of $5 \times (0.9389)$, can be accomplished by a FOR-loop in Pascal:

```
X   :=   0.0;
FOR   I   := 1 TO 5 DO
      X   := X + 0.9389;
```

Assuming that the machine allows four significant digits, let's calculate the absolute and relative error each time through the loop. Table 1.3 illustrates the example. The error compounds as the loop is executed. If the loop was executed 11 times with "FOR I := 1 TO 11 DO" to calculate 11×0.9389, we would get a relative error of 0.1733% (see Table 1.3). Or if 108×0.9389 was to be evaluated, we would have a relative error more than 20 times the original one.   □

TABLE 1.3
Calculated Loop Errors

Value of I	Correct value of X	Computer value of X	Absolute error	Relative error
1	0.9389	0.9389	0	0
2	0.9389 + 0.9389 —— 1.8778	0.9389 + 0.9389 —— 1.8778	1.8778 − 1.877 —— 0.0008	0.0008/1.8778 ≈ 0.000426 = 0.0426%
3	0.9389 + 1.8778 —— 2.8167	0.9389 + 1.877 —— 2.8159	2.8167 − 2.815 —— 0.0017	0.0017/2.8167 ≈ 0.0006035 = 0.06035%
4	0.9389 + 2.8167 —— 3.7556	0.9389 + 2.815 —— 3.7539	3.7556 − 3.753 —— 0.0026	0.0026/3.7556 ≈ 0.0006923 = 0.06923%
5	0.9389 + 3.7556 —— 4.6945	0.9389 + 3.753 —— 4.6919	4.6945 − 4.691 —— 0.0035	0.0035/4.6945 ≈ 0.0007456 = 0.07456%
11	10.3279	10.31	0.0179	0.0179/10.3279 ≈ 0.001733 = 0.1733%
108	101.4012	100.5	0.9012	0.9012/101.4012 ≈ 0.008887 = 0.8887%

EXERCISES 1.2

In Exercises 1–12 express the numbers without using a power of 10.

1. 0.701×10^4 **2.** 0.701×10^{-4} **3.** -0.701×10^4

4. 0.35×10^2 **5.** 0.35×10^{-2} **6.** 0.0061×10^6

7. 9807.6×10^{-7} **8.** 78.32×10^3

9. 78.32×10^{-3} **10.** $0.4 \times 10^{-1} \times 10^6$

11. $0.528 \cdot 10^{-5} \times 10^2$ **12.** $10^{-3} \times 10^3$

Write the numbers of Exercises 13–24 in normalized exponential notation.

13. 63,850   **14.** 29.748   **15.** 0.00032

16. 53.7×10^3 **17.** 0.496   **18.** 0.0000017

19. 0.009×10^{-5} **20.** 0.009×10^5 **21.** -0.82

22. -82 **23.** -0.00082 **24.** 4.4

25. Give the magnitude and precision of 0.743621×10^{25}.

26. Repeat Exercise 25 using 93.6×10^7.

27. Give the precision and the largest magnitude of numbers in a computer where the fractional part has 8 digits and the largest power of 10 is 125.

Give the number of significant digits and the most significant digit for the numbers in Exercises 28–33.

28. 63,850   **29.** 29.004   **30.** 0.00074

31. 10^3 **32.** 4×10^{-5} **33.** 0.0300500

34. Give the range of the normalized positive numbers where the mantissa has 3 digits and the exponent of 10 is from -5 to $+5$.

35. Give the range of the normalized positive numbers that can be expressed with 7 digits in the mantissa and an exponent of 10 from -78 to $+73$.

Perform the arithmetic indicated in Exercises 36–40, expressing the answers in normalized exponential notation.

36. $(0.134 \times 10^{-5})(0.28 \times 10^{-13})$ **37.** $(7 \times 10^3) - (8 \times 10^2)$

38. $(9.4 \times 10^{-5}) + (3.6 \times 10^4)$ **39.** $(7 \times 10^3)(8 \times 10^2)$

40. $(7 \times 10^3)/(8 \times 10^2)$

*For Exercises 41–44 find **a.** the absolute error and **b.** the relative error of each number as it is rounded to two decimal places. Then compute the **c.** absolute and **d.** relative error as each number is truncated to two decimal places.*

41. 6.239   **42.** 6.231

43. 6.235   **44.** 1/3 stored with 5 significant digits

45. Suppose the following is executed:

$$X := 6.239;$$
$$X := X + X;$$

If values are truncated to 3 significant digits, give the value stored for X after execution of each statement and the relative error for the value of X after the last statement. Compare this error with your answer in Exercise 41.

46. Recall that $1.0 = 0.9\overline{9} = 0.99\overline{9}$. Suppose this value is assigned to X and to Y as a series of 9's truncated to 4 significant digits.

a. If $X := X + Y$ is executed 4 times in a loop, give the value of X and the absolute and relative error for the original assignment and each iteration of the loop.

b. By observing the results of part a, give the value of X and the absolute and relative error after the tenth iteration of the loop. Notice how the error increases with each iteration.

1.3

Operator Precedence

LANGUAGE EXAMPLE

The arithmetic operations of addition, subtraction, multiplication, division, and exponentiation are standard in most computer languages. In a language context, unless we are discussing a specific language that is exceptional, we will use the symbols, $+$, $-$, $*$, $/$, and $**$, respectively, for these operators.

When several operations are present in an expression, some priority must be established as to which is calculated first. Most computer languages adopt the same priority used in mathematics. Standard Pascal, however, does not have the operation of exponentiation; and the language APL, as we will see later, has a completely different convention. □

Understanding this priority is important because an incorrectly written expression can cause errors in output. These errors can be difficult to find when the program is long. Worse yet, the error may go undetected while the company for which you are working makes decisions based on your incorrect information.

The convention for operator precedence in computer mathematics is as follows:

OPERATION PRECEDENCE IN MATHEMATICS

1. Parentheses can override the precedence below
2. $**$
3. Unary minus; i.e., negation of a number
4. $*$ and $/$
5. $+$ and $-$

When two operations of equal priority are encountered in an expression, such as $*$ and $/$ or two pluses, they are evaluated from left to right. For example, $10/5 * 4$ is $(10/5) * 4 = 2 * 4 = 8$. One exception occurs in mathematics when exponentiation follows itself as in $5 ** 2 ** 3$ or 5^{2^3}. Here, the expression is evaluated from right to left with the 2 cubed first to obtain $5^{2^3} = 5^8 = 390{,}625$. Of course, parentheses can override this priority because parentheses always indicate top precedence. To multiply before dividing in $10/5 * 4$, write $10/(5 * 4)$ which is $10/20 = 0.5$. To square 5 first in 5^{2^3}, use parentheses again: $(5^2)^3 = 25^3 = 15{,}625$.

EXAMPLE 1

Evaluate $14 - 10 \ / \ 2 + 8$.

Solution

The appearance is deceptive, but since division has a higher priority than addition or subtraction, $10/2$ is calculated first. The answer is $14 - 5 + 8 = 17$. ■

EXAMPLE 2

How would you write the expression in Example 1 to perform the subtraction first?

Solution

Write $(14 - 10)/2 + 8$ since this expression equals $4/2 + 8 = 2 + 8 = 10$. ■

EXAMPLE 3

How would you write the expression in Example 1 to perform the addition first?

Solution

Write $14 - 10/(2 + 8)$ which yields $14 - 10/10 = 14 - 1 = 13$. ■

EXAMPLE 4

LANGUAGE EXAMPLE

Write the expression

$$\frac{X + Y}{Z + W}$$

in a computer language format.

Solution

$(X + Y) / (Z + W)$

As illustrated in Examples 1, 2, and 3, the parentheses are essential. $(14 - 10)/(2 + 8) = 4/10 = 0.4$ is entirely different than the answers obtained in those problems. Imagine how difficult it would be for you to find an error in omission of parentheses if evaluation of this expression was just one step in a long series of calculations in your program. ☐ ■

EXAMPLE 5

LANGUAGE
EXAMPLE

Suppose A = 1, B = 10, C = 2, and D = 5. Evaluate the expression

A/B ** C * D − D ** B ** (C − 2)

giving the operation evaluated at each step. Also, write the expression as it is normally displayed in mathematics.

Solution

```
  1/10 ** 2 * 5 − 5 ** 10 ** (2 − 2)    (−)
= 1/(10 ** 2) * 5 − 5 ** 10 ** 0        (**)
= 1/100 * 5 − 5 ** (10 ** 0)           (**)
= 1/100 * 5 − (5 ** 1)                 (**)
= (1/100) * 5 − 5                      (/)
= (0.01 * 5)− 5                        (*)
= 0.05 − 5                             (−)
= −4.95
```

Written as normally seen in mathematics, the expression is

$$\left(\frac{A}{B^C}\right)D - D^{B(C-2)}$$ ☐ ■

EXAMPLE 6

Write $-X^2 + 8Y^3$ as in a computer language, and evaluate the expression for $X = 7$ and $Y = 6$.

Solution

LANGUAGE
EXAMPLE

-X ** 2 + 8 * Y ** 3.

In evaluating

-7 ** 2 + 8 * 6 ** 3

exponentiation is performed before the unary minus or multiplication to give

$-49 + 8 * 216 = -49 + 1728 = 1679.$

Parentheses around $-X$ as in $(-X)$ ** 2 + 8 * Y ** 3 would cause -7 to be squared giving $49 + 1728 = 1777.$ ☐ ■

The operator precedence is especially important when performing arithmetic where truncation can occur. Often in languages, when variables are of data type INTEGER, each evaluation of the expression as well as of the answer must be of the same type. Suppose, for example, you wanted to take two-thirds of I, which has been declared to be an INTEGER **identifier** (variable). In mathematics it would not matter if you wrote

$$\frac{2}{3}I \quad \text{or} \quad I\frac{2}{3} \quad \text{or} \quad \frac{2I}{3}.$$

LANGUAGE EXAMPLE

But when using integer arithmetic in a computer, the manner of writing the expression makes a big difference. With 2/3 * I, because of equal priority of / and *, the leftmost part of the expression, 2/3, is evaluated first. But the intermediate result must be an integer, so $2/3 = 0.6\overline{6}$ is truncated to 0. The evaluation of the expression will always be 0, regardless of the value of I. Multiplication must be performed first to avoid this difficulty: 2 * I/3 or I * 2/3. If I has the value of 7, the expression evaluates to 4, a truncated $14/3 = 4.6\overline{6}$.

There is, however, a comparable situation in which we want to divide first: there is a danger of the product in the numerator exceeding the largest allowable integer in the computer. Suppose MAXINT is 65,535. If we want to take 75 fourths of 1002, multiplication of 1002 by 75 first yields 75,150 which is larger than can be expressed in this computer. Unpredictable results happen, with truncation usually occurring on the left. But division before multiplication would yield the more reasonable, though still not exact, answer of

```
(1002/4) * 75 = 250 * 75 = 18,750.
```

Usually when you have the choice, however, write * before /. Even in real arithmetic, this rule is important because truncation can still cause errors. Suppose I is a real variable with value $0.1430 \times 10^3 = 143.0$, and the precision is 4 (that is, real numbers can have 4 significant digits). Calculating two-thirds of I with division first, we obtain

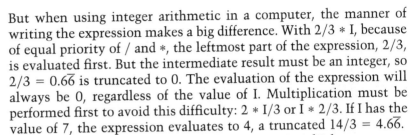

```
I/3 * 2 = 143.0/3 * 2
        = 47.66 * 2        Truncated from 47.66̄
        = 95.32.
```

But multiplying first we obtain

```
I * 2/3 = 143.0 * 2/3
        = 286.0/3
        = 95.33            Truncated from 95.33̄
```

Truncation occurs earlier in the first evaluation, compounding the error upon multiplication. In this example, the former problem has four times the relative error of the latter:

For $I/3 * 2$, since $95\frac{1}{3}$ is the exact answer

$$\frac{95\frac{1}{3} - 95\frac{32}{100}}{95\frac{1}{3}} = \frac{\frac{1}{3} - \frac{32}{100}}{\frac{286}{3}}$$

$$= 4(\tfrac{1}{300} \times \tfrac{3}{286}).$$

For $I * 2/3$

$$\frac{95\frac{1}{3} - 95\frac{33}{100}}{95\frac{1}{3}} = \frac{\frac{1}{300}}{\frac{286}{3}}$$

$$= (\tfrac{1}{300} \times \tfrac{3}{286}). \quad \square$$

Other operators such as relational operators $(=, \neq, \leq, \geq, >, <)$, Boolean operators (OR, AND, NOT), and set operators (union, intersection, difference) can occur in expressions. It is very important for you to understand the priority of operations in the language in which you are working. For instance, in some languages AND has a higher priority than $=$, while in others it does not. In the statement

IF I = 4 AND K = 3 THEN. . .

if AND has a higher priority, the computer will evaluate (4 AND K) first, yielding an error or unpredictable results. In this case parentheses are essential to override the evaluation of AND:

IF (I = 4) AND (K = 3) THEN. . .

We consider Boolean and set operations in the next chapter.

LANGUAGE EXAMPLE

It is interesting that the computer language APL has an order of priority that deviates significantly from that of mathematics. Namely, all operations have the same precedence, and expressions are evaluated from right to left, except when overridden by parentheses. Thus, with $\div$ as the symbol for real division and $\times$ as the symbol for multiplication, $10 \div 5 \times 4$ is evaluated as $10 \div (5 \times 4) = 0.5$. Example 1, $14 - 10 \div 2 + 8$, becomes $14 - (10 \div (2 + 8)) = 14 - (10 \div 10) = 13$. While $\times$ is used for multiplication, $*$ is used for exponentiation. Thus, Example 5 is evaluated as

$$
\begin{array}{ll}
1 \div 10 * 2 \times 5 - 5 * 10 * 0 & \text{Evaluate } 10^0 \\
= 1 \div 10 * 2 \times 5 - 5 * 1 & \text{Evaluate } 5^1 \\
= 1 \div 10 * 2 \times 5 - 5 & \\
= 1 \div 10 * 2 \times 0 & \\
= 1 \div 10 * 0 & \text{Evaluate } 10^0 \\
= 1 \div 1 = 1. \quad \square &
\end{array}
$$

EXERCISES 1.3

Calculate the real number answers for Exercises 1–24 using the usual order of priority of operations. Assume X = 5, Y = 9, Z = 2, W = 3, *and* V = 6.

1. X + Y/Z
2. (X + Y)/Z

3. X/Z + Y
4. Y + X/Z + W

5. Y + X/(Z + W)
6. (X + Y)/Z + W

7. Z ** W ** Z
8. Z ** Z ** W

9. (Z ** Z) ** W
10. (Z ** W) ** Z

11. V * X/Z
12. (X * V)/Z

13. V/Z * W
14. V/(Z * W)

15. X/Z * V
16. X/(Z * V)

17. 2 * Y ** 2
18. X ** Z * Z ** W

19. V/W * (Y − 4)
20. −Y ** 2

21. Z/V * W
22. W * Z/V

23. X − Y − Z
24. X − (Y − Z)

25. Assuming INTEGER arithmetic, reevaluate Exercises 2, 3, **4**, 6, **11**, 12, 13, 14, **15**, 19, and 22.

26. Assuming INTEGER arithmetic, reevaluate Exercises 1, 5, 16, and 21, and calculate the relative error.

27. Rewrite the following exercises in APL, then evaluate the expression in real arithmetic with the order of priority of APL: 1, 2, **3**, 4, 6, **7**, 11, **13**, 16, 17, 18, 19, 21, 22, and **23**. □

28. Write a programming statement to evaluate

$$\frac{X^2}{Y + 3}Z$$

with the operator precedence of mathematics to obtain the most accurate result. □

29. Repeat Exercise 28 for

$$\frac{Y}{Z^8}\frac{4Y^2}{Y − 1}.$$ □

For Exercises 30 and 31 suppose the computer holds only three significant digits and that X = 5.94, Y = 9.21, Z = 2.4, W = 3.44, *and* V = 6.82.

30. Reevaluate Exercises 1, 2, 4, 5, **6**, and 21.

31. Reevaluate Exercises **3**, 11, and 17, and calculate the relative error for each.

Polish Notation

HISTORICAL NOTE

The usual way to write arithmetic expressions, such as $A + B$, is called **infix notation** with the sign of operation between the two operands. However, the same expression could be writen with **prefix notation** as $+AB$ or with **postfix notation** or **suffix notation** as $AB+$. These forms are called **Polish notation** after the Polish mathematician, Jan Łukasiewicz (1878–1956), who suggested them. $\square$

Some calculators demand that the user enter expressions in Polish notation, and some computers accept the expression in infix notation but convert it to postfix form before execution. These are stack-oriented machines. A **stack** is a data structure in which data can only be added to or removed from one end, the **top,** of the sequence. The last item placed on the stack must be the first item taken from the stack. This structure is well named, for it is analogous to a stack, say, of trays in a cafeteria. If you put back a tray, it goes on the top of the stack; the next person will pick up that tray. (For the sake of our analogy, we must assume that no one will reach into the middle of the stack.) Thus, there are two operations on the stack: **PUSH X**, to place X on the top of the stack; and **POP X**, to pull X from the top.

When the expression is written in postfix notation, such as $AB+$, during execution the values of the variables are pushed on the stack as the expression is read from left to right. When an operator, such as $+$, is encountered, the top two pieces of data are popped from the stack, the operation is performed, and the result is pushed onto the stack. For example, suppose $A = 4$ and $B = 3$. Figure 1.2 illustrates what happens during execution of $AB+$.

Figure 1.2

The stack during execution of $AB+$ with $A = 4$ and $B = 3$. The left arrow indicates assignment.

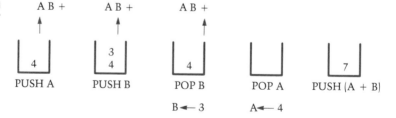

One of the advantages of Polish notation is that parentheses are unnecessary. Consider the infix expression $A * (B + C)$ with $A = 4$, $B = 3$, $C = 5$. Correctly evaluated, its value is $4(3 + 5) = 4 \times 8 = 32$. Here, with infix notation, parentheses are essential. With postfix notation, however, parentheses are unnecessary since $A * (B + C)$ would be written as $ABC+*$. Observing the stack illustrated in Figure 1.3, we find the operation is always between the top two operands.

Figure 1.3

The stack during execution
of A B C + * with A = 4,
B = 3, and C = 5.

C	5
B	3
A	4

1. PUSH A   (4)
2. PUSH B   (3)
3. PUSH C   (5)

(B + C)	8
A	4

4. POP   C   (5)
5. POP   B   (3)
6. PUSH (B + C) (8)

	32

7. POP   (B + C)   (8)
8. POP   A   (4)
9. PUSH A * (B + C) (32)

HISTORICAL NOTE

The stack-oriented language FORTH uses postfix notation, such as 4 3 5 + *, for expressions. The story goes that Charles Moore, its creator, wanted to call the new language FOURTH to indicate that it was a fourth-generation language. The computer system he was using, however, allowed only five-character names, hence the shortened, but meaningful, name FORTH. □

Algorithm to Convert from Postfix to Infix Notation

Do while there are at least two operands:

1. Read the expression from left to right.

2. When an operation is encountered, write it between the two operands immediately preceding. Place parentheses around this expression. The expression in parentheses is now treated as a single operand.

EXAMPLE 1

Convert A B C + * to infix notation.

Solution

Reading from left to right, stop when you get to +

    A   B   C   +   *
             ↑

Write A(B + C) * . Continuing to the right

    A (B + C) *
             ↑

the * operation will be applied to the two previous operands, A and (B + C), giving A * (B + C). Note that (B + C) is one operand and is not split. ■

EXAMPLE 2

Convert A B C + D ** − to infix notation, where ** stands for exponentiation. Also, write the result as you would normally see it expressed mathematically, with superscripts instead of **.

Solution

$$A\ B\ C\ +\ D\ **\ -$$
$$\uparrow$$

$$A\ (B\ +\ C)\ D\ **\ -$$
$$\uparrow$$

$$A\ (\ (B\ +\ C)\ **\ D)\ -$$
$$\uparrow$$

$$A\ -\ (\ (B\ +\ C)\ **\ D)\ \text{ or}$$
$$A\ -\ (B\ +\ C)\ **\ D,$$

which without ** is $A - (B + C)^D$. ∎

**Algorithm to Convert from Infix Notation
to Prefix or Postfix Notation**

1. Write the infix expression with a full set of parentheses for the appropriate order of priority of operations; there should be a pair of parentheses for each operator. For example, write $A + B$ as $(A + B)$.

2. a. For postfix notation replace each right parenthesis with the corresponding operation, and drop the matching left parenthesis. For example, $A\ B\ +$.

   b. For prefix notation replace the left parenthesis, and drop the matching right parenthesis. For example, $+\ A\ B$.

EXAMPLE 3

Convert

$$\frac{A + B}{B^2} + C$$

to infix notation as the expression would be written with operators +, /, and **; then write it in postfix and prefix notation.

Solution

Infix: $(A + B)/B ** 2 + C$
With full parentheses: $(((A + B)/(B ** 2)) + C)$
Postfix: $A\ B\ +\ B\ 2\ **\ /\ C\ +$
Prefix: $+\ /\ +\ A\ B\ **\ B\ 2\ C$ ∎

Notice that in the three notations the order of appearance of the numbers and variables does not change; only the operators move.

EXERCISES 1.4

1. Suppose A = 7, B = 4, C = 6, D = 2. Evaluate the expression A B + C *, written in postfix notation, with the given values of the variables. Also, show the stack before and after each step of processing the expression, and write the expression in infix notation.

2. Repeat Exercise 1 using A B C D + * − .

For the expressions of Exercises 3–7 given in postfix notation, write the corresponding infix representation.

  3. A B C * /   **4.** A B * C D ** +   **5.** A B C * D + /

  6. X Y + Z − W /   **7.** X Y Z W + − **

Write each of the expressions in Exercises 8–20 in postfix and prefix notation.

  8. X * Y   **9.** X * Y + Z   **10.** X + Y * Z
  11. A / B + C * D   **12.** A + B / C + D   **13.** (A + B)/C + D
  14. A + B/(C + D)   **15.** (A + B)/(C + D) **16.** A ** B ** C
  17. A + B ** C   **18.** (A + B) ** C   **19.** A − B + C
  20. A * (B − C)

Write the expressions of Exercises 21–28 in infix and postfix notation.

  21. $2X^3$ **22.** $(2X)^3$ **23.** $\dfrac{3X}{Y}$

  24. $3\dfrac{X}{Y}$ **25.** $\dfrac{Y - 6Z}{5 + X^4}$ **26.** $\dfrac{A^3}{B + \frac{1}{C}}$

  27. $\dfrac{Y + Z^3}{5X}$ **28.** $B^2 - 4AC$

Note: The Advanced Placement (AP) Computer Science Examination seeks to test "students' abilities to use computing in powerful, intelligent, and responsible ways." After taking a one-year AP computer science course in high school which is designed as a college freshman-year, majors' course, a student may take the AP examination. If the student scores high enough, many universities and colleges will give him or her credit for a first-year computer science course. Several sample questions from the *1984 AP Course Description in Computer Science* will be given throughout the book, illustrating the close relationship between computer science and mathematics.*

---

*AP questions selected from *AP Course Description in Computer Science*. College Entrance Examination Board, 1984. Reprinted by permission of Educational Testing Service, the copyright owner of the sample questions.

LANGUAGE
EXERCISE

29. Consider the following sequence of procedure calls:

  Push(*x*);
  Push(*y*);
  Add;
  Push(*z*);
  Push(*w*);
  Mult;
  Add;

Invoking *Push* causes its argument to be pushed onto a stack. Invoking the procedures *Add* or *Mult* causes (1) the stack to be popped twice, (2) the two popped items to be added or multiplied, and (3) the result to be pushed onto the stack. If $x = 10$, $y = 20$, $z = 30$, and $w = 40$, and the stack is empty initially, then at the end of the sequence of procedure calls above, the stack contains

(A) Nothing   **(B)** 0   **(C)** 940   **(D)** 1230   **(E)** 1410   □

Chapter 1   REVIEW

Note: The chapter reviews at the end of each chapter outline the most important terms, notation, and methods in the chapter. The comments are not meant to contain exact definitions or methods, but only to indicate the essence of the terms, to give examples, or to note algorithms.

Terminology/Algorithms	Pages	Comments
Set	2	Collection of things
Set of integers, $\mathbb{I}$	2	$\{\ldots, -3, -2, -1, 0, 1, 2, 3 \ldots\}$
Set of computer integers, $\mathbb{I}_c$	2	No decimals, finite set
Set of natural numbers, $\mathbb{N}$	2	$\{0, 1, 2, 3, \ldots\}$
Set of positive integers, $\mathbb{I}^+$	3	$\{1, 2, 3, \ldots\}$
Binary set, $\mathbb{B}$	3	$\{0, 1\}$
Hexadecimal set, $\mathbb{H}$	3	$\{0, 1, 2, \ldots, 9, A, B, C, D, E, F\}$
Set of rationals, $\mathbb{Q}$	3	Ratio of integers; repeating or terminating decimal expansion
Set-builder notation	3	$\{x \mid \text{property } x \text{ satisfies}\}$
Set of computer reals, $\mathbb{R}_c$	4	Terminating decimal expansion, finite set
Set of reals, $\mathbb{R}$	4	Has decimal expansion

Terminology/Algorithms	Pages	Comments
Set of irrationals, $\mathbb{IR}$	4	Not rational
Set of primes, $\mathbb{P}$	5	$p > 1$, 1 and p only positive integer divisors
Bit	5	0, 1
Properties of exponents	6	Table 1.1
Byte	6	Number of bits to encode character, often 8 bits
K	6	2^{10} bytes
Metric system	7	Table 1.2
Giga, g	7	$1\text{ g}\_\_ = 10^9\_\_$
Mega, M	7	$1\text{ M}\_\_ = 10^6\_\_$
Kilo, K	7	$1\text{ K}\_\_ = 10^3\_\_$
Milli, m	7	$10^3\text{ m}\_\_ = 1\_\_$
Micro, μ	7	$10^6\ \mu\_\_ = 1\_\_$
Nano, n	7.	$10^9\text{ n}\_\_ = 1\_\_$
Pico, p	7	$10^{12}\text{ p}\_\_ = 1\_\_$
Normalized exponential notation	13	$r \times 10^n$, $0.1 \le r < 1$, Example: 0.358×10^7
Fractional part or mantissa	13	Example: 358 in 0.358×10^7
Algorithm	13	Method
Number $\rightarrow$ normalized exponential notation	13	Algorithm
Significant digit	14	Not leading zeros; without decimal point, not trailing zeros either
Most significant digit	14	Leftmost
Precision	14	Number significant digits
Magnitude	14	Power of 10
Truncate	15	Chop off digits
Absolute error	15	(correct value) − result
Relative error	15	(absolute error)/(correct value)

Terminology/Algorithms	Pages	Comments
Round to n positions	15	Check $(n + 1)$st position
Operator precedence	18	See list
Integer arithmetic	21	Truncate to integers
Infix notation	24	Example: A + B
Prefix notation	24	Example: +A B
Postfix or suffix notation	24	Example: A B +
Stack	24	Data added to and removed from one end (top)
POP	24	Take data from stack
PUSH	24	Put data on stack
Postfix → infix	25	Algorithm
Infix → prefix or postfix	26	Algorithm

REFERENCES

Bloomenthal, J. "Modeling the Mighty Maple." *SIGGRAPH '85 Proceedings* 19 (3): 305–311, July 1985.

Educational Testing Service. *1984 AP Course Description in Computer Science.* Princeton, N.J.: College Board Publications, 1984.

Knuth, D. E. "Mathematics and Computer Science: Coping with Finiteness." *Science* 194 (4271): 1235–1242, December 1976.

Larson, H. "Reversible Monotonic Primes." *Tentacle* 3 (8): 44, August 1983.

McCabe, C. K. *FORTH Fundamentals*, Vol. I, *Language Usage.* Beaverton, Ore.: Dilithium Press, 1983.

Max, N. L. "Shadows for Bump Mapped Surfaces." *Advanced Computer Graphics: Proceedings of Computer Graphics, Tokyo '86*, T. L. Kunii, ed., 145–156, Tokyo: Springer Verlag, 1986.

The Pioneer Computers Comparative Statistics. *The Computer Museum Report.* 14–15, 1983.

Spencer, D. D. *The Illustrated Computer Dictionary*, 3rd ed. Columbus: Merrill, 1986.

Strecker, G. E. "Round Numbers: An Introduction to Numerical Expression." *UMAP* 3 (4): 425–447, 1982.

2 SETS AND LOGIC

Mandelbrot Set. Membership of a point (x, y) in the set is decided by an algorithm that is implemented on a computer. If (x, y) is in the set, the corresponding point is black in the picture.

2.1

Properties of Sets

The concept of a set, which is a collection of things, was introduced in Section 1.1. There, we considered sets of numbers, but sets can be composed of other things as well. For example, you might talk about the set of programs on your diskette. The set of student records for a school might be stored on paper in file cabinets or on a tape file or in a data base. The collection of invoices a company sends out in a day is also a set.

LANGUAGE EXAMPLE

As we will see, the concept of a set is part of the foundation of a relational data base. Some computer languages even allow the programmer to define his or her own sets. In Pascal, for example, WEEKDAYS and WEEKEND could each be declared to be a SET OF DAYS. DAYS must be an appropriately defined type, such as (SUN, MON, TUE, WED, THU, FRI, SAT).Then values could be assigned, such as

WEEKDAYS : = [MON, TUE, WED, THU, FRI]

and

WEEKEND : = [SAT, SUN].

There is even a way of asking whether the value of a variable is in a set. Suppose we wish to find out if the value of FIELDDAY is on the weekend. We would have in Pascal

IF FIELDDAY IN WEEKEND THEN . . .

In mathematics we would express the same concept by defining WEEKEND = {SAT, SUN} and writing, "if FIELDDAY $\in$ WEEKEND then. . . ." The symbol, $\in$, means "belongs to." Should the condition not be met, we would write

FIELDDAY $\notin$ WEEKEND.

Additionally, since the number 2 is in the set $\mathbb{I}$ but not in $\mathbb{B}$, we write

$2 \in \mathbb{I}$ and $2 \notin \mathbb{B}$.

To indicate that the number of elements in $\mathbb{B}$ is 2, we write $n(\mathbb{B}) = 2$. Thus, $n(\text{WEEKDAYS}) = 5$, and $n(\mathbb{I})$ is infinite. □

NOTATION

The notation for the number of elements in set S is $n(S)$.

The language Pascal, which we are using for several exercises involving set operations, is named after Blaise Pascal, the great scientist and mathematician of the 17th century. During his late teens and twenties, he designed and built the first workable, automatic calculator, to help his father who was overwhelmed with work as a tax assessor. Pascal suffered from violent headaches all his life and alternated between periods of great scientific achievement and intense religious fervor. Intrigued by the questions of a gambling friend, he helped to establish the mathematical theory of probability. During a period in which he was writing many religious essays, to take his mind off of a painful toothache he immersed himself in the study of geometry, working many previously unsolved problems. Blaise Pascal died at the age of 39 in 1662 after suffering a brain hemorrhage. □

In the language Pascal, [ ] is the symbol used for one special set, the **empty** or **null set.** □ This set with no elements is written as ϕ or $\{\}$ in mathematics. On your blank diskette, the set of programs is certainly empty.

Be careful you do not confuse $\{0\}$ or $\{\phi\}$ with ϕ. The first two sets contain one element each while the third has no elements; the number $0 \in \{0\}$ and the set $\phi \in \{\phi\}$. Also, though in a computer science context 0 is often written with a slash through it, we will use the Greek letter phi, ϕ, to stand for the empty set.

Another special set is a **universal set** U or a set from which all the elements come in a particular context. While discussing WEEKDAYS, WEEKEND, and FIELDDAY, ALLDAYS := [SUN, MON, TUE, WED, THU, FRI, SAT] (in the notation of Pascal) would be an appropriate universal set. The set of students at your school, S, would be an appropriate universal set in discussing various class rolls.

Notice that every student in your class is a student in your school. We say the set of students in your class, C, is a subset of S and write $C \subseteq S$.

DEFINITION

Let A and B be sets. A is a **subset** of B if every element of A is an element of B. Write $A \subseteq B$.

Another way of expressing this definition, using mathematical notation and an arbitrary element x of A, is helpful when trying to prove one set is a subset of another.

REPHRASED DEFINITION ▬▬▬▬▬▬▬▬▬▬▬▬▬▬▬

> Let A and B be sets. A is a **subset** of B if
>
> $$x \in A \quad \text{implies} \quad x \in B$$
>
> written
>
> $$x \in A \Rightarrow x \in B.$$

EXAMPLE 1 $\mathbb{B} \subseteq \mathbb{I}$. ■

EXAMPLE 2 $\{1, 3, 5\} \subseteq \{5, 4, 1, 2, 3\}$. ■

As with $=$ and $\in$, to negate $\subseteq$, draw a slash through it. Thus, $\{5, 7\} \not\subseteq \{1, 3, 5\}$.

Notice that in a discussion, every set A is a subset of the universal set and of itself: $A \subseteq U$ and $A \subseteq A$. The empty set is always a subset of every set S. Since ϕ contains no elements, by default every element of ϕ is in S. A formal proof of the statement $\phi \subseteq S$ is discussed in Section 2.6.

Be careful about the symbols $\in$ and $\subseteq$. For $A \in B$, A must itself be a member of B. For $A \subseteq B$, A must be a set, and each of its elements must be a member of B. Thus, for $\mathbb{B} = \{0, 1\}$ and $C = \{0, 1, 2\}$

$$\mathbb{B} \subseteq C \qquad \mathbb{B} \subseteq \mathbb{B} \qquad C \subseteq C \qquad \phi \subseteq \mathbb{B} \quad \text{and} \quad \phi \subseteq C$$

while

$$0 \in C \qquad 1 \in C \qquad 2 \in C \quad \text{or} \quad 0, 1, 2, \in C.$$

EXAMPLE 3   For the set, $\mathbb{B} = \{0, 1\}$, the following are true:

$0 \in \mathbb{B}$	The number 0 is listed in $\mathbb{B}$.
$\{0\} \subseteq \mathbb{B}$	The set $\{0\}$ has one element, zero. That element is also in $\mathbb{B}$.
$\phi \notin \mathbb{B}$	The set $\mathbb{B}$ has two elements, the numbers zero and one; ϕ is not listed as one of the elements of $\mathbb{B}$.
$\phi \subseteq \mathbb{B}$	The empty set is a subset of every set.

$\phi \in \{\alpha, \phi, \beta\}$	ϕ is one of the elements in this set.
$\{\phi\} \subseteq \{\alpha, \phi, \beta\}$	The set $\{\phi\}$ has one element, ϕ; that element is listed in $\{\alpha, \phi, \beta\}$.
$\{\phi\} \notin \{\alpha, \phi, \beta\}$	ϕ is an element of $\{\alpha, \phi, \beta\}$, but $\{\phi\}$, braces and all, is not listed in $\{\alpha, \phi, \beta\}$.
$\phi \subseteq \{\alpha, \phi, \beta\}$	The empty set is always a subset of any given set. ■

If two sets contain exactly the same elements, they are equal. It is possible that the set of students taking Pascal, P, is the same as the set of students taking Discrete Mathematics, D. Then, every student in P is in $D (P \subseteq D)$, and everyone in D is also in $P (D \subseteq P)$.

DEFINITION

The sets A and B are **equal**, $A = B$, if each is a subset of the other.

If you are trying to prove two sets are equal, you often use this definition, which is expressed in symbols as follows.

REPHRASED DEFINITION

Let A and B be sets. $A = B$ if $A \subseteq B$ and $B \subseteq A$.

In Section 1.1 we indicated two definitions for the set of rational numbers $\mathbb{Q}$:

$$\{p/q \mid p \text{ and } q \text{ are integers; } q \neq 0\}$$

and

$$\{x \mid x \text{ has a repeating or terminating decimal expansion}\}.$$

These sets are equal because every number that is written as a fraction can be expressed with a repeating or terminating decimal expansion and vice versa; that is, each set is a subset of the other.

EXAMPLE 4

For the set $\mathbb{B} = \{0, 1\}$, the following are true:

$\mathbb{B} = \{1, 0\}$	The order in which we write the elements is unimportant since $\mathbb{B} \subseteq \{1, 0\}$ and $\{1, 0\} \subseteq \mathbb{B}$.
$\mathbb{B} = \{0, 1, 0\}$	Duplicates are unimportant; 0 and 1 are the only elements in the set. Thus, $n(\{0, 1, 0\}) = 2$. ■

EXERCISES 2.1 *Let* $X = \{4, \{5, 6\}, 7\}$. *Indicate true or false for each of the statements in Exercises 1–8.*

1. $7 \in X$ **2.** $5 \in X$ **3.** $\{5, 6\} \in X$ **4.** $\{5, 6\} \subseteq X$

5. $\{\{5, 6\}\} \subseteq X$ **6.** $X \subseteq \mathbb{I}$ **7.** $n(X) = 3$ **8.** $4 \subseteq X$

For sets defined in Section 1.1, indicate true or false for each of the statements in Exercises 9–42.

9. $0 \in \mathbb{I}$ **10.** $0 \in \mathbb{Q}$ **11.** $0 \in \mathbb{R}$ **12.** $0 \in \mathbb{B}$

13. $0 \in \mathbb{P}$ **14.** $34 \in \mathbb{I}$ **15.** $\phi \in \mathbb{I}_c$ **16.** $\sqrt{5} \in \mathbb{Q}$

17. $\sqrt{5} \in \mathbb{IR}$ **18.** $\sqrt{5} \in \mathbb{R}$ **19.** $\sqrt{5} \in \mathbb{R}_c$ **20.** MAXINT $+ 1 \in \mathbb{I}_c$

21. $\pi \in \mathbb{IR}$ **22.** $\mathbb{H} \subseteq \mathbb{R}$ **23.** $16.35\overline{35} \in \mathbb{Q}$

24. $14 \in \mathbb{R}$ **25.** $-2/9 \in \mathbb{I}_c$ **26.** $-2/9 \in \mathbb{IR}$

27. $-2/9 \in \mathbb{R}$ **28.** $\phi = \{0\}$ **29.** $\phi \subseteq \{\phi\}$

30. $\phi = \{\phi\}$ **31.** $\phi \in \{\phi\}$ **32.** $\phi = \{\}$

33. $\{\phi\} = \{\{\phi\}\}$ **34.** $49.3 \in \mathbb{Q}$ **35.** $49.5 \in \mathbb{R}_c$ **36.** $6 \in \mathbb{H}$

37. $\mathbb{Q} \not\subseteq \mathbb{IR}$ **38.** $\mathbb{IR} \not\subseteq \mathbb{Q}$ **39.** $n(\mathbb{H}) = 15$ **40.** $\mathbb{N} = \mathbb{I}^+$

41. $\mathbb{P} \subseteq \mathbb{I}^+$ **42.** $\mathbb{B} \subseteq \mathbb{I}^+$

In Exercises 43–53 write the sets enumerated, without set-builder notation.

43. $\{x \mid x \in \mathbb{I} \text{ and } 3 \leq x \leq 7\}$

44. $\{x \mid x \in \mathbb{I} \text{ and } x > 3\}$

45. $\{m \mid m \text{ is a month beginning with the letter "M"}\}$

46. $\{y \mid y \text{ is an even integer}\}$

47. $\{y \mid y = 2^n, n \in \mathbb{N}\}$

48. $\{y \mid y = 2^n, n \in \mathbb{I}\}$

49. $\{y \mid y = 3x \text{ for } x \in \mathbb{N}\}$

50. $\{x \mid x = 2 \text{ or } x = 3\}$

51. $\{x \mid (x - 2)(x - 3) = 0\}$

52. $\{x \mid x^2 - 5x + 6 = 0\}$

53. $\{x \mid x \text{ is a vowel in the word "computer"}\}$

Write the sets in Exercises 54–58 using set-builder notation.

54. $\{4, 5, 6, 7, 8, \ldots\}$

55. $\{4, 5, 6, 7, 8\}$

56. $\{-4, -5, -6, -7, -8\}$

57. $\{0, 5, 10, 15, 20, \ldots\}$

58. $\{\ldots, -15, -10, -5, 0, 5, 10, 15, \ldots\}$

59. Write the five smallest elements of the set
$$A = \{m + (m - 1)(m - 2)(m - 3) \,|\, m \in \mathbb{I}^+\}.$$
Note: Notice that $\{1, 2, 3, \ldots\}$ does not adequately describe A. In this case, as with many infinite sets, set-builder notation is preferable.

60. Write every possible subset relation involving the following sets: $\mathbb{I}$, $\mathbb{N}$, $\mathbb{I}^+$, $\mathbb{Q}$, $\mathbb{IR}$, $\mathbb{R}$, $\mathbb{P}$, $\mathbb{B}$, $\mathbb{H}$, ϕ. For example, $\phi \subseteq \mathbb{IR} \subseteq \mathbb{R}$. *Note:* See Section 1.1 for definitions.

61. Give the set of all the subsets of $\mathbb{B}$.

62. Give the set of all the subsets of $\{x, y, z\}$.

*Indicate whether each statement in Exercises 63–72 is true or false. If false, give a **counterexample**; that is, give specific sets that do not satisfy the statement. For example, the statement*
$$\text{If } A \subseteq B, \text{ then } B \subseteq A$$
is false. One counterexample is to let $A = \{1, 2\}$ and $B = \{1, 2, 3, 4\}$. $A \subseteq B$ but $B \not\subseteq A$.

63. If $A \subseteq B$ and $B \subseteq C$, then $A \subseteq C$.

64. If $A \subseteq B$ and $B \neq C$, then $A \neq C$.

65. If $A \subseteq B$ and $B \subseteq C$ and $B \neq C$, then $A \neq C$.

66. If $A \neq B$ and $B \neq C$, then $A \neq C$.

67. If $A \in B$ and $B \subseteq C$, then $A \in C$.

68. If $A \in B$ and $B \not\subseteq C$, then $A \notin C$.

69. If $A \notin B$ and $B \subseteq C$, then $A \notin C$.

70. If $A \notin B$ and $B \subseteq C$, then $A \in C$.

71. If $A \subseteq B$ and $B \not\subseteq C$, then $A \not\subseteq C$.

72. If $A = B$, then $A \subseteq B$.

LANGUAGE EXERCISE

73. In Pascal, $A \subseteq B$ is written as A <= B and $A \neq B$ as A < > B. Suppose WEEKDAYS, WEEKEND, ALLDAYS, ACLASSDAYS, and BCLASSDAYS, each is declared to be a SET OF DAYS. Let
$$\text{DAYS} = (\text{SUN, MON, TUE, WED, THU, FRI, SAT})$$
Let PLAYDAY be a variable which has as its value one of the DAYS of the week. Suppose
$$\text{ACLASSDAYS} := [\text{MON, WED, FRI}]$$
and
$$\text{BCLASSDAYS} := [\text{TUE, THU}].$$
Indicate whether each of the following conditions is TRUE or FALSE.

a. ACLASSDAYS <= WEEKDAYS

b. MON IN BCLASSDAYS

c. [ ] IN WEEKEND

d. [ ] <= WEEKDAYS

e. WEEKDAYS = BCLASSDAYS

f. (WEEKEND <= ALLDAYS) AND (WEEKEND < > ALLDAYS)

g. (PLAYDAY IN WEEKDAYS) OR (PLAYDAY IN WEEKEND)   □

Operations on Sets

Just as you can perform operations on numbers, such as addition, subtraction, and multiplication, you can also perform operations on sets. In fact, we will see that operations between two sets are as much a part of many computing environments as operations on numbers.

One way of combining two sets is to put all the elements together to make a new set. For example, suppose there is an outside speaker coming to your school, and all the mathematics and computer science classes are required to attend. The set of names of those to attend would come from a combination of the rolls of the mathematics and computer science classes. A number of students may be taking both classes, but, of course, they can only attend the talk once. This new set is the union of the two original sets.

EXAMPLE 1

Let $A = \{1, 2, 3\}$ and $B = \{2, 4, 6, 8\}$. The union of A and B, $A \cup B$, is $\{1, 2, 3, 4, 6, 8\}$. ∎

DEFINITION

Let A and B be sets. The **union** of A and B is

$$A \cup B = \{x \mid x \in A \text{ or } x \in B\}.$$

In this definition, "or" means x belongs to either A or B or both.

A **Venn diagram** gives a pictorial representation of sets. The interior of the large rectangle represents the universal set, while the interiors of circles and other closed figures stand for subsets. Shading is used to emphasize particular sets. In Figure 2.1, $A \cup B$ is shaded for sets A and B that have elements in common.

Figure 2.1

Venn diagram of A union B.
Shaded area indicates $A \cup B$

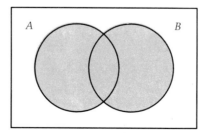

EXAMPLE 2

LANGUAGE EXAMPLE

In Pascal, + is the notation for union in the context of sets. Thus, if WEEK is also a SET OF DAYS,

WEEK := WEEKEND + WEEKDAYS

assigns to WEEK the value [SAT, SUN, MON, TUE, WED, THU, FRI]. (See Section 2.1 for variable definitions.) □ ■

If only students taking both mathematics and computer science were eligible for a scholarship, the set of candidates would be the intersection of the set of mathematics students and the set of computer science students (see Figure 2.2).

Figure 2.2

Venn diagram of the intersection of A and B. $A \cap B$ is the darker shaded portion.

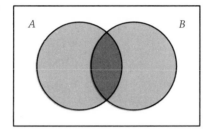

EXAMPLE 3

Let $A = \{1, 2, 3\}$ and $B = \{2, 4, 6, 8\}$. The intersection of A and B, $A \cap B$, is $\{2\}$. ■

DEFINITION

Let A and B be sets. The **intersection** of A and B is

$$A \cap B = \{x \,|\, x \in A \text{ and } x \in B\}.$$

EXAMPLE 4

LANGUAGE EXAMPLE

When using sets in Pascal, * is the notation for intersection. Consequently, the expression

WEEKEND * WEEKDAYS

has the value [ ], the empty set. Since these sets have no elements in common, we say that they are disjoint (see Figure 2.3). □ ■

Figure 2.3

Venn diagram of disjoint
sets A and B

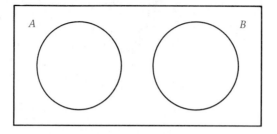

DEFINITION

Let A and B be sets. A and B are **disjoint** if $A \cap B = \phi$.

Another operation, set difference, has a similar concept and notation to that of the difference of numbers. Suppose in a mathematics class the professor wants to cover a computer-related topic for which the students need some computer science concepts. The instructor might ask the mathematics students who are not in a computer science course to attend an extra lecture on these topics. The resulting set of students is the set of mathematics students minus the set of computer science students (see Figures 2.4a and b).

Figure 2.4

(a) $A - B$ is shaded;
(b) $B - A$ is shaded

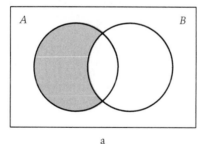
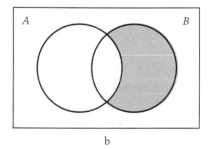

a   b

EXAMPLE 5

Let $A = \{1, 2, 3\}$ and $B = \{2, 4, 6, 8\}$. The set difference

$$A - B = \{1, 3\}$$

while

$$B - A = \{4, 6, 8\}. \quad \blacksquare$$

DEFINITION

Let A and B be sets. The **difference** of A and B is
$$A - B = \{x \mid x \in A \text{ and } x \notin B\}.$$

EXAMPLE 6

In Pascal if ACLASSDAYS is [MON, WED, FRI] and BCLASSDAYS is [TUE, THU], then the logical expression

   (WEEKDAYS − ACLASSDAYS) = BCLASSDAYS

is TRUE. The same notation is used in mathematics and Pascal for set difference. ☐ ■

If the difference is between the universal set U and another set A, the result is called the complement of A, written A' (see Figure 2.5). The set of all students from a university not taking a mathematics class is the complement of the set of all students enrolled in a mathematics class.

Figure 2.5

Venn diagram of A', the complement of A. A' is shaded

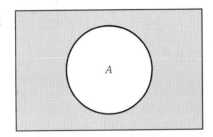

EXAMPLE 7

If the universal set is $\mathbb{N}$ and $A = \{1, 2, 3\}$, the complement of A in $\mathbb{N}$ is

   $A' = \mathbb{N} - A = \{0, 4, 5, 6, \ldots\}.$

If the universal set is $\mathbb{I}$, then

   $A' = \{\ldots, -2, -1, 0, 4, 5, 6, \ldots\}.$ ■

DEFINITION

Let A be a subset of universal set U. The **complement** of A in U is

   $A' = U - A$

or $A' = \{x \mid x \in U \text{ and } x \notin A\}.$

The union of two sets A and B is the set of elements in A or B or both. If we exclude those elements that belong to both, we have A "exclusive or" B or the symmetric difference of A and B, $A \oplus B$

(see Figure 2.6). For example, if a school advises students to take mathematics and computer science in the same semester, during registration, the registrar would want to talk to those students enrolled in one of the subjects but not both.

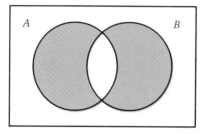

EXAMPLE 8

Let $A = \{1, 2, 3\}$ and $B = \{2, 4, 6, 8\}$. The symmetric difference, $A \oplus B$, is $\{1, 3, 4, 6, 8\}$. ■

DEFINITION

> Let A and B be sets. The **symmetric difference** of A and B is
>
> $$A \oplus B = \{x \mid x \in A \text{ or } x \in B \text{ but not both}\}.$$
>
> Thus,
>
> $$A \oplus B = (A \cup B) - (A \cap B).$$

Set operations are used in solid modeling, a graphics technique where solid objects such as cubes, spheres, cylinders, and cones are combined to form 3-D shapes as illustrated in Figure 2.7. An engineer might use the union, intersection, difference, and symmetric difference of these various primitive solids in the computer-aided design of machine parts. A terminal screen during such a design is shown in Figure 2.8. To verify that two parts fit together properly, the engineer must check that, when placed close together, the parts are still disjoint. After the design is complete, by taking the complement of each part in a cube that contains it, a mold to make a cast of the part is designed.

Figure 2.7

Primitives and set
operations. From J. W. Boyse
and J. E. Gilchrist.
"GMSolid: Interactive
Modeling for Design and
Analysis of Solids." *IEEE
Computer Graphics and
Applications*, p. 28, March
1982 (© 1982 IEEE)

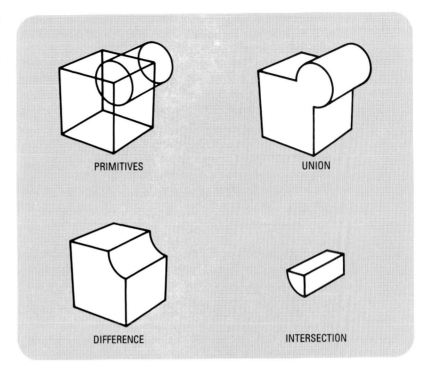

Figure 2.8

Combining solids with set
operations. From J. W. Boyse
and J. E. Gilchrist.
"GMSolid: Interactive
Modeling for Design and
Analysis of Solids." *IEEE
Computer Graphics and
Applications*, p. 36, March
1982 (© 1982 IEEE)

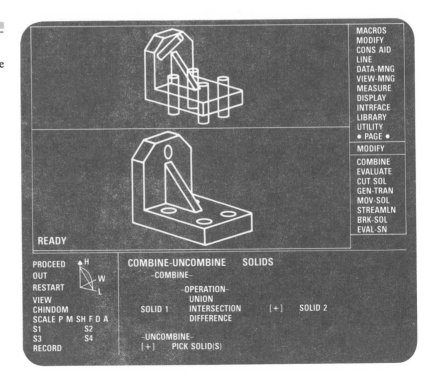

EXERCISES 2.2

Let A = {6, 7, 8, 9}, B = {3, 6, 9}, C = {2, 3, 4, 6, 8, 10}, D = {7, 9} *with universal set* U = {1, 2, 3, . . . , 10}. *Evaluate the sets in Exercises 1–40, using symbols, such as* A, U, *or* ϕ, *if appropriate.*

1. $A \cap D$ **2.** $A \cup D$ **3.** $D - A$ **4.** $A - D$

5. $A \cap D'$ **6.** C' **7.** $C \cup C'$ **8.** $C \cap C'$

9. $A \cap B$ **10.** $B \cap A$ **11.** $A \cup B$ **12.** $B \cup A$

13. $(A \cup B) \cup C$ **14.** $A \cup (B \cup C)$ **15.** $(A \cap B) \cap C$

16. $A \cap (B \cap C)$ **17.** $A \cup (B \cap C)$ **18.** $(A \cup B) \cap (A \cup C)$

19. $A \cap (B \cup C)$ **20.** $(A \cap B) \cup (A \cap C)$

21. $(A \cap B) - (A \cup B)$ **22.** $A \cup \phi$

23. $(A \cup B)'$ **24.** $A' \cap B'$ **25.** $A \cap U$ **26.** $(A \cap B)'$

27. $A' \cup B'$ **28.** $A \cap \phi$ **29.** $A \cup U$ **30.** $C \cup U$

31. $C \cap \phi$ **32.** $A \cup A$ **33.** $A \cap A$ **34.** A''

35. $A \oplus B$ **36.** $(A \cup B) - (A \cap B)$

37. $(A \cap B') \cup (A' \cap B)$ **38.** $(A - B) \cup (B - A)$

39. $A \oplus D$ **40.** $C \oplus D$

Let the universal set be ℝ. *Find the sets in Exercises 41–46.*

41. $\mathbb{Q}'$ **42.** $\mathbb{N}'$ **43.** ϕ' **44.** $\{0\}'$ **45.** $\mathbb{Q} \cup \mathbb{R}$ **46.** $\mathbb{Q} \oplus \mathbb{R}$

Let A = {x | x *is an odd integer*}, B = {y | y = 10x, x ∈ 𝕀}, C = {z | 5 ≤ z ≤ 11, z ∈ 𝕀}, D = {i | −2 ≤ i ≤ 8, i ∈ 𝕀} *with universal set* 𝕀. *Evaluate the sets in Exercises 47–60.*

47. $A' \cap C$ **48.** $C - A$ **49.** $D \cap C$

50. $A - C$ written without set-builder notation.

51. $C \cup D$ **52.** A' **53.** $B \cap C$

54. $𝕀'_c$, where MAXINT = 32,767 and the smallest integer in $𝕀_c$ is −32,768.

55. $C - B$ **56.** $A \cap B$ **57.** $A - B$

58. $A' \cap D$ **59.** $C \oplus D$ **60.** $B \oplus C$

Suppose A, B, C, *and* D *are subsets of the universal set* U *such that* C *and* D *are disjoint and* A *is a subset of* B. *Simplify each set in Exercises 61–74.*

Note: *Reference to particular examples in Exercises 1–40 may be helpful. Observations of the specific can lead to ideas about the general.*

61. $A \cup \phi$ **62.** $A \cap \phi$ **63.** $A \cup B$ **64.** $A \cap B$ **65.** $A \cup U$

66. $A \cap U$ **67.** $A - A$ **68.** $U - A$ **69.** $A \cup A'$ **70.** $A \cap A'$

71. $C \cap D$ **72.** $C - D$ **73.** $A \oplus B$ **74.** $C \oplus D$

75. Give a counterexample to the following statement:

Let A, B, and C be sets with $A \neq \phi$.

If $A \cap B = A \cap C$, then $B = C$.

76. Give a counterexample to the following statement:

If $A \cup B = A \cup C$, then $B = C$.

Note: Based on Exercises 75 and 76 we see that the cancellation law does not hold in the algebra of sets. This law does hold for multiplication and addition in the real number system where

if $a \neq 0$ and $ax = ay$, then $x = y$;

and if $x + a = y + a$, then $x = y$

77. Write all possible subset relationships for the following where A and B are any sets:

A, $A \cap B$, $A - B$, $B - A$, A', $A \oplus B$.

78. Give the set in terms of A, B, and C illustrated by the Venn diagram in the Figure #78 below.

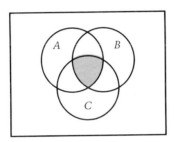

79. Repeat Exercise 78 using the Figure #79 below.

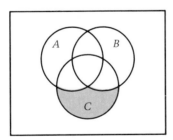

80. Repeat Exercise 78 using the Figure #80 below.

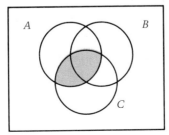

Suppose A, B, and C are distinct sets where any two of them, and in fact, all three of them have elements in common. Draw a Venn diagram to represent each of the sets in Exercises 81–87.

81. $A \cap B'$ **82.** $(A \cup B) - (A \cap B)$ **83.** $A \cap (B \cup C)$

84. $(A \cap B) \cup (A \cap C)$ **85.** $(A \cup B)'$ **86.** $A' \cap B'$

87. $(A \oplus B) \oplus C$

88. Recall in Pascal that +, *, and − stand for set union, intersection, and difference, respectively. Let MIDWEEK be a SET OF DAYS with value [TUE, WED, THU]. Give values for the following expressions in terms of WEEKDAYS, WEEKEND, ALLDAYS, ACLASSDAYS, BCLASSDAYS, [ ], and other sets. (Refer to Exercise 73 of Section 2.1.)

    a. ACLASSDAYS + MIDWEEK

    b. ACLASSDAYS * MIDWEEK

    c. ACLASSDAYS − MIDWEEK

    d. (ALLDAYS − MIDWEEK) * BCLASSDAYS

    e. ALLDAYS − (MIDWEEK * BCLASSDAYS) ☐

2.3

Algebra of Sets

In this section we take time to reflect on properties of set operations already observed in specific examples. As will be seen in a later chapter, a collection of all the subsets of a universal set along with union, intersection, complement, and certain properties forms a Boolean algebra, a structure fundamental to computer science.

The power set of U, $\mathcal{P}(U)$, is the special name given to the set of subsets of the set U.

EXAMPLE 1

For $\mathbb{B} = \{0, 1\}$, find $\mathcal{P}(\mathbb{B})$.

Solution

$\mathcal{P}(\mathbb{B}) = \{\phi, \{0\}, \{1\}, \mathbb{B}\}$. ■

EXAMPLE 2

For $U = \{a, b, c\}$, find $\mathcal{P}(U)$.

Solution

$\mathcal{P}(U) = \{\phi, \{a\}, \{b\}, \{c\}, \{a, b\}, \{a, c\}, \{b, c\}, U\}$. ■

As in these examples, the empty and universal sets are always in the power set.

DEFINITION

The collection of all subsets of a set U is called the **power set** of U, written $\mathcal{P}(U)$.

Notice that if the number of elements of U is $n(U) = 2$, then $n(\mathcal{P}(U)) = 4 = 2^2$; and for $n(U) = 3$, $n(\mathcal{P}(U)) = 8 = 2^3$. In general, if U has n elements, $\mathcal{P}(U)$ has 2^n elements. The reason the number of elements in the power set is a power of 2 is based on the Fundamental Counting Principle which is covered in Section 3.1.

THEOREM

Let U be a set with n elements. The power set of U has 2^n elements.

Now, taking a universal set U with operations of union, intersection, and complement, the properties in the box below are true for all $A, B, C \in \mathcal{P}(U).$*

Properties that hold In the algebra of sets. A, B, and C are any subsets of the universal set U.

1. **Commutative** properties for union and intersection
   a. $A \cup B = B \cup A$ b. $A \cap B = B \cap A$

2. **Associative** properties for union and intersection
   a. $A \cup (B \cup C) = (A \cup B) \cup C$
   b. $A \cap (B \cap C) = (A \cap B) \cap C$

3. **Distributive** properties
   a. $A \cup (B \cap C) = (A \cup B) \cap (A \cup C)$
   b. $A \cap (B \cup C) = (A \cap B) \cup (A \cap C)$

4. **Identities** for union (ϕ) and intersection (U)
   a. $A \cup \phi = A$ b. $A \cap U = A$

5. **Complement**
   a. $A \cup A' = U$ b. $A \cap A' = \phi$

---

* For particular examples of each property refer to the exercises in Section 2.2.

Property	Exercises	Property	Exercises
1a	11 and 12	4a	22
1b	9 and 10	4b	25
2a	13 and 14	5a	7
2b	15 and 16	5b	8
3a	17 and 18		
3b	19 and 20		

Properties, however, are not justified by examples or even Venn diagrams. Pictures and specific examples only suggest general properties and aid in understanding them. As presented in the rephrased definitions of Section 2.1, to show two sets are equal using definitions, follow the hierarchy chart in Figure 2.9.

Figure 2.9

Hierarchy chart for proving equality of sets. The open arrow "⇒" means "implies"

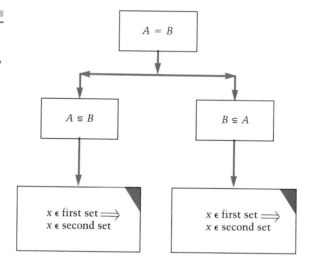

The shaded corner in the lower boxes indicates that the module occurs more than once in the chart. Recall also from the definitions that union, ∪, means "or" and intersection, ∩, means "and."

Devising a proof is very much like designing a computer program. For a proof you need to know what you are assuming, which is similar to the input, and what you want to show, much like the output. Just as a program should be designed from the top down, so should a proof. Keep breaking the problem down into smaller and smaller parts, asking at each step, "What does this mean? What is another way to express this?" In a computer program at the statement level you must make sure you are using legal coding for the language. In mathematical proofs you must also use only what has been given or derived previously. Consequently, list your reasons at each step.

Mathematical techniques can be used to prove that a program is correct. John W. Verity points out in "Bridging the Software Gap" that while software costs were about one-tenth of hardware costs in 1955, today the situation is reversed. He writes that a body of prominent computer scientists feels that application of the mathematical method to programming, including proof of program correctness, will improve the productivity of programmers and the dependability of the software they produce.

EXAMPLE 3

Prove 4b: $A \cap U = A$.

Proof

Following the hierarchy chart, we prove the set on each side is a subset of the other.

1. Show $A \cap U \subseteq A$.
   Take any $x \in A \cap U$. Show that this assumption implies $x \in A$.

   $x \in A \cap U \Rightarrow x \in A$ and $x \in U$ (definition of $\cap$)

           $\Rightarrow x \in A$ (definition of "and").

2. Show $A \subseteq A \cap U$.
   Take any $x \in A$. Show $x \in A \cap U$.

   $x \in A \Rightarrow x \in U$ (The universal set U contains
                                     all elements under discussion)

        $\Rightarrow x \in A$ and $x \in U$ ($x \in A$, given)

        $\Rightarrow x \in A \cap U$ (definition of $\cap$). ▪

EXAMPLE 4

Prove that if $A \subseteq B$, then $A \cup B = B$.

Proof

Here, we have some extra information with which to work, $A \subseteq B$. When proving equality from definitions, we will always follow the hierarchy chart. In the justifications for the steps, "def." is used as an abbreviation for definition.

1. Show $A \cup B \subseteq B$.
   Take $x \in A \cup B$. Then $x \in A$ or $x \in B$ def. $\cup$.
   If $x \in B$, we are through.
   If $x \in A$, then $x \in B$ $A \subseteq B$, def. $\subseteq$.
   Thus, in either case $x \in B$.

2. Show $B \subseteq A \cup B$.

   $x \in B \Rightarrow x \in A$ or $x \in B$ def. "or"

           $\Rightarrow x \in A \cup B$ def. $\cup$. ▪

    The properties on the left and right in the box on page 47 are called **duals** of each other. One is formed from the other by exchanging union and intersection as well as the universal and empty sets. If a statement is true, its dual must also hold. To prove the dual statement take the proof of the original statement and simply interchange each step of justification with the dual property of the step.

EXAMPLE 5

Write the dual of $A \cap (B \cup \phi) = A \cap B$.

Solution

$A \cup (B \cap U) = A \cup B$. ■

Other properties that also hold in the algebra of sets are listed in the box below.

Properties that hold in the algebra of sets. A and B are any subsets of the universal set U.

6. **Idempotent** properties
   - a. $A \cup A = A$ b. $A \cap A = A$

7. **DeMorgan's laws**
   - a. $(A \cup B)' = A' \cap B'$ b. $(A \cap B)' = A' \cup B'$

8. **Intersection with ϕ; union with U**
   - a. $A \cap \phi = \phi$ b. $A \cup U = U$

9. **Double complement**
   - a. $A'' = A$

Look at Section 2.2 (Exercises 32, 33, 23 and 24, 26 and 27, 28, 29, 34) for numeric examples of the above. Using these properties we can prove other statements without resorting to definitions.

EXAMPLE 6

Prove $A \cap B = A \cap (A' \cup B)$.

Solution

Start with the more involved side, and with a series of justified statements arrive at the other side.

$$A \cap (A' \cup B) = (A \cap A') \cup (A \cap B) \qquad \text{distribution}$$
$$= \phi \cup (A \cap B) \qquad \text{complement}$$
$$= A \cap B \qquad \text{identity.} ■$$

HISTORICAL NOTE

Not only are proof techniques applied to show that programs are correct, but computers can be programmed to prove some simple theorems. In the mid 1960s a geometry theorem-proving program was instructed to prove the following: If two sides of a triangle are equal, then the triangle is isosceles, i.e., the angles opposite the equal sides are equal. Surprisingly, the computer came up with an elegant, one-line proof unknown to the program's designer (see Figure 2.10 for more detail).   □

Figure 2.10

Proof of if $AB = AC$ in triangle ABC ($\triangle ABC$), then angles B and C are equal. ($\angle B = \angle C$)

Proof.
Since triangles $\triangle ABC$ and $\triangle ACB$ are similar,
$\angle B = \angle C$

EXERCISES 2.3

1. Find $\mathscr{P}(U)$ for $U = \{1, 2, 3, 4\}$.

2. Find the number of elements in $\mathscr{P}(U)$ if U has the following number of elements:

   a. 5   **b.** 8   **c.** 10   **d.** 32

Let A and B be sets. Fill in the blanks for the steps or the reasons to prove each statement in Exercises 3–5. The statements are proved entirely from definitions instead of from the properties of the algebra of sets.

3. Prove $A \cup \phi = A$.

Proof

1. Show $A \cup \phi \subseteq A$.
   Show $x \in A \cup \phi \Rightarrow x \in A$:

   $x \in A \cup \phi \Rightarrow x \in A$ __a__ $x \in \phi$ def. $\cup$

   $\Rightarrow$ __b__ ϕ has no elements, so x must belong to A

2. Show __c__ $\subseteq$ __d__ .
   Show __e__ $\Rightarrow$ __f__ :

   $x \in A$ $\Rightarrow$ __g__   def. "or"

   $\Rightarrow x \in A \cup \phi$ def. __h__

4. Prove $A \cup A' = U$.

Proof

1. Show $A \cup A' \subseteq U$.
   Show $x \in A \cup A' \Rightarrow$ __a__ :

   $x \in A \cup A' \Rightarrow$ __b__   def. U

2. Show __c__ .
   Show $x \in U \Rightarrow x \in A \cup A'$:

   $x \in U \Rightarrow x \in A$ or $x \notin A$ __d__

   $\Rightarrow x \in A$ or __e__   def. A'

   $\Rightarrow$ __f__   __g__

5. If A and B are disjoint, then $A \cup B = A \oplus B$.

Proof

    1. Show $A \cup B \subseteq A \oplus B$.

        Show ___**a**___ :

            $x \in A \cup B \Rightarrow x \in A$ or $x \in B$ or both.   ___**b**___

            But x cannot belong to both A and B.   ___**c**___

            Thus, $x \in$ ___**d**___   ___**e**___

    2. Show ___**f**___ $\subseteq$ ___**g**___ .

        Show ___**h**___ $\Rightarrow$ ___**i**___ :

            $x \in A \oplus B \Rightarrow x \in A$ or $x \in B$ but not both   ___**j**___

                      $\Rightarrow x \in A \cup B$ ___**k**___

Let A and B be sets. Prove the statements in Exercises 6–15 using only the definitions and not properties.

6. $A \subseteq A \cup B$ **7.** $A \cap B \subseteq A$

8. $A \cup U = U$ **9.** $A - B = A \cap B'$

10. $A \oplus B \subseteq A \cup B$ **11.** $A - B = A - (A \cap B)$

12. If $A \subseteq B$, then $A \cap B = A$.

13. If $A \subseteq B$ and $B \subseteq C$, then $A \subseteq C$.

14. If A and B are disjoint, then $A - B = A$.

15. If $A \subseteq B$, then $A \oplus B = B - A$.

Let A, B, and C be sets. Fill in the blanks for the steps or the reasons to prove each statement in Exercises 16–18. The statements are proved using the properties of the algebra of sets.

16. Prove $A \cup (B \cap A')' = A \cup B'$.

Proof

$$
\begin{aligned}
A \cup (B \cap A')' &= A \cup (\ \underline{\quad a \quad}\ ) &&\text{DeMorgan} \\
&= A \cup (B' \cup A) &&\underline{\quad b \quad} \\
&= A \cup (A \cup B') &&\underline{\quad c \quad} \\
&= (\underline{\quad d \quad}) \cup B' &&\text{associative} \\
&= A \cup B' &&\underline{\quad e \quad}
\end{aligned}
$$

17. Prove $(A \cap B) \cup (A \cap B') = A$.

Proof

$$
\begin{aligned}
(A \cap B) \cup (A \cap B') &= A \cap (B \cup B') &&\underline{\quad a \quad} \\
&= A \cap \underline{\quad b \quad} &&\text{complement} \\
&= A &&\underline{\quad c \quad}
\end{aligned}
$$

18. Prove $A - (A - B) = A \cap B$.

Proof
$$A - (A - B) = A - (A \cap B') \qquad\qquad \text{Exercise 9}$$
$$= A \cap \underline{\quad\textbf{a}\quad} \qquad\qquad\qquad \text{Exercise 9}$$
$$= A \cap (A' \cup B'') \qquad\qquad\qquad \underline{\quad\textbf{b}\quad}$$
$$= A \cap (A' \cup \underline{\quad\textbf{c}\quad}) \qquad\qquad \underline{\quad\textbf{d}\quad}$$
$$= (\underline{\quad\textbf{e}\quad} A') \cup (\underline{\quad\textbf{f}\quad} B) \qquad \text{distributive}$$
$$= \underline{\quad\textbf{g}\quad} \cup (A \cap B) \qquad\qquad \text{complement}$$
$$= A \cap B \qquad\qquad\qquad\qquad \underline{\quad\textbf{h}\quad}$$

Let A, B, *and* C *be sets. Prove the statements in Exercises 19–29 using any of the nine properties of the algebra of sets listed in this section.*

19. $(A' \cup B')' = A \cap B$ **20.** $(A \cup B) \cup B = A \cup B$

21. $(A \cup \phi) \cap (A \cup B) = A$ **22.** $A = (A \cap B) \cup (A \cap B')$

23. $U = A \cup (A \cap B)'$ **24.** $A \cap (B \cup A') = A \cap B$

25. $((A \cap B)' \cup B)' = \phi$ **26.** $(A \cap B')' \cup B = A' \cup B$

27. $A - (B \cap C) = (A - B) \cup (A - C)$

28. $(((A \cup B) \cap C)' \cup B')' = B \cap C$

29. $A \cap B' = (A' \cup B)'$

30. Prove the identity for union is unique in an algebra of sets.
    Note: Property 4a says there exists at least one identity for union; it doesn't explicitly say there is exactly one. To show there is only one identity for union, assume there are two different ones, ϕ and Z. Then show $\phi = Z$ to prove we really could not have two different identities. An identity for union must satisfy the identity property 4a, so
$$X \cup \phi = X \quad \text{and} \quad X \cup Z = X \qquad \text{for any set } X.$$
    Since these are true for any X, substitute $X = \phi$ in the second equation and $X = Z$ in the first. What do you have? Why are the two expressions equal?

31. With a similar proof to that in Exercise 30, prove the identity for intersection is unique.

2.4

Algebra of Propositions

In the last section we studied the algebra of sets. For a universal set U with n elements it was noted that the collection of all subsets of U, the power set $\mathcal{P}(U)$, always has 2^n elements. This set $\mathcal{P}(U)$ along with the operations of union, intersection, complementation, and the empty and universal sets was shown to satisfy ten basic properties. A number of other properties also were derived from these.

The algebra of sets is but one example of a general structure, called a Boolean algebra, which is covered in Chapter 9. Still another example of a Boolean algebra is the algebra of propositions. Though the algebra of sets and the algebra of propositions have vastly different applications and even different notations and terminology, they behave in exactly the same way. The study of the algebra of propositions is called **propositional calculus** or **logic.** Logic provides the basis for reasoning; it is fundamental to how we write computer programs, how we build computers, and how we develop mathematics.

An element of the algebra of propositions is a proposition or statement that can be determined to be true or false. For example, in a computer program you might have the instruction

    IF   (X > 3)   THEN. . .

A condition having the value TRUE or FALSE, such as (X > 3), is called a **logical** or **Boolean expression** in a programming context. Often these expressions are combined using the logical operators, OR, AND, NOT, to form compound statements that themselves are TRUE or FALSE. For instance, in

    IF   (X > 3)   AND   (Y = 7)   THEN. . .

the instruction after THEN will be executed only in the situation where the expression "(X > 3) AND (Y = 7)" has the value TRUE. The condition (X > 3) is either TRUE (T) or FALSE (F) as is (Y = 7). If we were to test the flow of the program thoroughly through this segment, we would need to check four possibilities—for (X > 3) being F with (Y = 7) being F or T, and for (X > 3) being T with again (Y = 7) being F or T. A diagram called a **tree** is often helpful in picturing the possible outcomes. Notice in Figure 2.11 that there are four terminating branches in the tree.

Figure 2.11 ▨▨▨▨▨▨▨▨▨▨

Tree of possible T–F values
for two statements

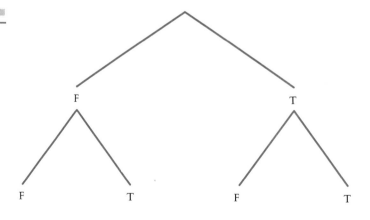

These same possibilities can be enumerated in another type of diagram called a truth table. A partial truth table for the expression "$(X > 3)$ AND $(Y = 7)$" is provided in Table 2.1.

TABLE 2.1
Partial truth table for
$(X > 3)$ AND $(Y = 7)$

$X > 3$	$Y = 7$
F	F
F	T
T	F
T	T

Information in a truth table can be more extensive, however, giving the resulting T–F values for the compound statement. For instance, the only way the expression "$(X > 3)$ AND $(Y = 7)$" can be true is for each of the component statements, $(X > 3)$ and $(Y = 7)$, to be true. As you can see in Table 2.2 a **truth table** provides a scheme for displaying all possible T–F values for conditions and the resulting values for the compound statement.

TABLE 2.2
Truth table for $(X > 3)$ AND $(Y = 7)$

$X > 3$	$Y = 7$	$(X > 3)$ AND $(Y = 7)$
F	F	F
F	T	F
T	F	F
T	T	T

Boolean expressions may also be connected using the logical operators of OR and NOT. The compound statement

$(X > 3)$ OR $(Y = 7)$

is true as long as at least one of the component statements is true. Only when both $(X > 3)$ and $(Y = 7)$ are false is the compound proposition false. The statement

NOT $(Y = 7)$

negates the value of $(Y = 7)$. If $(Y = 7)$ is TRUE, NOT $(Y = 7)$ is FALSE and vice versa. ☐ The truth tables for the three basic Boolean operations can be seen in Table 2.3 where A and B are any Boolean expressions.

TABLE 2.3
Truth tables for OR, AND, and NOT

A	B	A OR B		A	B	A AND B
F	F	F		F	F	F
F	T	T		F	T	F
T	F	T		T	F	F
T	T	T		T	T	T

A	NOT A
F	T
T	F

Suppose there are three conditions, A, B, and C, instead of two, linked together with logical operators. How many rows are necessary then? There are three statements, each having two possible values (T or F). As pictured in the tree and partial truth table of Figure 2.12 and Table 2.4, respectively, $2^3 = 8$ situations must be considered. To be consistent we always consider the situation of all F's in the first row of the truth table. In listing the other possibilities, we have the value of the right condition change the fastest while the value of the left one varies the slowest.

Figure 2.12

Tree of all possible values for statements A, B, C

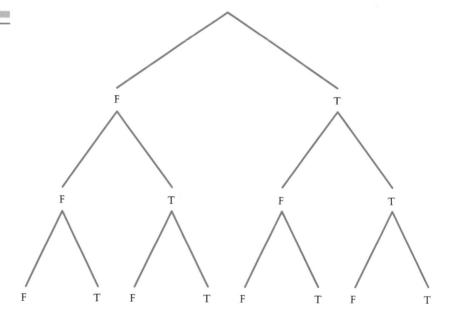

TABLE 2.4
Truth table of all possible values
for statements A, B, C

A	B	C
F	F	F
F	F	T
F	T	F
F	T	T
T	F	F
T	F	T
T	T	F
T	T	T

Recall that in the algebra of sets for a universal set with three elements, the power set has $2^3 = 8$ elements. Moreover, just as for $n(U) = n$, $n(\mathscr{P}(U)) = 2^n$, n Boolean expressions require 2^n rows in the truth table. Justification for both these facts is presented in the next chapter in a discussion of the Fundamental Counting Principle.

In the language of logic, statements are often represented by the small letters p, q, r. Where we use the connectors "or," "and," and "not" in English or their equivalent in a computer language, the notation for these **logical connectives** is $\vee$, $\wedge$, $\sim$, respectively, though the meaning is precisely as interpreted above.

DEFINITION

A **statement** or **proposition** is a declarative sentence that is either true or false.

DEFINITION

The following are some of the **logical connectives:**

    disjunction   denoted $\vee$: $p \vee q$ means "p or q."

    conjunction denoted $\wedge$: $p \wedge q$ means "p and q."

    negation   denoted $\sim$: $\sim p$ means "not p."

DEFINITION

A **compound statement** or **compound proposition** is a statement or statements combined with logical connectives to form a sentence which is either true or false.

EXAMPLE 1

The following are examples of statements: p: "George Boole developed an algebraic foundation for logic." q: "The electronic computer was invented in 1500 A.D." Both sentences are statements though the second is clearly false. Express the compound statement $p \wedge \sim q$ using the above sentences.

Solution

"George Boole developed an algebraic foundation for logic, and the electronic computer was not invented in 1500 A.D." ∎

Once again we will see that truth tables provide a concise way of indicating under what conditions a compound statement is true. Using Table 2.5 we note that the basic truth tables for conjunction, disjunction, and negation are identical to those for OR, AND, and NOT, respectively.

TABLE 2.5
Truth tables for $p \vee q$, $p \wedge q$, and $\sim p$

p	q	$p \vee q$		p	q	$p \wedge q$		p	$\sim p$
F	F	F		F	F	F		F	T
F	T	T		F	T	F		T	F
T	F	T		T	F	F			
T	T	T		T	T	T			

EXAMPLE 2

Draw the truth table for $p \wedge \sim q$. Under what condition(s) is the proposition $p \wedge \sim q$ true? Relate the answer to Example 1.

Solution

To help us complete the table we have a column for the intermediate result $\sim q$ as well.

p	q	$\sim q$	$p \wedge \sim q$
F	F	T	F
F	T	F	F
T	F	T	T
T	T	F	F

The proposition $p \wedge \sim q$ is true only when p is true (T) and q is false (F). The sentence of Example 1 represented by $p \wedge \sim q$ is true because statement p, "George Boole developed an algebraic foundation for logic," is true (T) and sentence q, "The electronic computer was invented in 1500 A.D.," is false (F). ∎

EXAMPLE 3

Find the truth table for $\sim(\sim p \lor q)$, and compose an English sentence using the corresponding statements from Example 1.

Solution

p	q	$\sim p$	$\sim p \lor q$	$\sim(\sim p \lor q)$
F	F	T	T	F
F	T	T	T	F
T	F	F	F	T
T	T	F	T	F

The compound statement derived from Example 1 is as follows: "It is not true that George Boole did not develop an algebraic foundation for logic and/or the electronic computer was invented in 1500 A.D." To emphasize that $\lor$ indicates one or the other or both statements are true, we sometimes use "and/or" instead of "or" for the English equivalent of $\lor$. ∎

Notice that $p \land \sim q$ of Example 2 and $\sim(\sim p \lor q)$ of Example 3 have the same values in their truth tables. Hence, the two statements are equivalent or mean the same thing. We write

$$p \land \sim q \equiv \sim(\sim p \lor q).$$

DEFINITION

Two propositions p and q are **logically equivalent** or **equivalent** or **equal,** written $p \equiv q$, provided they have identical values in their truth tables.

Actually, in the algebra of sets you established the equality $A \cap B' = (A' \cup B)'$ in Exercise 29 of Section 2.3. With a change of notation of $\cap$ for $\land$, $\cup$ for $\lor$, $'$ for $\sim$, $=$ for $\equiv$, and a change of letters, we arrive at a statement for sets that behaves the same as the statement $p \land \sim q \equiv \sim(\sim p \lor q)$ in logic. In fact, the basic properties of the algebra of propositions as listed in the box below are just a restatement of the definition of the algebra of sets in the notation of logic. Notice that F, a statement that is always FALSE, takes the place of ϕ and T, a statement that is always TRUE, replaces U as the identities in this switch to the new definition.

For all statements *p*, *q*, and *r* the following properties are true:

1. **Commutative** properties for disjunction and conjunction
   a. $p \lor q \equiv q \lor p$ b. $p \land q \equiv q \land p$

2. **Associative** properties for disjunction and conjunction
   a. $p \lor (q \lor r) \equiv (p \lor q) \lor r$
   b. $p \land (q \land r) \equiv (p \land q) \land r$

3. **Distributive** properties
   a. $p \lor (q \land r) \equiv (p \lor q) \land (p \lor r)$
   b. $p \land (q \lor r) \equiv (p \land q) \lor (p \land r)$

4. **Identities** for disjunction (F) and conjunction (T)
   a. $p \lor F \equiv p$ b. $p \land T \equiv p$

5. **Complement**
   a. $p \lor \sim p \equiv T$ b. $p \land \sim p \equiv F$

Several other significant properties were covered in the algebra of sets as listed in the box on page 50. Each of these can be proved in one of two ways, from the definitions or from the properties. Similarly, additional properties in the algebra of propositions can be verified in one of two ways, from truth tables or from the properties. The next box displays several properties that are important to logic and that parallel those discussed in set theory.

For all statements *p*, *q*, and *r* the following properties are true:

6. **Idempotent** properties
   a. $p \lor p \equiv p$ b. $p \land p \equiv p$

7. **DeMorgan's laws**
   a. $\sim(p \lor q) \equiv \sim p \land \sim q$ b. $\sim(p \land q) \equiv \sim p \lor \sim q$

8. **Conjunction with F; disjunction with T**
   a. $p \land F \equiv F$ b. $p \lor T \equiv T$

9. **Double complement**
   a. $\sim\sim p \equiv p$

Each of these properties can be used in programming situations where an equivalent way of expressing an instruction might be more understandable and, hence, easier to maintain. For instance, employing DeMorgan's law (7a) for Boolean expressions X and Y, we see that the instruction

LANGUAGE
EXAMPLE

IF NOT (X OR Y) THEN...

is equivalent to the statement

IF (NOT X) AND (NOT Y) THEN....

In plain English you might emphatically state, "I did not lie or cheat! To repeat, I did not lie, and I did not cheat!" Notice the use of "not," "or," and "and." □

EXERCISES 2.4

Let p be "It is cloudy" and q be "It is rainy." Assume the opposite of "cloudy" is "clear." Write each of the statements of Exercises 1–6 in symbolic logic.

1. It is cloudy and rainy.
2. It is cloudy but not rainy.
3. It is neither cloudy nor rainy.
4. It is false that it is clear or rainy.
5. It is clear or cloudy.
6. It is clear, or it is cloudy and rainy.

Consider the following statements: p, I go to dinner; q, I eat spaghetti; r, I eat salad. Write English equivalents to the compound propositions in Exercises 7–13. Then notice the relationship between your two answers in each exercise, and give the name of the property that is illustrated.

7. $q \wedge r, r \wedge q$
8. $p \wedge (q \vee r), (p \wedge q) \vee (p \wedge r)$
9. $p \wedge \sim p$. Is this situation possible?
10. $p \vee \sim p$. Can this situation ever be false?
11. $\sim(q \vee r), \sim q \wedge \sim r$
12. $\sim(q \wedge r), \sim q \vee \sim r$
13. $q \wedge q, q$

Consider the following statements: p, I like pizza. q, I dislike liver. r, I like chicken. Write English equivalents to the compound propositions in Exercises 14–17.

14. $\sim(p \vee r)$
15. $\sim p \wedge \sim r$
16. $p \wedge \sim q \wedge r$
17. $r \vee (\sim p \wedge q)$

Negate the statements in Exercises 18–21, then by DeMorgan's laws simplify each using the symbols $<, \leq, >, \geq, =, \neq$, and the connectives "or," "and."

18. $x \leq y$ and $x \leq z$
19. $x > y$ or $y = z$
20. $x \neq y$ and $x \geq y$
21. $x \neq y$ and $(y \geq z$ or $z < u)$

22. In COBOL there are no symbols or terms for $\leq$ and $\geq$. Thus, "$\leq$" is written "NOT GREATER THAN," and "$\geq$" is written "NOT LESS THAN." Using DeMorgan's laws rewrite the COBOL condition so that only one "NOT" is used. Determine if the condition is true or false for the given values of the variables.

(RATE NOT GREATER THAN 2) AND (COST NOT LESS THAN 6)

	RATE	COST		RATE	COST
a.	0	0	**b.**	0	8
c.	8	0	**d.**	8	8
e.	2	2	**f.**	2	6

23. Repeat Exercise 22 using the following condition and values of variables:

((X NOT LESS THAN 3) OR (Y NOT LESS THAN 7)) AND
(Z NOT GREATER THAN 8)

	X	Y	Z		X	Y	Z
a.	0	0	0	**b.**	5	5	5
c.	8	8	8	**d.**	9	9	9

24. Negate the following statement, then simplify the resulting sentence, omitting the word "not." Assume x and y are real numbers: "x is rational; and y is negative; but x does not exceed y." *Note:* The word "but" has the same effect as "and."

25. Prove the idempotent properties using truth tables.

26. Fill in the lettered blanks for the steps and reasons in the proof from the properties that
$$(p \lor \sim q) \equiv ((p \land q) \lor \sim q)$$
holds in the algebra of propositions.

Proof Starting with the more involved side we have

$((p \land q) \lor \sim q) \equiv \sim q \lor (p \land q)$		__a__
$\equiv$ __b__ $\land$ __c__		distributive
$\equiv (\sim q \lor p) \land T$		__d__
$\equiv$ __e__		identity
$\equiv p \lor \sim q$		__f__

27. a. Prove $p \lor T \equiv T$ from the other properties. *Note:* Use the complement property (5a) to substitute for T first.
   b. In a similar proof show that $p \land F \equiv F$ is true.

28. a. Prove $p \lor T \equiv T$ using truth tables.
   b. Prove $p \land F \equiv F$ using truth tables.

29. Prove DeMorgan's laws using truth tables.

30. Prove the double complement property using a truth table.

Using a truth table, verify each equivalence in Exercises 31–33.

31. $p \land (q \land r) \equiv (p \land q) \land r.$ associative property

32. $p \lor (q \land r) \equiv (p \lor q) \land (p \lor r)$.   distributive property

33. $p \lor (p \land q) \equiv p$.   absorption law

34. Prove the absorption law of Exercise 33 from the properties of the algebra of propositions.

   Note: You could start the proof by substituting $(p \land T)$ for p on the left side and then applying the distributive property to the resulting compound statement.

2.5

Logic

**LANGUAGE
EXAMPLES**

In this section we continue our discussion of the algebra of propositions by covering additional logical connectives and studying valid arguments.

We should, however, express a note of caution before considering any other logical connectives. When programming, check the operator precedence of the language that you are using. In the majority of languages, as in mathematics, NOT has the highest priority, followed by AND and then OR. Hence, the statement

   X OR NOT Y AND Z

written with parentheses is

   X OR ((NOT Y) AND Z)

One exception to this priority is the language Ada where "and" and "or" have the same precedence but cannot be used in the same expression without parentheses.

Languages vary, however, in whether the logic operators have higher or lower priority than the arithmetic operators, such as addition and multiplication, or the relational operators, such as < and =. Liberal use of parentheses can override precedence as well as confusion. Some of the variety in precedence is described in Table 2.6.

TABLE 2.6
Order of precedence in several languages

Ada	COBOL	FORTRAN	Pascal	PL/I
** not	not	**	not	** $\neg$
/ *	/ *	/ *	/ * and	/ *
+ −	+ −	+ −	+ − or	+ −
\<relational\>	\<relational\>	\<relational\>	\<relational\>	\<relational\>
and or	and	.NOT.		&
	or	.AND.		|
		.OR.		

   Note: In PL/I $\neg$ means "not," and $|$ means "or."

EXAMPLE 1

Which of the languages listed in Table 2.6 would require parentheses to interpret the following statement as indicated?

If $(x < (y + 5))$ or $(\text{not } (z))$ then ...

Solution

Only Pascal requires parentheses since "or" has a higher priority than $<$. □ ■

HISTORICAL
NOTE

The language Ada, mentioned in Table 2.6, like COBOL, was developed under the direction of the Department of Defense. The language derives its name from Lady Ada Augusta Lovelace, considered to be the first programmer. Born in the early 1800s, she was the daughter of the poet Lord Byron. Lovelace became fascinated with the work of Charles Babbage who conceived the idea of the analytical engine, a machine to solve general arithmetic problems. Babbage even thought of using punched cards to enter data and operations. His engine contained most of the major features of modern digital computers. And the writings of Ada Lovelace on Babbage's work reveal that she envisioned the concepts of loop and subroutine. Ada died at an early age, and Babbage's analytical engine, called "Babbage's folly" by his critics, was never completed. Unfortunately, his ideas were far beyond the technology of the times. □

We continue our discussion of logic by considering another logical connective that will be used extensively in exercises, theorems, and program examples, "if ..., then ...." The compound statement $p \Rightarrow q$, if p then q, is true in every situation except where p is true (T) and q is false (F) as indicated in the truth table of Table 2.7.

TABLE 2.7

Truth table for $p \Rightarrow q$

p	q	$p \Rightarrow q$
F	F	T
F	T	T
T	F	F
T	T	T

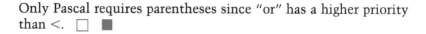

DEFINITION

The following is another logical connective:

implication denoted $\Rightarrow$:　　$p \Rightarrow q$ means "if p, then q"
　　　　　　　　　　　　　　　　or "p implies q"

For the implication $p \Rightarrow q$ to hold, when p is true, we must have q likewise being true. For instance, the compound statement, "If you are reading this sentence, then you are swimming," is false. Clearly, if the statement "you are reading this sentence" is true, an assertion about your swimming is ridiculous.

Should p be false, the value of q is irrelevant; in this situation, the statement $p \Rightarrow q$ is true regardless. Consider the proposition, "If you can memorize this book in a minute, I will give you a million dollars." This entire compound statement is true. Who can dispute it? Since obviously no one can memorize that fast, I am safe in making my offer. The compound statement is still true whether I write that million dollar check or not.

Notice that the truth tables for $p \Rightarrow q$ in Table 2.7 and for $\sim p \vee q$ of Example 3 in Section 2.4 are identical. Thus, those two propositions are equivalent, and we can write

$$(p \Rightarrow q) \equiv (\sim p \vee q).$$

In fact, all compound statements may be expressed in terms of the basic connectives $\vee$, $\wedge$, and $\sim$.

EXAMPLE 2

The double arrow $\Leftrightarrow$, indicating "if and only if," is often used to express equivalence. Draw the truth table for $p \Leftrightarrow q$, which can be interpreted as "if p, then q; and if q, then p," $(p \Rightarrow q) \wedge (q \Rightarrow p)$.

Solution

p	q	$p \Rightarrow q$	$q \Rightarrow p$	$(p \Rightarrow q) \wedge (q \Rightarrow p)$
F	F	T	T	T
F	T	T	F	F
T	F	F	T	F
T	T	T	T	T

Notice that $p \Leftrightarrow q$ is true only in the situation when p and q have the same values, that is, $p \Leftrightarrow q$ means the same as $p \equiv q$. ■

DEFINITION

The following is another logical connective:

equivalence denoted $\Leftrightarrow$ or $\equiv$: $p \Leftrightarrow q$ means "p if and only if q" or "if p, then q; and if q, then p"

We should note the order of priority is as mentioned above with implication and equivalence evaluated last:

    1. $\sim$ 2. $\wedge$ 3. $\vee$ 4. $\Rightarrow$ 5. $\Leftrightarrow$

Thus, in forming the truth table for $p \vee q \Rightarrow r \Leftrightarrow r \wedge \sim q$, the columns would be evaluated in the order dictated by the parentheses as follows:

$$((p \vee q) \Rightarrow r) \Leftrightarrow (r \wedge (\sim q))$$

EXAMPLE 3

Find the truth table for

$$p \wedge q \Rightarrow p \vee q \quad \text{and} \quad {\sim}p \wedge q \Leftrightarrow p \vee {\sim}q.$$

Solution

p	q	$p \wedge q$	$p \vee q$	$p \wedge q \Rightarrow p \vee q$
F	F	F	F	T
F	T	F	T	T
T	F	F	T	T
T	T	T	T	T

${\sim}p$	${\sim}q$	${\sim}p \wedge q$	$p \vee {\sim}q$	${\sim}p \wedge q \Leftrightarrow p \vee {\sim}q$
T	T	F	T	F
T	F	T	F	F
F	T	F	T	F
F	F	F	T	F

The first statement of Example 3, $p \wedge q \Rightarrow p \vee q$, is always true, regardless of the truth or falsity of p and q. Such a compound statement is called a tautology; it always holds; it could be stated as a theorem. The second statement, ${\sim}p \wedge q \Leftrightarrow p \vee {\sim}q$, however, is called a contradiction since it is never true.

DEFINITION

> A statement is a **tautology** if it is true for all possible cases. A statement is a **contradiction** if it is false for all possible cases.

EXAMPLE 4

Show that the empty set ϕ is a subset of any set S.

Solution

Recall, to show one set is a subset of another as in $\phi \subseteq S$, we need to show the following:

$$x \in \phi \Rightarrow x \in S.$$

We have two statements combined by the implication symbol:

$$p: \quad x \in \phi \quad \text{and} \quad q: \quad x \in S.$$

Can the statement "$x \in \phi$" ever be true? No, ϕ, being empty, has no elements. Since in this situation statement p is always false, as Table 2.7 indicates, the implication $p \Rightarrow q$ is always true. Thus, ϕ is a subset of any set S. ∎

Symbolic logic was first applied to business in 1936 by Edmund Berkeley. In examination of certain rearrangements of premium payments for insurance policy holders, he discovered that there were contradictory company rules. The maze of special situations was quite complicated, but by converting statements to symbols and applying Boolean algebra, he discovered conflicts that no one else had detected. ☐ The following example illustrates the process employed.

EXAMPLE 5

By translating the following statements into symbols and using a truth table, show the argument is valid:

The language Pascal allows recursion. Early computer languages did not allow recursion. Therefore, Pascal was not an early computer language.

Solution

The argument can be rephrased using implications as follows:

If the language is Pascal, then the language allows recursion. If the language was an early one, then the language does not allow recursion. The language is Pascal. Therefore, the language was not an early one.

Assign symbols for the statements as follows:

p: The language is Pascal.

r: The language allows recursion.

e: The language was an early one.

Thus, for the individual sentences we have,

$p \Rightarrow r$: If the language is Pascal, then the language allows recursion.

$e \Rightarrow \sim r$: If the language was an early one, then the language does not allow recursion.

p: The language is Pascal.

$\sim e$: The language was not an early one.

The argument asserts the conjunction of the first three compound statements implies the last:

$$(p \Rightarrow r) \wedge (e \Rightarrow \sim r) \wedge p \Rightarrow \sim e.$$

Since we have three basic statements, p, r, e, there are $2^3 = 8$ rows in the truth table. Notice that the last column verifies that the argument is a tautology.

p	r	e	$p \Rightarrow r$	$\sim r$	$e \Rightarrow \sim r$	$\wedge$	$\sim e$	$\Rightarrow$
F	F	F	T	T	T	F	T	T
F	F	T	T	T	T	F	F	T
F	T	F	T	F	T	F	T	T
F	T	T	T	F	F	F	F	T
T	F	F	F	T	T	F	T	T
T	F	T	F	T	T	F	F	T
T	T	F	T	F	T	T	T	T
T	T	T	T	F	F	F	F	T

The column labeled $\wedge$ stands for $(p \Rightarrow r) \wedge (e \Rightarrow \sim r) \wedge p$, while the last column, labeled $\Rightarrow$, gives the values for the implication, $(p \Rightarrow r) \wedge (e \Rightarrow \sim r) \wedge p \Rightarrow \sim e$. Often such compound statements are written with the component statements in a column and the conclusion at the bottom:

$$p \Rightarrow r$$
$$e \Rightarrow \sim r$$
$$\underline{p}$$
$$\sim e$$

Any **argument** or implication that is a tautology, is called a **valid argument.** Even if the component statements are ridiculous, tautology indicates validity. For example, the following argument would be written in symbolic notation in exactly the same way as the above:

> Pigs eat corn. Animals that roll in the mud do not eat corn. Therefore, pigs do not roll in the mud.

Though this argument is valid, it certainly is not sound. The trouble arises with the second statement about animals that roll in the mud not eating corn. It simply is not true. Faulty assumptions can lead to faulty conclusions even though the process of the argument is valid. The situation is similar in programming. The logic of your program can be perfect; but if the data is wrong, your output will be meaningless—"Garbage in, garbage out!"

EXERCISES 2.5

1. Place parentheses in "If not $x < y$" to correspond to the order of priority in each of the languages in Table 2.6.

2. Repeat Exercise 1 for "If $x + y$ and $z * w$."

Draw parentheses to indicate the order of priority in Exercises 3–5.

3. $p \Rightarrow q \wedge r$ **4.** $\sim p \vee q \wedge r \Leftrightarrow \sim\sim r$ **5.** $\sim p \Rightarrow q \Leftrightarrow p \vee q \Rightarrow r$

6. a. Write $p \Rightarrow q$ using only p, q, and the three basic logical connectives, $\vee$, $\wedge$, $\sim$.
   b. Write $p \Leftrightarrow q$ using only p, q, and the three basic logical connectives.

7. Using a truth table show that the following is a tautology:
$$((p_1 \vee p_2) \Rightarrow q) \Longleftrightarrow ((p_1 \Rightarrow q) \wedge (p_2 \Rightarrow q))$$

8. Using a truth table show that the following is a tautology:
$$((p_1 \Rightarrow p_2) \wedge (p_2 \Rightarrow q)) \Rightarrow (p_1 \Rightarrow q)$$
Note: This basic principle of logic is called the law of syllogism.

Using a truth table verify each equivalence in Exercises 9–12.

9. $(p \Rightarrow q) \equiv (\sim q \Rightarrow \sim p)$.   contrapositive

10. $(p \Rightarrow q) \equiv (p \wedge \sim q \Rightarrow F)$.   proof by contradiction

11. $((p \Rightarrow q) \wedge (p \Rightarrow r)) \equiv (p \Rightarrow (q \wedge r))$

12. $(p \wedge q) \Rightarrow r \equiv p \Rightarrow (q \Rightarrow r)$

13. Using a truth table show that each of the following is a tautology:
   a. $(p \wedge (p \Rightarrow q)) \Rightarrow q$.   *modus ponens*
   b. $p \wedge q \Rightarrow p$

Construct the truth tables for Exercises 14–17. State if each is a tautology, contradiction, or neither.

14. $(p \Rightarrow q) \Leftrightarrow (q \Rightarrow p)$ **15.** $\sim p \vee (p \wedge q) \Leftrightarrow \sim p \vee q$

16. $(p \vee q \Rightarrow r) \Leftrightarrow r \wedge \sim q$ **17.** $\sim p \Rightarrow (q \vee r)$

18. The "exclusive-or" connective $\oplus$ is defined such that $p \oplus q$ means "either p or q is true but not both."
   a. Construct the truth table for $p \oplus q$.
   b. Show $p \oplus q \equiv \sim(p \Leftrightarrow q)$.
   c. Construct a truth table for $p \oplus p$.
   d. Show $p \oplus T \equiv \sim p$.
   e. Construct the truth table for $(p \oplus q) \oplus r$.
   f. Show $p \oplus q \equiv (p \vee q) \wedge \sim(p \wedge q)$.
   Note: Compare the logic definition of $\oplus$ with the set exclusive-or definition and Figure 2.6, the exclusive-or Venn diagram, of Section 2.2.

Write each argument of Exercises 19–24 in symbolic notation and test its validity.

19. If John is married or older than 25, then he pays a lower insurance rate. If John pays a lower insurance rate, then he can afford more insurance. John cannot afford more insurance. Thus, he is not married and is not older than 25.

20. Cats can fly. Animals that can eat at restaurants do not wear shoes. Animals that fly wear shoes. Therefore, cats cannot eat at restaurants.

21. Anyone who knew how to work all the homework problems passed discrete mathematics. R. H. Smith passed discrete mathematics. Therefore, R. H. Smith knew how to work all the homework problems.

22. If a value is read for X, then X is not assigned a value. If X is nonnegative, then a value is assigned to X. If X is negative, then X is printed. A value is read for X. Therefore, X is printed.

23. I have some money if and only if I will eat out. If I have some money, then I will buy a tape. I will go to the movies or I will not eat out. Therefore, I will go to the movies or I will buy a tape.

24. If I go to the dance, then if I will buy a corsage, the corsage will be roses. I will buy a corsage. Therefore, if I go to the dance, the corsage will be roses.

25. Use the equivalence of Exercise 12,
$$(p \wedge q) \Rightarrow r \equiv p \Rightarrow (q \Rightarrow r),$$
to rephrase the statement, "If I go to the dance, then if I will buy a corsage, the corsage will be roses."
Note: A similar rephrasing could be used to simplify the following program segment with r being an instruction:

```
IF p THEN
    IF q THEN r  □
```

26. Show that $p \Rightarrow (q \Rightarrow r)$ is not equivalent to $(p \Rightarrow q) \Rightarrow r$.
Note: You are showing that the logical connective implication, $\Rightarrow$, is not associative.

27. Write the following in symbolic logic:

```
IF p THEN q ELSE r
```

28. Show that $p \Rightarrow q$ is not equivalent to $q \Rightarrow p$. *Note:* You are showing that the logical connective implication, $\Rightarrow$, is not commutative. The implication $q \Rightarrow p$ is called the **converse** of $p \Rightarrow q$.

2.6

Proof Techniques

Throughout this book we are justifying statements. In this section we consider various techniques of proof, presenting their foundation in logic and organizing the various approaches.

Most **theorems** or mathematical statements can be written in the form "$p \Rightarrow q$." For instance, in set theory the statement "$A \subseteq A \cup B$" can be rephrased as "If $x \in A$, then $x \in A \cup B$" or "$x \in A \Rightarrow x \in A \cup B$." The first proposition, "$x \in A$," is called the antecedent or hypothesis, while the second statement, "$x \in A \cup B$," is called the consequent or conclusion.

DEFINITION

In the implication $p \Rightarrow q$, p is called the **antecedent** and q is called the **consequent.** If the implication is a theorem, p is called the **hypothesis** and q is called the **conclusion.**

There are a number of ways an implication of the form "$p \Rightarrow q$" can be phrased in English and mathematics. Certainly, we need to know what we are proving before we can prove it. The following are several equivalent ways of stating an implication:

1. $p \Rightarrow q$
2. If p, then q.
3. p implies q.
4. p is a sufficient condition for q.
5. p only if q.
6. $q \Leftarrow p$.
7. q is implied by p.
8. q is a necessary condition for p.
9. q follows from p.
10. q if p.

EXAMPLE 1

Restate "If $x \in A$, then $x \in A \cup B$" using statements of type 4, 5, and 8 above.

Solution

4. $x \in A$ is a sufficient condition for $x \in A \cup B$.
5. $x \in A$ only if $x \in A \cup B$.
8. $x \in A \cup B$ is a necessary condition for $x \in A$. ∎

In proving "$x \in A \Rightarrow x \in A \cup B$" we can use a **direct method** where a sequence of implications starts at the hypothesis and leads to the conclusion. The method is based upon the **law of syllogism,** shown to be true using the truth table in Exercise 8 of the last section and in Table 2.8 below:

$$((p_1 \Rightarrow p_2) \wedge (p_2 \Rightarrow q)) \Rightarrow (p_1 \Rightarrow q).$$

Tracing the proof of the above theorem, we have

$$x \in A \Rightarrow x \in A \ \text{ or } \ x \in B \qquad \text{def. "or"}$$
$$\Rightarrow x \in A \cup B \qquad \qquad \text{def. } \cup$$

TABLE 2.8

Truth table for the law of syllogism: $((p_1 \Rightarrow p_2) \wedge (p_2 \Rightarrow q)) \Rightarrow (p_1 \Rightarrow q)$

p_1	p_2	q	$p_1 \Rightarrow p_2$	$p_2 \Rightarrow q$	$(p_1 \Rightarrow p_2) \wedge (p_2 \Rightarrow q)$	$p_1 \Rightarrow q$	$(p_1 \Rightarrow p_2) \wedge (p_2 \Rightarrow q) \Rightarrow (p_1 \Rightarrow q)$
F	F	F	T	T	T	T	T
F	F	T	T	T	T	T	T
F	T	F	T	F	F	T	T
F	T	T	T	T	T	T	T
T	F	F	F	T	F	F	T
T	F	T	F	T	F	T	T
T	T	F	T	F	F	F	T
T	T	T	T	T	T	T	T

Here, the statement symbols and implications could be assigned as

p_1: $x \in A$

p_2: $x \in A$ or $x \in B$

q: $x \in A \cup B$.

Thus,

$$p_1 \Rightarrow p_2: \quad x \in A \Rightarrow x \in A \quad \text{or} \quad x \in B$$

and $p_2 \Rightarrow q$: $x \in A$ or $x \in B \Rightarrow x \in A \cup B$

imply $p_1 \Rightarrow q$: $x \in A \Rightarrow x \in A \cup B$.

Of course, direct proofs usually involve a longer sequence of implications, but the concept is the same.

Another technique, **proof by contradiction,** was used in the uniqueness proofs of Exercises 30 and 31 in Section 2.3. The theorem in Exercise 30 of that section is, "The identity for union is unique in an algebra of sets." In the proof you assume there are two distinct identities for union, ϕ and Z, and show in fact that $\phi = Z$; two different identities for union are impossible. In symbols, we want to show "$p \Rightarrow q$," meaning, "If union takes place in an algebra of sets, then the identity for union is unique." The proof by contradiction, however, consists of using p, "union is in an algebra of sets," and $\sim q$, "the identity for union is not unique," and finding a contradiction. Symbolically, we show $p \wedge \sim q \Rightarrow F$. Table 2.9 shows the equivalence of the truth values of $p \Rightarrow q$ and $p \wedge \sim q \Rightarrow F$.

TABLE 2.9

Truth table showing the equivalence of $p \Rightarrow q$ and
$p \wedge \sim q \Rightarrow F$

p	q	$p \Rightarrow q$	$\sim q$	$p \wedge \sim q$	F	$p \wedge \sim q \Rightarrow F$
F	F	T	T	F	F	T
F	T	T	F	F	F	T
T	F	F	T	T	F	F
T	T	T	F	F	F	T

Another way to rephrase "$p \Rightarrow q$" is by its **contrapositive,** $\sim q \Rightarrow \sim p$. For instance, Example 4 of Section 2.3 states, "If $A \subseteq B$, then $A \cup B = B$." The contrapositive reads, "If $A \cup B \neq B$, then $A \not\subseteq B$." Negate and switch the hypothesis and the conclusion. The contrapositive restatement is in fact equivalent to the original implication as shown by the truth table in Table 2.10.

TABLE 2.10
Truth table showing the equivalence of $p \Rightarrow q$ and $\sim q \Rightarrow \sim p$

p	q	$p \Rightarrow q$	$\sim q$	$\sim p$	$\sim q \Rightarrow \sim p$
F	F	T	T	T	T
F	T	T	F	T	T
T	F	F	T	F	F
T	T	T	F	F	T

Sometimes the contrapositive, $\sim q \Rightarrow \sim p$, is easier to prove than the implication $p \Rightarrow q$. Proof of the contrapositive of the theorem, "If $A \subseteq B$, then $A \cup B = B$" is given in Example 2.

EXAMPLE 2

Proof

Prove if $A \cup B$ is not equal to B, then A is not a subset of B.

Suppose $A \cup B \neq B$. Now, by the definition of union $A \cup B = \{x \mid x \in A \text{ or } x \in B\}$. Since $A \cup B$ is not equal to B and $B \subseteq A \cup B$, there must be at least one element y of $A \cup B$, which is not in B. Since $y \in A \cup B$ and $y \notin B$, y must belong to A. Because there is an element y in A but not in B, A is not a subset of B. ■

Frequently, a computer problem or part of such a problem is solved by considering a number of cases or situations. Moreover, in verifying that a program is correct, you need to check that each case of input produces correct output. Similarly, in mathematics some proofs are handled by considering various cases. Suppose you are asked to simplify $\sqrt{x^2}$. The correct answer is the absolute value of x, $|x|$, not just x. For instance,

$$\sqrt{3^2} = \sqrt{9} = 3 = |3|$$
$$\text{and} \quad \sqrt{(-3)^2} = \sqrt{9} = 3 = |-3|.$$

The absolute value of x is itself defined by cases:

$$|x| = \begin{cases} x, & \text{if } x \geq 0 \\ -x, & \text{if } x < 0. \end{cases}$$

Thus, for $x = -3 < 0$, $|x| = -x = -(-3) = 3$. The proof that follows of the statement that $\sqrt{x^2} = |x|$ illustrates a **proof by cases.**

EXAMPLE 3   Prove if $x \in \mathbb{R}$, then $\sqrt{x^2} = |x|$.

Proof   **Case 1** $x \geq 0$.

Since x is nonnegative, $\sqrt{x^2} = x = |x|$.

Case 2 $x < 0$.

Since x is negative, $-x$ is the corresponding positive number. Thus, $\sqrt{x^2} = -x = |x|$. ■

In Example 3 the hypothesis "$x \in \mathbb{R}$" can be broken into the disjunction of two statements, $x \geq 0$ or $x < 0$. Let p_1 be $x \geq 0$ and p_2 be $x < 0$. Taking the conclusion to be q, $\sqrt{x^2} = |x|$, the implication can be given symbolically as $p_1 \vee p_2 \Rightarrow q$. The proof by cases consists, however, of showing "$x \geq 0 \Rightarrow \sqrt{x^2} = |x|$" and "$x < 0 \Rightarrow \sqrt{x^2} = |x|$," which in the symbolism of logic is

$$(p_1 \Rightarrow q) \wedge (p_2 \Rightarrow q)$$

Table 2.11 shows with truth tables the equivalence of $p_1 \vee p_2 \Rightarrow q$ and $(p_1 \Rightarrow q) \wedge (p_2 \Rightarrow q)$.

TABLE 2.11
Truth table showing the equivalence of $p_1 \vee p_2 \Rightarrow q$ and $(p_1 \Rightarrow q) \wedge (p_2 \Rightarrow q)$

p_1	p_2	q	$p_1 \vee p_2$	$p_1 \vee p_2 \Rightarrow q$	$p_1 \Rightarrow q$	$p_2 \Rightarrow q$	$(p_1 \Rightarrow q) \wedge (p_2 \Rightarrow q)$
F	F	F	F	T	T	T	T
F	F	T	F	T	T	T	T
F	T	F	T	F	T	F	F
F	T	T	T	T	T	T	T
T	F	F	T	F	F	T	F
T	F	T	T	T	T	T	T
T	T	F	T	F	F	F	F
T	T	T	T	T	T	T	T

A number of times in exercises you will be asked to provide a counterexample to a statement. For instance, in Exercise 75 of Section 2.2 you were asked to find sets A, B, C to show for $A \neq \phi$ the implication $A \cap B = A \cap C \Rightarrow B = C$ is false. Certainly, the implication does hold in some situations, but to prove $p \Rightarrow q$ is not a tautology, we need to find an example where p is true and q is false. By referring to the truth table of $p \Rightarrow q$, we see that $p \equiv T$ and $q \equiv F$ are the only values of p and q that yield an F for $p \Rightarrow q$. When one counterexample has been found, we are through—the implication is false. Thus, in conjecturing a theorem we either prove it is false by finding a counterexample or prove it is true.

Table 2.12 summarizes the proof techniques covered in this section. The principle of mathematical induction, mentioned in the table, is a method for proving a statement is true for all positive integers. Discussion of this method is postponed until Section 6.4 where it will be covered in depth.

TABLE 2.12
Proof techniques

Name	Prove	Method
Direct	$p \Rightarrow q$	Law of syllogism, $(p \Rightarrow p_2) \wedge (p_2 \Rightarrow p_q)$
By contradiction	$p \Rightarrow q$	$p \wedge \sim q \Rightarrow F$
By contrapositive	$p \Rightarrow q$	$\sim q \Rightarrow \sim p$
By cases	$p_1 \vee p_2 \Rightarrow q$	$(p_1 \Rightarrow q) \wedge (p_2 \Rightarrow q)$
Counterexample	$p \Rightarrow q$	Example where p is T and q is F
Mathematical induction	$P(n), n \in \mathbb{I}^+$	See Section 6.4

We can use the same techniques of theorem proving in verifying that arguments are valid. Consider Example 5 of the last section, which was written in symbolic logic as

$$(p \Rightarrow r) \wedge (e \Rightarrow \sim r) \wedge p \Rightarrow \sim e.$$

Instead of using truth tables as we did then, we can show the conjunction of the first three statements implies the conclusion using the properties of the algebra of propositions.

The property of Exercise 13, Section 2.5, called *modus ponens*, is often useful in completing a proof of validity. *Modus ponens* asserts that $(p \wedge (p \Rightarrow q)) \Rightarrow q$. In other words, should the implication $p \Rightarrow q$ hold and should the antecedent p be true, then the consequent q is also true. The truth of the implication $p \Rightarrow q$ is not enough. For instance, consider the statement used earlier as an example of an implication, "If you can memorize this book in a minute, I will give you a million dollars." Suppose the implication is true, and I have every intention of following through on the promise. Truth of the implication, however, is not enough to make me write that check. But truth of the implication *and* your doing the fast job of memorizing would make you a million dollars richer. We now return to the reality of using *modus ponens* to verify an argument.

EXAMPLE 4

Prove $((p \Rightarrow r) \wedge (e \Rightarrow {\sim}r) \wedge p) \Rightarrow {\sim}e$ using the properties of the algebra of propositions.

Proof

$(p \Rightarrow r) \wedge (e \Rightarrow {\sim}r) \wedge p$

$\equiv (p \Rightarrow r) \wedge (r \Rightarrow {\sim}e) \wedge p$ contrapositive and double complement

$\equiv ((p \Rightarrow r) \wedge (r \Rightarrow {\sim}e)) \wedge p$ associative

$\Rightarrow (p \Rightarrow {\sim}e) \wedge p$ syllogism

$\equiv p \wedge (p \Rightarrow {\sim}e)$ commutative

$\Rightarrow {\sim}e$ *modus ponens* ■

EXERCISES 2.6

For each sentence in Exercises 1–4 give the antecedent and consequent.

1. I will be thrilled if I make an "A" in this class.

2. Eating a whole pizza is a sufficient condition for my being full.

3. Let $n \in \mathbb{N}$. $n^2 \leq 2^n$ is a necessary condition for $n \geq 4$.

4. Let $n \in \mathbb{N}$. $3n$ is even only if n is even.

5. State the following implications in the ten different ways indicated at the first of this section.
$$x \text{ even} \Rightarrow x^2 \text{ even.}$$

6. Prove if x is an even integer then so is x^2.
Note: An even integer x can be written as $x = 2n$ for some integer n.

7. Prove if x is an odd integer then so is x^2.
Note: An odd integer x can be written as $x = 2n + 1$ for some integer n.

8. Two numbers are said to have the same **parity** if they are both even or both odd. List the two cases you would need to verify the following: For an integer x, x and x^2 have the same parity.

9. Consider the theorem that $\sqrt{2}$ is irrational. Supply the justification for each step of the following proof by contradiction: Assume $\sqrt{2}$ is rational $\Rightarrow \sqrt{2} = a/b$ for some integers a and b, $b > 0$, where a/b is completely reduced. (**a.** Why?)

$\Rightarrow \sqrt{2}\, b = a$ (**b.** Why?)

$\Rightarrow 2b^2 = a^2$ (**c.** Why?)

$\Rightarrow a^2$ is even   (**d.** Why?)

$\Rightarrow a$ is even   (**e.** Why?)

$\Rightarrow a = 2c$ for some integer c (**f.** Why?)

$\Rightarrow b^2 = 2c^2$ (**g.** Why?)

$\Rightarrow b^2$ is even   (**h.** Why?)

$\Rightarrow b$ is even   (**i.** Why?)

$\Rightarrow b = 2d$ for some integer d (**j.** Why?)

Since $a = 2c$ and $b = 2d$, we have a contradiction. (**k.** Why?)

State and prove the contrapositive of each statement in Exercises 10–13.

10. Let $x, y \in \mathbb{N}$. If $x + y \geq 83$, then $x \geq 41$ or $y \geq 41$.

11. Let $x, y, z \in \mathbb{N}$. If $xy = z$, then $x \leq \sqrt{z}$ or $y \leq \sqrt{z}$.

12. Let $x \in \mathbb{I}$. x being positive is a sufficient condition for $x + 1$ to be positive.

13. Let $x, y \in \mathbb{I}$. x and y are odd only if $x + y$ is even.

14. Prove the following statement is false:

$$\text{If } x \in \mathbb{N} \qquad \text{then } x^2 + x + 11 \in \mathbb{P}.$$

15. Prove the following statement is false:

$$\text{If } x \in \mathbb{P} \quad \text{and} \quad x > 3 \qquad \text{then } x^2 - 8 \in \mathbb{P}.$$

LANGUAGE EXERCISE

16. The following Pascal program segment does *not* find the minimum value in a data set of N integers. Find a counterexample.

```
MINN := 0;
FOR I := 1 TO N DO
    BEGIN
        READLN (A);
        IF A < MINN THEN MINN := A
    END  □
```

17. Consider the theorem, $|xy| = |x| \cdot |y|$.
    a. In a proof by cases we consider two possibilities for each number, the number is nonnegative or negative. How many cases must be considered in the proof?
    b. Prove the theorem.

18. Prove directly that the sum of three consecutive integers is divisible by 3.
    Note: Recall, for a number x to be divisible by 3, there must exist a $y \in \mathbb{I}$ such that $x = 3y$.

19. Prove the following by cases using definitions: For sets A, B, and C, $A \cup B \subseteq A \cup (B \oplus C) \cup C$.

20. Prove the following statement is false: The sum of two prime numbers is never a prime.

21. Prove by cases that for $x \in \mathbb{I}$, $x^2 + x$ is even.

Using symbolic notation and the properties of the algebra of propositions, not truth tables, show each argument in Exercises 22–25 is valid.

22. Use the argument in Exercise 19 of Section 2.5.

23. Use the argument in Exercise 20 of Section 2.5.

24. Use the argument in Exercise 22 of Section 2.5.

25. If X is nonnegative, then X is printed. If a value is read for X, then X is not assigned a value. If X is less than or equal to 0, then X is assigned a value. A value is read for X. Thus, X is printed.

Chapter 2   REVIEW

Terminology/Algorithms	Pages	Comments
$\in$	32	Belongs to
$n(S)$	32	Number of elements in set
Empty set, ϕ	33	Set with no elements
Universal set, U	33	Set of all elements
Subset, $A \subseteq B$	33	$x \in A \Rightarrow x \in B$
Equal, $A = B$	35	$A \subseteq B$ and $B \subseteq A$
Counterexample	37	Specific example, contradicting statement
Venn diagram	38	Pictorial representation of sets
Union, $A \cup B$	38	$\{x \mid x \in A \text{ or } x \in B \text{ or both}\}$
Intersection, $A \cap B$	39	$\{x \mid x \in A \text{ and } x \in B\}$
A and B disjoint	40	$A \cap B = \phi$
Difference $A - B$	40	$\{x \mid x \in A \text{ and } x \notin B\}$
Complement, A'	41	$U - A$
Symmetric difference, $A \oplus B$	42	$\{x \mid \text{either } x \in A \text{ or } x \in B \text{ but not both}\}$
Power set, $\mathcal{P}(U)$	47	Collection of all subsets of U
Theorem: $n(U) = n \Rightarrow n(\mathcal{P}(U)) = 2^n$	47	Example: $n(U) = 3 \Rightarrow n(\mathcal{P}(U)) = 2^3 = 8$
Dual	49	$U \leftrightarrow \phi$ and $\cup \leftrightarrow \cap$
Commutative	47	$A \cup B = B \cup A$ $A \cap B = B \cap A$
Associative	47	$A \cup (B \cup C) = (A \cup B) \cup C$ $A \cap (B \cap C) = (A \cap B) \cap C$
Distributive	47	$A \cup (B \cap C) = (A \cup B) \cap (A \cup C)$ $A \cap (B \cup C) = (A \cap B) \cup (A \cap C)$

Terminology/Algorithms	Pages	Comments
Identities, ϕ, U	47	$A \cup \phi = A,$ $\qquad A \cap U = A$
Complement	47	$A \cup A' = U,$ $\qquad A \cap A' = \phi$
Idempotent	50	$A \cup A = A,$ $\qquad A \cap A = A$
DeMorgan	50	$(A \cup B)' = A' \cap B'$ $\quad (A \cap B)' = A' \cup B'$
$\cap \phi, \cup U$	50	$A \cap \phi = \phi,$ $\qquad A \cup U = U$
Double complement, $(A')'$	50	$A'' = A$
Tree	54	Type of diagram
Truth table	55	Table of T–F values of conditions and results
Statement, proposition	57	True or false
Logical connectives	57	$\vee, \wedge, \sim, \Rightarrow, \Leftrightarrow$
Disjunction, $p \vee q$	57	p or q
Conjunction, $p \wedge q$	57	p and q
Negation, $\sim p$	57	Not p
Compound statement	57	Statement(s) combined with $\vee, \wedge, \sim, \Rightarrow, \Leftrightarrow$
Logically equivalent, equal, $p \equiv q$	59	Same truth tables
Commutative	60	$p \vee q \equiv q \vee p,$ $\;\; p \wedge q \equiv q \wedge p$
Associative	60	$p \vee (q \vee r) \equiv (p \vee q) \vee r$ $p \wedge (q \wedge r) \equiv (p \wedge q) \wedge r$
Distributive	60	$p \vee (q \wedge r) \equiv (p \vee q) \wedge (p \vee r)$ $p \wedge (q \vee r) \equiv (p \wedge q) \vee (p \wedge r)$
Identities, F and T	60	$p \vee F \equiv p,$ $\qquad p \wedge T \equiv p$
Complement	60	$p \vee \sim p \equiv T,$ $\qquad p \wedge \sim p \equiv F$
Idempotent	60	$p \vee p \equiv p,$ $\qquad p \wedge p \equiv p$

Terminology/Algorithms	Pages	Comments
DeMorgan	60	$\sim(p \vee q) \equiv \sim p \wedge \sim q, \quad \sim(p \wedge q) \equiv \sim p \vee \sim q$
$\wedge$ F, $\vee$ T	60	$p \wedge F \equiv F, \qquad\qquad p \vee T \equiv T$
Double complement, $\sim(\sim p)$	60	$\sim\sim p \equiv p$
Order of precedence	63	Varies with language
Implication, $p \Rightarrow q$	64	If p, then q
Equivalence, $p \Leftrightarrow q$	65	p if and only if q
Tautology	66	All truth values T
Contradiction	66	All truth values F
Argument	68	Implication, $p \Rightarrow q$
Valid argument	68	Implication is a tautology
Law of syllogism	69	$((p_1 \Rightarrow p_2) \wedge (p_2 \Rightarrow q)) \Rightarrow (p_1 \Rightarrow q)$
Theorem	70	Mathematical statement
Antecedent in $p \Rightarrow q$	71	p
Consequent in $p \Rightarrow q$	71	q
Hypothesis in $p \Rightarrow q$	71	p in theorem
Conclusion in $p \Rightarrow q$	71	q in theorem
Direct method of proof	71	Sequence of implications from hypothesis to conclusion
Proof by contradiction	72	$p \wedge \sim q \Rightarrow F$
Proof by contrapositive	73	$\sim q \Rightarrow \sim p$
Proof by cases	73	$(p_1 \Rightarrow q) \wedge (p_2 \Rightarrow q)$
Modus ponens	75	$(p \wedge (p \Rightarrow q)) \Rightarrow q$

REFERENCES

Boyse, J. W. and J. E. Gilchrist. "GMSolid: Interactive Modeling for Design and Analysis of Solids." *IEEE Computer Graphics and Applications* 2(2): 27–40, March 1982.

Cralle, R. K. "Computer Thinking." *Tentacle* 2(2): 22–23, February 1982.

Dewdney, A. K. "Computer Recreations: A Computer Microscope Zooms in for a Look at the Most Complex Object in Mathematics." *Scientific American* 253:16–20, August 1985.

Fleury, A. "The Discrete Structures Course: Making Its Purpose Visible." Presented at the ACM Annual Conference, 1985.

Levine, S. R. *Introduction to Computer Graphics.* Tutorial at SIGGRAPH 85, 1985.

Mortimer, E. "Blaise Pascal." In *Mathematics: An Introduction to Its Spirit and Use*, p. 17. San Francisco: Freeman, 1979.

Peitgen, H. O. and P. H. Richter. *The Beauty of Fractals.* New York: Springer Verlag, 1986.

Verity, J. W. "Bridging the Software Gap," *Datamation* 31(4): 84–88, February 15, 1985.

3 COUNTING

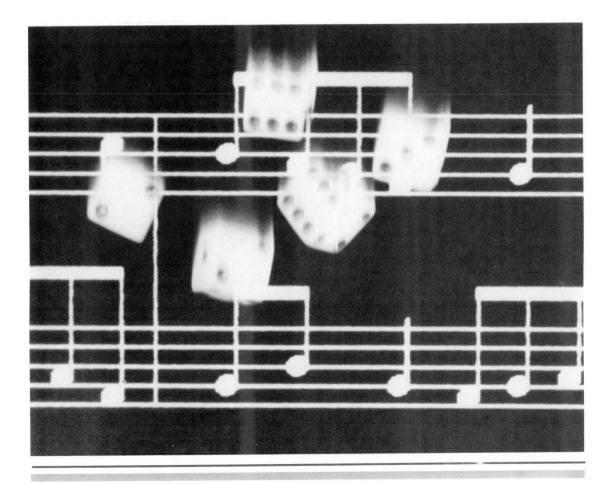

Mathematics by the Measure. The music was written with the aid of a computer. A much older piece of music, Mozart's dice waltz, illustrates one of the counting principles (see Exercise 40 of Section 3.1).

Counting Principles

There are many counting questions that arise in the study of computer science, and a branch of mathematics called **combinatorics** provides the means to answer them.

Let us consider one counting problem you might find in programming. Suppose input data contains students' names, classes (freshman, sophomore, junior, senior), and dorms (A, B, C). Output data will be in the form of several reports, one report for each class/dorm combination, such as freshmen in Dorm B or juniors in Dorm A. How many reports are necessary? For the freshmen there are three reports, one for each dorm; for sophomores, 3 dorm reports, as well as three reports each for the junior and senior classes. Thus, the number of reports is

$$3 + 3 + 3 + 3 = 4 \times 3 = 12$$

The Fundamental Counting Principle states the general idea.

FUNDAMENTAL COUNTING PRINCIPLE

> If there are m possibilities for one event and n for another independent event, then there are $m \times n$ possibilities for the sequence of the two events.

In the above example there are four possible classes and three possible dorms. Hence, there are 4×3 possible combinations of both. A tree diagram is often helpful in picturing the possible outcomes. Notice in Figure 3.1 that there are 12 terminating branches.

Figure 3.1

Tree associating dorms A, B, C with each class

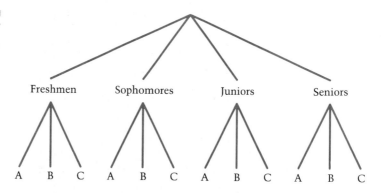

Situations in which there are two possibilities for each event are prevalent in computer science because of the importance of logic and the binary number system in computers. In Section 2.4 we considered the statement

IF $(X > 3)$ AND $(Y = 7)$ THEN...

As noted, the condition (X > 3) is either TRUE (T) or FALSE (F) as is the condition (Y = 7). Consequently, there are 2 × 2 = 4 possibilities that can arise, and we need 4 rows in the truth table of (X > 3) AND (Y = 7).

In a compound statement with three conditions, A, B, and C, instead of two, 8 rows are necessary for the truth table. Since there are three events (statements) and two possibilities for each statement (T or F), we see by an extension of the counting principle to three events that there are $2 \times 2 \times 2 = 2^3 = 8$ rows. Do not fall into the trap of multiplying 3 by 2. Each event has two possible outcomes, two for the first, two for the second, and two for the third. Extending the argument further, we see that for n component statements, the truth table for the compound statement requires 2^n rows.

EXAMPLE 1

The most common method of encoding characters in the computer, the ASCII system, uses 7 bits to represent each character. For example, the semicolon, ";", is encoded with the bit string, 0111011. How many different characters can be encoded by the ASCII system? In other words, how many different 7-bit strings are possible?

Solution

With two choices (0 or 1) for each of the seven positions, there are $2 \times 2 \times 2 \times 2 \times 2 \times 2 \times 2 = 2^7 = 128$ different possibilities. ◼

EXAMPLE 2

Let us now return to the justification for the theorem in Section 2.3:

Let U be a set with n elements. The power set $\mathscr{P}(U)$ has 2^n elements.

Justification

We are really asking how many possible choices are there for an arbitrary subset S of U. Or looking at the problem in a different way, there are two possibilities for each element of U—the element can either be in S or not. Since there are two possibilities for each of the n elements, in all there are

$$\underbrace{2 \times 2 \times \cdots \times 2}_{n \text{ factors}} = 2^n$$

choices for an arbitrary subset S. ◼

Unlike in Examples 1 and 2 where there are two possibilities for each event, the next application of the Fundamental Counting Principle has a different number of choices for each event.

EXAMPLE 3

Suppose four programs have been submitted to the computer for execution almost simultaneously. In how many ways can they be ordered in the job queue (line) to establish which program is to be executed first, which second, etc.?

Solution

For the event of executing first, any one of the four programs is a possible choice. But once that program has executed, it is not available, so there are only three programs that could run second. There are then two possibilities for third place in line; and once the other positions in the queue are decided, the program left over must be last. Thus, the number of arrangements in the queue is

$$4 \cdot 3 \cdot 2 \cdot 1 = 24. \quad \blacksquare$$

HISTORICAL NOTE

A notation for the product of these successive positive integers is 4!, called 4-factorial. In the early 1800s, ⌐n was the notation used for n-factorial, but $n!$ evolved because it was easier to use for printing. Some even suggested changing the name to "n-admiration" since "!" is a "note of admiration." ☐

DEFINITION

Let n be a positive integer. Then **n-factorial** is

$$n! = n(n-1)\cdots 2 \cdot 1 = n(n-1)!.$$

Also, define $0! = 1$.

EXAMPLE 4

Evaluate 7!, 7!/6!, 7!/7, and 7!/(2! 5!).

Solution

$$7! = 7 \cdot 6 \cdot 5 \cdot 4 \cdot 3 \cdot 2 \cdot 1 = 5040$$

$$\frac{7!}{6!} = \frac{7 \cdot \cancel{6} \cdot \cancel{5} \cdot \cancel{4} \cdot \cancel{3} \cdot \cancel{2} \cdot \cancel{1}}{\cancel{6} \cdot \cancel{5} \cdot \cancel{4} \cdot \cancel{3} \cdot \cancel{2} \cdot \cancel{1}} = 7.$$

Or writing more succinctly, using the above definition we have

$$\frac{7 \cdot 6!}{6!} = \frac{7 \cdot \cancel{6!}}{\cancel{6!}} = 7$$

$$\frac{7!}{7} = \frac{\cancel{7} \cdot 6!}{\cancel{7}} = 6! = 6 \cdot 5 \cdot 4 \cdot 3 \cdot 2 \cdot 1 = 720$$

7!/(2! 5!): Expanding 7! down to the largest factorial in the denominator and then canceling, we have

$$\frac{7 \cdot \overset{3}{\cancel{6}} \cdot \cancel{5!}}{\cancel{2} \cdot \cancel{5!}} = 21 \qquad \text{since } 2! = 2 \cdot 1 = 2. \quad \blacksquare$$

Often we combine events with OR instead of AND. For example, suppose you have two favorite restaurants, one a sandwich shop and the other an Italian restaurant. At the sandwich shop you have a choice of five meals, while at the Italian restaurant you have a choice of eight. When deciding to go out to eat at a favorite place, how many possible meals do you have from which to choose? You do not multiply $5 \times 8 = 40$ since you will not eat both a sandwich *and* an Italian dinner. There are $5 + 8 = 13$ possible meals to order: one of 5 sandwiches *or* one of 8 Italian dishes.

DISJOINT-OR COUNTING PRINCIPLE

> If there are m possibilities for one event and n for another event and all possibilities are distinct, then there are $m + n$ possibilities for one event or the other, but not both, to occur.

We must be careful that there is no possibility that occurs in both events, or using this formula would result in counting that possibility twice. For example, if a meatball sandwich is on both menus, then there are only 12 different meals you can order. The two events must be **mutually exclusive** or **disjoint;** that is, they must have no possibilities in common. Thus, the set of possibilities for one event E_1 and the set of possibilities for the other E_2 must be disjoint, so that $E_1 \cap E_2 = \phi$.

In the situation where there is overlap, as in the case of the meatball sandwich being on both menus, we must alter the formula. In this situation we subtract the number of elements in the intersection $E_1 \cap E_2$ since those elements have been counted once in E_1 and again in E_2.

INCLUSION–EXCLUSION PRINCIPLE

> If there are m possibilities for one event, n for another event, and b possibilities for both events, then there are $m + n - b$ possibilities for one event or the other to occur.

Expressed in the language of sets, the inclusion–exclusion principle says that for the sets E_1 and E_2

$$n(E_1 \cup E_2) = n(E_1) + n(E_2) - n(E_1 \cap E_2).$$

Thus, the disjoint-OR counting principle is just a special case of the inclusion–exclusion principle with $E_1 \cap E_2 = \phi$ so that $n(E_1 \cap E_2) = 0$.

EXERCISES 3.1   *Evaluate Exercises 1–15, simplifying as much as possible before multi-plication or division.*

1. $3!$ 2. $5!$ 3. $3! \cdot 4 \cdot 5$ 4. $3 \cdot 3!$ 5. $(3 \cdot 3)!$

6. $3! \, 3!$ 7. $0 \cdot 5!$ 8. $(0 \cdot 5)!$ 9. $0! \, 5!$ 10. $3 \cdot 4!$

11. $\dfrac{10!}{3! \, 7!}$ 12. $\dfrac{29!}{0! \, 29!}$ 13. $\dfrac{6!}{2 \cdot 3!}$ 14. $\dfrac{71!}{70! \, 1!}$ 15. $\dfrac{9!}{8! \, 9}$

16. Suppose an argument is composed of six statements. To prove that it is valid using a truth table, how many rows will be needed?

17. Mainframe IBM computers use the EBCDIC system with 8 bits to encode a character. In this scheme how many different characters could be encoded?

18. Suppose there are 326 students in CS 101, 52 in CS 499, and no students taking both. How many students are taking
    a. CS 101 and CS 499?
    b. CS 101 or CS 499?

19. Suppose there are 52 students in CS 499, 35 in CS 350, and 12 students taking both courses. How many students are taking
    a. CS 350 and CS 499?
    b. CS 350 or CS 499?
    c. CS 499 but not CS 350?

20. How many different social security numbers are possible?

21. For disjoint sets E_1 and E_2 by the disjoint-OR counting principle, we have

$$n(E_1 \cup E_2) = n(E_1) + n(E_2).$$

    a. Solve for $n(E_2)$.
    b. Of the 39 students working at a computer terminal, 17 are programming in Pascal. How many of those at terminals work with a language other than Pascal?

22. Of the 200 computer science majors who can program at a college, 103 have programmed in only one computer language; 58 have programmed in exactly two; and 24 in exactly three. How many have programmed in
    a. At least two languages?
    b. More than three languages?

23. Suppose there are four statements, A, B, C, and D, each of which can be true or false.
    a. In constructing a truth table, how many rows would be needed?
    b. How many rows have T as the value for all four statements?
    c. How many rows have at least one F as a value for A, B, C, or D?

24. The menu at a steak restaurant in town gives the following options for a dinner: soup or salad; bread; baked potato, French fries, hash browns, or onion rings; one of seven steak entrees; and apple pie and/or ice cream.
    a. How many different dinners are possible?
    b. How many different dinners are possible if you definitely want onion rings?

    c. How many different dinners are possible if you have decided against onion rings?

25. Suppose a password to use a computer system must have six letters in it.
    a. How many different passwords are possible on the system? How many different passwords are possible
    b. If no letter can be repeated?
    c. If the first letter is "U?"
    d. If the first letter is "U" or "D?"
    e. If the first letter is "U" and the last letter is "X?"
    f. If the first letter is "U" or the last letter is "X?"

26. a. In how many ways can ten books be arranged in a row on a bookshelf?
    b. Suppose three are computer science and seven are mathematics books. In how many ways can they be arranged so that the math books are on the left?
    c. In how many ways can they be arranged such that all the books of a subject are together?

27. Suppose you plan a great vacation and decide to pack carefully with seven shirts, three pairs of slacks, and two pairs of shorts, all color-coordinated. How many days can you go without repeating the same outfit?

28. Suppose you will be traveling to six European cities on this great vacation.
    a. In how many different ways can you do this without repeating a city?
    b. In how many ways can you do this without repetition if you must land in Rome and leave from Paris?

29. In certain versions of the language BASIC, only the first two characters of a variable name are recognized. The first character is a letter which is followed by a letter, digit, or nothing. How many different variables are possible?

30. A variable name in the language FORTRAN is from one to six characters long. The first character must be a letter, while the others can be letters or digits. How many different variables are possible?

31. a. How many 5-digit positive integers are there without a leading zero? How many 5-digit positive integers are there
    b. That have no digit repeated?
    c. That end in the number 3?
    d. That are odd?
    e. That begin and end with 3?
    f. That begin or end with 3?
    g. That are odd or end in 3?
    h. That are odd or begin with 3?

32. Suppose a 9-bit string is transmitted from a terminal to a computer.
    a. How many different strings are possible? How many of these have
    b. An even number of 1's?
    c. An odd number of 1's?
    d. All 1's?
    e. At least one 0?

33. Suppose in filling out anonymous questionnaires, people are asked to indicate their age in one of five categories (1–17, 18–30, 31–50, 51–65, 66+), sex (M, F), race (Caucasian, Negroid, American Indian, Oriental), marital status (married, divorced, single), political preference (Democrat, Republican, Independent).
   a. How many different questionnaire responses are possible?
   b. Repeat part a assuming the questionnaire was given to fifth graders in a girls' boarding school.
   c. Draw a tree to illustrate part b.

34. Suppose you had two children, and they each had two children as did each of their children.
   a. How many grandchildren would you have?
   b. Great-grandchildren?
   c. Draw a "family" tree to illustrate this situation.

35. Suppose you and your descendants had three children each.
   a. What would be the number of your grandchildren?
   b. Of your great-grandchildren?
   c. Draw a "family" tree to illustrate this situation.

36. In filling in a crossword puzzle suppose you are looking for a 6-letter word where the first letter is p, the third is r, and the sixth is t.
   a. How many possibilities are there for filling in the puzzle? How many possibilities are there
   b. If you assume the second letter is a vowel? (Include y.)
   c. If you assume there is at least one vowel? (Include y.)

37. Louis Braille in 1829 designed the code, which is named after him, for the blind. To encode each of 40 characters he had at most six dots arranged in two columns and three rows. For example, to indicate 0 he used ⸫ and for Y ⸪. What is the maximum number of characters he could have encoded? ☐

38. How many different times can be displayed on a 12-hr digital clock that shows
   a. Hours and minutes?
   b. Hours, minutes, and seconds?
   c. Hours, minutes, and seconds, and A.M. and P.M.?

39. a. What is the minimum number of bits you would need to construct an encoding scheme for 40 characters?
   b. For 26 characters?   **c.** For 92 characters?

40. Mozart's dice waltz has 16 bars. He has 11 different selections for each of 14 bars, but only one possibility (11 identical choices) for bar 8 and two possibilities for bar 16. Before being performed, the audience would roll dice 16 times to obtain 16 numbers between 2 and 12, inclusively; the music would be played based on their rolls. How many different ways can the dice waltz be performed? ☐

41. In the game of twenty questions one person thinks of an object and then states whether it is animal, vegetable, or mineral. The other player asks at most 20 yes/no questions, trying to guess the object. How many possible outcomes are there for an object and set of answers assuming exactly 20 questions are asked?

42. In a computer network suppose a message is sent from a central host to ten other computer centers. Then each of these centers relays the message to five other centers, and in turn those centers send the message to three centers each.
  a. What is the maximum number of computer centers that will receive the message at the third step?
  b. Why is the actual number probably much less?

3.2

Permutations and Combinations

In Example 3 of the last section the Fundamental Counting Principle was used to calculate that there are 4! ways of arranging four programs in a job queue. In mathematical terms we say there are 4! = 24 permutations of four programs, where a **permutation** is an ordered arrangement.

As another example of ordered arrangements, suppose five friends walk over to the computer center together to enter their programs. They find only a cathode-ray tube (CRT) terminal and teletype terminal available. In how many ways can the two input/output (I/O) devices to the computer be used? (The friends that do not get on a terminal, impatient and hungry, will leave and go out for a hamburger.) By the counting principle, any of the five could get the CRT, leaving four from whom to choose for the other terminal. We still have an ordered arrangement of five people, but we are taking only two at a time. We are not ordering the remaining three people, and hence we cut out the $3 \cdot 2 \cdot 1$ in the $5! = 5 \cdot 4 \cdot 3 \cdot 2 \cdot 1$, stopping after multiplying the first two numbers. We can view the number of permutations of five things taken two at a time as

$$5 \cdot 4 = \frac{5 \cdot 4 \cdot 3 \cdot 2 \cdot 1}{3 \cdot 2 \cdot 1} = \frac{5!}{3!} = \frac{5!}{(5-2)!}.$$

This ratio can be generalized into a formula. Since many calculators have a factorial key, the number of permutations is often easier to compute using this formula than using the Fundamental Counting Principle.

DEFINITION

The **number of permutations** or arrangements of r distinct objects chosen from n distinct objects is

$$P(n, r) = \frac{n!}{(n-r)!} \qquad 0 \le r \le n.$$

EXAMPLE 1

Suppose users of a large computer system are assigned a 5-letter user-identification (user-id), where no letter is repeated. Someone trying to gain illegal access to the computer overhears a password. (A user-id and a password are necessary to access this system.) What is the largest number of user-ids that the unauthorized person will need to try to get into the system?

Solution

This is a permutation problem. There are 26 letters from which to choose 5 and then arrange in a user-id:

$$P(26, 5) = \frac{26!}{(26 - 5)!} = \frac{26!}{21!} = \frac{26 \cdot 25 \cdot 24 \cdot 23 \cdot 22 \cdot \cancel{21!}}{\cancel{21!}}$$

$$= 7,893,600. \blacksquare$$

HISTORICAL NOTE

The person of Example 1 and the five friends are all trying to access a mainframe using terminals. Since a number of people can use the computer system at the same time, it is a **time-sharing** system. The computer's time is shared so that each person gets the impression he or she is the sole user of the system. In the early days of computers before terminology was established, "time sharing" was called "time stealing." □

Suppose we do not care about the number of ways of arranging the items but just about choosing some of them. Consider again the five friends going to the computer center: perhaps they are stopped at the door with the news that only two terminals are available. In how many ways can they decide who works on the program and who goes out to eat? The five friends form a set, and we are interested in how many different two-element subsets there are. Here, we are concerned with what is called the number of combinations of two people chosen from five, usually denoted by $C(5, 2)$. In the original problem, two friends were chosen and then arranged, one getting the CRT and one unfortunate soul getting the teletype terminal. Once two are picked in $C(5, 2)$ number of ways, there are $2! = 2 \times 1$ ways of arranging them. Thus, by the counting principle

$$C(5, 2) \times 2! = P(5, 2)$$

or $\quad C(5, 2) \times 2! = \dfrac{5!}{(5 - 2)!}$

or dividing, $C(5, 2) = \dfrac{5!}{2! \, (5 - 2)!}$

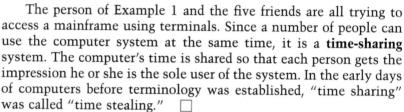

$$C(5, 2) = \frac{5 \cdot 4 \cdot 3!}{2 \cdot 3!} = \frac{5 \cdot \overset{2}{\cancel{4}}}{\cancel{2}} = 10.$$

There are 10 ways to pick a subset of two friends from five. There are more ways, 20, to pick the subset and then arrange the two. This formula for $C(n, r)$ was known by 1100 A.D., and the one for $P(n, r)$ was discovered as long as 2500 years ago. ☐

HISTORICAL NOTE

DEFINITION

The number of **combinations** or subsets of r distinct objects chosen from n objects is

$$C(n, r) = \frac{P(n, r)}{r!} = \frac{n!}{r! \, (n - r)!}, \qquad 0 \le r \le n.$$

EXAMPLE 2

Suppose a manufacturer of microcomputers wishes to do extensive testing of a sampling of 5 computers picked from every 100 produced. In how many ways can this be done?

Solution

Here, we are not concerned with ordering, only with subsets. Therefore, we need to find the number of combinations of 5 microcomputers chosen from 100:

$$C(100, 5) = \frac{100!}{5! \, 95!}$$

$$= \frac{\overset{5}{\cancel{100}} \cdot \overset{33}{\cancel{99}} \cdot \overset{49}{\cancel{98}} \cdot 97 \cdot 96 \cdot \cancel{95!}}{\cancel{5} \cdot \cancel{4} \cdot \cancel{3} \cdot \cancel{2} \cdot 1 \cdot \cancel{95!}} = 75{,}287{,}520.$$

By all means, cancel the denominator completely before multiplying so that computations will be easier. ■

Combinations, permutations, and the counting principle may all be needed in solving some interesting problems.

EXAMPLE 3

How many different 5-card hands can be dealt that contain 2 diamonds and 3 spades? *Note:* There are 52 cards in a deck. They are divided into two colors (black and red), and four suits (clubs (black), diamonds (red), hearts (red), and spades (black)). Each suit has 13 cards labeled 2 through 10, jack, queen, king, and ace.

Solution

Arrangement does not matter; you do not really care in which order you are dealt the cards, so combinations are required. There are two events, drawing 2 of the 13 diamonds and drawing 3 of the 13 spades. We use the Fundamental Counting Principle to handle the overall situation and combinations to calculate the number of subsets in each set:

$$C(13, 2) \times C(13, 3) = \frac{13!}{2!\ 11!} \times \frac{13!}{3!\ 10!}$$

$$= \frac{13 \cdot 12}{\cancel{2}} \times \frac{13 \cdot \cancel{12} \cdot 11}{\cancel{3} \cdot \cancel{2}} = 22{,}308. \quad \blacksquare$$

The Fundamental and Disjoint-OR Counting Principles are both needed in the following example.

EXAMPLE 4

A combinational circuit in a computer has several input lines and one output line. Each line carries a value of 0 or 1 and the combined input determines the output. (Such circuits are considered in greater detail in Chapter 9.) The action of this circuit can be illustrated in a truth table with 0 and 1 instead of FALSE and TRUE. Suppose there are three inputs, x_1, x_2, x_3, and one output, y. The number of possible input values for all three lines is $2^3 = 8$, so there are 8 rows in the table as in Table 3.1. Answer the following questions:

a. In how many ways can exactly two input lines be activated to carry a 1?

Solution

We are not arranging, only picking a subset, so the answer is $C(3, 2) = 3!/(2!\ 1!) = 3$. Notice that Table 3.1 is filled in such a way that y is 1 on the 3 rows where exactly two of the x's have values of 1.

TABLE 3.1
Truth table for
Example 4

x_1	x_2	x_3	y
0	0	0	0
0	0	1	0
0	1	0	0
0	1	1	1
1	0	0	0
1	0	1	1
1	1	0	1
1	1	1	0

b. Suppose the output y is 1 if and only if one or two x's are 0. In how many ways can this be accomplished?

Solution

Here we have an OR situation where exactly one event can occur, so summation should be used:

$$C(3, 1) + C(3, 2) = 3 + 3 = 6.$$

c. In how many ways can y be 1 if at least one input line is active?

Solution

There are two main ways to calculate the answer:

1. "At least one" here means 1, 2, or 3 lines are active; and we have an OR situation:

$$C(3, 1) + C(3, 2) + C(3, 3) = 3 + 3 + 1 = 7.$$

2. "At least one" is the opposite of none. There are 2^3 possibilities in all, but we are excluding those situations with all inactive lines. Since there is $C(3, 0) = 1$ way to obtain all 0's on input, we have

$$2^3 - C(3, 0) = 8 - 1 = 7. \quad \blacksquare$$

Another important application of combinations is found in the Binomial Theorem. Undoubtedly, you have already expanded many expressions such as $(x + 5)^2 = x^2 + 10x + 25$ and $(x + 5)^3 = x^3 + 15x^2 + 75x + 125$. The Binomial Theorem gives us a general formula for expanding $(x + y)^n$ with $n \in \mathbb{I}^+$. For $n > 2$ this formula will be less susceptible to errors and be easier to use than repeated multiplication.

THEOREM

> **Binomial Theorem.** If x and y are real numbers, and n is a positive integer then
> $$(x + y)^n = C(n, 0)x^n y^0 + C(n, 1)x^{n-1}y^1$$
> $$+ \cdots + C(n, i)x^{n-i}y^i + \cdots$$
> $$+ C(n, n - 1)x^1 y^{n-1} + C(n, n)x^0 y^n.$$

EXAMPLE 5

Expand $(x + y)^4$ using the Binomial Theorem.

Solution

By the theorem we have,

$$(x + y)^4 = C(4, 0)x^4y^0 + C(4, 1)x^3y^1 + C(4, 2)x^2y^2$$
$$+ C(4, 3)x^1y^3 + C(4, 4)x^0y^4.$$

Since

$$C(4, 0) = C(4, 4) = \frac{4!}{0!\, 4!} = 1$$

$$C(4, 1) = C(4, 3) = \frac{4!}{1!\, 3!} = 4$$

$$\text{and } C(4, 2) = \frac{4!}{2!\, 2!} = 6$$

we have

$$(x + y)^4 = x^4 + 4x^3y + 6x^2y^2 + 4xy^3 + y^4. \quad \blacksquare$$

Justification for the Binomial Theorem is covered in the exercises (see Exercise 34), as is Pascal's triangle, a diagram containing the coefficients in binomial expansions.

EXERCISES 3.2

Evaluate Exercises 1–9.

1. $P(6, 0)$ **2.** $C(6, 0)$ **3.** $P(6, 6)$ **4.** $C(6, 6)$ **5.** $P(6, 2)$

6. $C(6, 2)$ **7.** $P(35, 32)$ **8.** $C(35, 32)$ **9.** $C(63, 58)$

10. a. The computer club must elect a president, vice president, treasurer, and secretary from 35 members. In how many ways can this be done?
   b. In how many ways can the president then appoint a committee of four not using any of the officers?

11. If there are 50 statements in a program, in how many ways can they be arranged whether the arrangement makes sense or not?

12. a. In how many ways can fourteen programs, awaiting execution on a mainframe computer, be run?
   b. If five of the programs are written in Pascal, seven in COBOL, and two in FORTRAN, in how many ways can all the Pascal programs execute first?
   c. In how many ways can all the Pascal programs run followed by all the COBOL programs and then all the FORTRAN codes?
   d. In how many ways can all the programs of one language be run, then all those of another language, then all the remaining programs?

13. On a test you are asked to answer five out of eight essay questions. In how many ways can this be done?

14. If a professional baseball team has thirty players, in how many ways can the batting order of nine people be established for a particular game?

15. **a.** A traveling salesperson plans to drive to five cities away from home and needs to choose the shortest route. Suppose there is exactly one direct road between each pair of cities, and the salesperson wants to drive through each city once. How many possible routes are there?
    b. If the salesperson travels to ten cities, how many routes are possible?
    c. Would it be feasible for the salesperson to check all possible routes by hand?

16. Suppose a simple cryptography code permutes every four characters in the same way. For example, perhaps the first and third characters are switched as are the second and fourth. Thus, the message, "GET HELP," becomes "T GELPHE."
    a. How many different such 4-element encoding schemes are there?
    b. How many different such 8-element encoding schemes are there?

17. **a.** In a combinational circuit with four input lines, how many possible input values for all four lines are there?
    b. In how many ways are exactly two lines active?
    c. Two or three lines?   **d.** At least three lines?
    e. All lines?   **f.** None?

18. **a.** How many ways are there to seat 12 people in a row?
    b. At a round table?

19. Give the number of possible ways to be dealt each of the following type of 13-card hands (use notation; do not multiply):
    a. Any 13 cards   **b.** All spades
    c. All of one suit   **d.** Six spades and seven diamonds
    e. Six spades and seven of anything else
    f. Eight diamonds   **g.** All black cards
    h. At least one red card **i.** Four aces
    j. No aces

20. 16,200 factors into primes as $2^3 \cdot 3^4 \cdot 5^2$. Any divisor of 16,200 (a number that divides into 16,200 with a remainder of 0) must contain only the prime factors 2, 3, or 5. Thus, a positive divisor has the form $2^n \cdot 3^m \cdot 5^r$, where $0 \le n \le 3$, $0 \le m \le 4$, and $0 \le r \le 2$, $n, m, r \in \mathbb{N}$. How many positive divisors of 16,200 are there?

21. By factoring first as in Exercise 20, find the number of positive divisors of 784.

22. How many numbers n are there such that $0 \le n < 10^6$, where n does not contain a 2 or a 4 as a digit?

23. Suppose an organization has 25 men and 21 women.
    a. In how many ways can a committee of six be chosen?
    How can the committee be chosen such that it contains
    b. Exactly three of each sex?   **c.** All men?
    d. At least one woman?   **e.** All women?
    f. At least one man?   **g.** Exactly two or three women?
    h. Exactly one or five men?

24. Suppose a combinational circuit has eight input lines.
  a. How many rows are in the corresponding truth table?
  b. In how many ways can exactly one line be activated to carry a 1?
  c. Exactly seven lines?   **d.** All but one line?
  e. At least one line?   **f.** Two or three lines?
  g. Which is more likely to occur, four 1's or five 1's?

25. In sorting the elements of an array A of 50 elements into ascending order, the values of two elements A[I] and A[J] are out of order if I < J and A[I] > A[J]. Find the number of possible pairs that are out of order.

26. a. How many 7-bit strings are there?
  b. How many of these have an even number of 1's?
  c. Exactly three 1's?   **d.** No 1's?   **e.** At least one 0?

27. How many different arrangements of the letters of the word "below" are there?

28. Suppose in a graphics application 8 bits are used to store the intensity of red, 8 bits for green, and 8 for blue. Various intensities of these three primary colors are combined to create other colors. How many different colors can be generated?

29. Using the Binomial Theorem verify that

$$(x + y)^2 = x^2 + 2xy + y^2.$$

30. Expand $(x + y)^3$ using the Binomial Theorem.

31. Expand $(3a - 1)^4$ using the Binomial Theorem.

32. Using the Binomial Theorem show

$$2^n = C(n, 0) + C(n, 1) + C(n, 2) + \cdots + C(n, n - 1) + C(n, n).$$

33. Prove the following using the Binomial Theorem:

$$\text{If } xy = 0, \text{ then } (x + y)^n = x^n + y^n \text{ for } n \in \mathbb{I}^+.$$

34. Where asked, provide reasons in the following justification of the Binomial Theorem:

Proof

$$(x + y)^n = \underbrace{(x + y)(x + y) \cdots (x + y)}_{n \text{ factors}}.$$

The expansion is a sum. Each summand is a product of n variables, where x or y is chosen from each factor $(x + y)$. Thus, each summand has the form $x^{n-i}y^i$, where $i = 0, 1, 2, \ldots, n$. (**a.** Why?) There are $C(n, i)$ ways that the summand $x^{n-i}y^i$ can arise in the expansion. (**b.** Why?) Thus, the equality in the theorem holds.

  Pascal's triangle *is shown in Figure 3.2. For* $n \in \mathbb{N}$*, row n contains the coefficients of the variables in the expansion of* $(x + y)^n$*. These numbers are called the* **binomial coefficients.**

Figure 3.2

Pascal's triangle

Row

```
0                              1
1                          1       1
2                      1       2       1
3                  1       3       3       1
4              1       4       6       4       1
5          1       5       10      10      5       1
.          . . . . . . . . . . . . . . . . . . . . . . . . . . .
```

35. Verify that row 3 of Pascal's triangle contains the coefficients in the expansion of $(x + y)^3$.

36. Verify that row 5 contains the binomial coefficients, $C(5, i)$, $i = 0, 1, 2, 3, 4, 5$.

37. Prove $C(n + 1, j) = C(n, j - 1) + C(n, j)$ using the definition of the number of combinations.

38. a. Show $C(5, 2) = C(4, 1) + C(4, 2)$ using the equality of Exercise 37.
b. Find $C(5, 2)$, $C(4, 1)$, and $C(4, 2)$ in Pascal's triangle.

39. The statement in Exercise 37 shows that any number in the triangle can be obtained by adding the two numbers directly above it. Using this fact, find row 6 of Pascal's triangle.

40. After finding row 7 of Pascal's triangle (see Exercise 39), expand $(a^2 + 2b)^7$.

41. In Pascal's triangle give the sum of the elements in
a. row 3   **b.** row 4.   **c.** row n.
Note: Use Exercise 32.

3.3

Loop Computations

One of the advantages of computers is that they can do monotonous, repetitive tasks without complaint or error. The loop or repetitive control structure found in most languages is so prevalent in programs that a study of its meaning and efficiency is important. In part, this section touches the surface of a significant area of computer science called analysis of algorithms. This area is reconsidered in Chapter 11 after the study of some needed, additional topics.

An **algorithm** is a precise method for doing something in a finite number of steps. The term itself is derived from the last name of a Persian mathematician with quite a name—Abu Ja'far Mohammed ibn Mûsâ al-Khowârizmî—who wrote an important arithmetic textbook about 825 A.D. □

HISTORICAL NOTE

First, let us consider an example that illustrates some manipulations with loops as well as some efficiency considerations. The problem consists of reading in exactly 10,000 integers and printing the sum of the first five numbers, then the sum of the sixth through tenth numbers, the sum of the next five, and so forth. Finally, the total of all the numbers is to be printed.

To accomplish this task we need to initialize the overall total (TOTAL) to zero. Then before each group of five numbers, we must initialize the sum for the five (SUBTOTAL) to be zero. As a number is read it should be added to this subtotal. After all five are read, the SUBTOTAL needs to be printed and added to the TOTAL. The algorithm is stated more precisely as follows:

Algorithm to produce a subtotal every five numbers and a final total

```
   I. [Initialize] TOTAL ← 0
  II. [Outer loop] While there is data do the
      following:
      A. [Initialize] SUBTOTAL ← 0
      B. [Inner loop] Do the following 5 times:
         a. [Read data value] Read NUM
         b. [Add to SUBTOTAL]
                         SUBTOTAL ← SUBTOTAL + NUM
      C. [Write]       Print SUBTOTAL
      D. [Add to TOTAL] TOTAL ← TOTAL + SUBTOTAL
 III. [Write]       Print TOTAL
```

Note that we have a **nesting of loops.** The inner loop at B is charged with processing a block of five numbers. The outer loop starting at II is driven through all the data. We know the B loop is executed five times whenever encountered, but how many times is loop II executed? The Fundamental Counting Principle can help us here. There are 10,000 data items in all, and for each time through the outer loop, five numbers are processed by the inner loop. Thus,

$$10{,}000 = X \cdot 5 \quad \text{or} \quad 2000 = X.$$

Translating this algorithm into Pascal we have the program in Program 3.1. The algorithmic step names are also given for clarification:

PROGRAM 3.1

Pascal program to produce a subtotal every five numbers and a final total

```
         PROGRAM LOOPEX ( INPUT, OUTPUT );
         VAR
               I, J, NUM, TOTAL, SUBTOTAL: INTEGER;
         BEGIN
  I.     TOTAL := 0;
  II.    FOR I:= 1 TO 2000 DO
            BEGIN
  A.          SUBTOTAL := 0;
  B.          FOR J := 1 TO 5 DO
                 BEGIN
  a.                 READ (NUM);
  b.                 SUBTOTAL := SUBTOTAL + NUM
                 END;
  C.          WRITELN ("THE SUM OF FIVE NUMBERS: ",
                       SUBTOTAL);
  D.          TOTAL   := TOTAL + SUBTOTAL
            END;
  III.   WRITELN ("OVERALL TOTAL IS ", TOTAL)
         END.
```

The first line says that the program called LOOPEX will accomplish reading and writing. The line after VAR declares all the variables to be integers. The actual implementation of the algorithm, however, starts after the first BEGIN. Since we know precisely how many times we want to go through each loop, we can use FOR loops.

Actually, I and J are employed only as counters and their values are not used within the loop. Thus, if we wanted to be convoluted, we could have J going from 2 to 6. The inner loop would still be executed $6 - 2 + 1 = 5$ times. Notice in calculating the number of times through the loop, we subtracted the smaller number from the larger but add 1 since the loop is executed for both initial and terminal values. Consequently, the loop defined with

```
FOR  I :=  -4000  TO  -2001  DO
```

is executed $-2001 - (-4000) + 1 = 2000$ times. There are other examples where a shift in the range of the index is far more natural than in this program.

Notice that statements b and D of the Pascal program 3.1 are executed $2000 \times 5 = 10,000$ and 2000 times, respectively.   □

Suppose we change the algorithm a little so that within the inner loop we add each number to the TOTAL as well as to the SUBTOTAL. Thus, Step D is eliminated and the algorithm for the inner loop becomes

Algorithm for Inner Loop

```
B.  [Inner loop]   Do the following 5 times:
    a.  [Read data]        Read NUM
    b.  [Add to subtotal]  SUBTOTAL ← SUBTOTAL
                                      + NUM
    c.  [Add to total]     TOTAL    ← TOTAL
                                      + NUM
```

Would the output be the same? Yes. There are no more coded statements, but would the algorithm be as efficient? No, Step c would be executed 10,000 times instead of the 2000 times that Step D was executed, a fivefold increase.

EXAMPLE 1

LANGUAGE
EXAMPLE

The iterated loop in FORTRAN uses the word DO instead of FOR and is followed by the number of the statement that closes the loop. Also, instead of writing "I = 1 TO N", "I = 1, N" is used. How many times is the following FORTRAN loop executed?

```
    DO   7   I = 1.3
```

(7 is a statement number and should be ignored for the purpose of this example.)

Solution

HISTORICAL
NOTE

Be careful! The loop is *not* executed three times. In error, the comma was changed to a period. So the FORTRAN compiler, ignoring blanks and not requiring variables to be declared as in Pascal, would consider this statement to be an assignment:

```
    DO7I = 1.3   □
```

This very error is what caused the first American probe to Venus to be lost. Such a small error—only a multimillion dollar one!   □ ■

Though efficiency considerations usually indicate otherwise, sometimes we do want to pad loops with extra statements. Perhaps a delay is desired in an interactive program so that what is on the screen is displayed for a certain amount of time. Or maybe we are writing a program to control the stoplights at an intersection with each colored light remaining on for a specified period of time.

Because we are dealing with time, we must consider the internal clock of the computer. The fastest microoperation in the computer takes one **clock cycle,** and all operations are executed in some integer number of cycles. To accomplish exact delays we must know the number of cycles per second (frequency) for our computer and how many clock cycles each statement takes. (See Section 3.4 for additional discussion of clock cycle and frequency.)

LANGUAGE EXAMPLE

Generally, this type of routine is written in assembler language, but for simplicity let us suppose we are programming a delay in BASIC on a machine that has a clock frequency of 1,023,000 cycles/sec. Suppose the empty loop

    FOR   I = 1   TO   N: NEXT I

takes 2 cycles to initialize I to 1, 5 cycles each for the N times the loop is executed, and 4 cycles when I achieves its final value of N + 1. What value must N have to cause a delay of 0.05 sec? First, calculating the number of cycles, we have

   0.05 sec × 1,023,000 cycles/sec = 51,150 cycles.

The number of cycles in the loop is

   $2 + 5 \times N + 4 = 51,150.$

Solving for N we obtain $N = 10{,}228.8$, which we would round to 10,229. For longer delays we could increase N, place some "dummy" statements within the loop, or nest this loop within another. □

EXERCISES 3.3

LANGUAGE EXERCISES

How many times would each of the Pascal loops in Exercises 1–7 be executed if it terminated normally?

1. FOR I := 7 TO 17 DO

2. FOR I := −4 TO 30 DO

3. FOR I := −60 TO −8 DO

4. FOR I := 800 DOWNTO 0 DO (This loop decrements.)

5. FOR K := 0 TO 50 STEP 2 DO (STEP 2 means K takes on every other integer, starting with 0.)

6. FOR I := 0 TO 49 STEP 2 DO

7. FOR I := 1 TO 700 STEP 2 DO □

8. a. Suppose for the example in the algorithm (on page 100) we wish to print the subtotals for groups of 25 numbers each. How should the inner loop be changed?

   b. How many times is the outer loop executed?

c. If there are 2214 numbers to be subtotaled in groups of 18, how should the inner loop be changed?

d. How many times is the outer loop executed?

9. Consider the following nested loops in Pascal:

```
FOR  I  := 5 TO 436 DO
      FOR J  := -7 TO 8 DO
            FOR K := 1 TO 100 DO
                  WRITELN(I,J,K);
```

How many lines are written? (A line is written every time WRITELN is executed.) ☐

10. Consider the following nested loops in Pascal:

```
FOR  I  := 0 TO N DO
      FOR J  := 0 TO M DO
            WRITELN(I,J);
```

How many times is the WRITELN statement executed?   ☐

11. Suppose you wish to execute a delay of 0.13 sec in BASIC using the machine mentioned at the end of this section. To accomplish this delay, what should N be in the following loop?

```
FOR  I = 1 TO N:  NEXT I  ☐
```

12. Consider the following algorithm for doing a **sequential search.** We are searching for the first occurrence of X in an array of 16 elements, A[1], A[2], . . . , A[16]. Each element, starting with A[1] and going up, is examined. If, say, element A[11] is X, then the index 11 is printed. If X is not found in the list, 0 is printed.

I. [Go through list]   For J from 1 to 16 do the following:
    A. [Found?]   If X = A[J], print J and stop
II. [Not found]   Print 0

a. In the best case, where X is found immediately, how many times is the loop executed?

b. In the worst case, where X is not found or is the last element in the list, how many times is the loop executed?

c. If we know X is in the list, on the average, how many times is the loop executed?

d. Repeat parts a, b, and c for a list of 16,000,000 elements.

e. Repeat parts a, b, and c for a list of N elements.

Note: We see later that a sequential search is not the most efficient technique when the number of elements in the list is more than 20 or 25.

13. Suppose the BASIC used on the machine of this section can only have integer values up to 32,767.

a. How many cycles are needed for a delay of 1 sec?

b. What approximately must M be in the next statement to accomplish this delay?

```
FOR J = 1 TO M: FOR I = 1 TO 10229: NEXT I: NEXT J
```

Note: As discussed in the section, the inner loop results in a delay of approximately 0.05 sec.   ☐

14. Suppose the computer has a clock frequency of 7,833,600 cycles per second. Find N for a delay of 0.04 sec.

FOR I = 1 TO N: NEXT I   ☐

15. If a routine executes a delay of 0.25 sec, how many times must it be invoked in a loop to get a delay of approximately 10 sec?

16. The following is one of the sample questions in the *1984 AP Course Description in Computer Science**:
Suppose that the following program segment is used to approximate a zero of the real-valued function

$$f(x) = x * x - 2$$

starting with *Left* = 1 and *Right* = 2.

```
var
      Left, Right, x, Epsilon: real;
repeat
      x := (Left + Right)/2;
      if f (x) < 0 then
            Left := x
      else
            Right := x
until (Right - Left) < Epsilon
```

How many times must the loop be executed to produce an *x* which is guaranteed to be within *Epsilon* of a zero of *f* when *Epsilon* is 0.001?

(A) Once
(B) 10 times
(C) 100 times
(D) 1,000 times
(E) It cannot be determined from the information given.   ☐

3.4

Computer Measurements

Various measurements arise in the study of computer science and facility in their manipulation can be quite helpful. A few calculations involved with time and space considerations are covered in this section. The purpose is not to have you memorize formulas, but to make you comfortable with some of the computations needed for insight into topics to be covered in other courses.

As mentioned in Section 3.3, signals in the circuitry of a computer are regulated by an internal clock. The **cycle time** is the unit of time for measuring all actions in the **central processing unit**

---

* AP questions selected from *AP Course Description in Computer Science.* College Entrance Examination Board, 1984. Reprinted by permission of Educational Testing Service, the copyright owner of the sample questions.

(CPU), the division that controls the computer and performs arithmetic and logic operations. Cycle time measures how long it takes to perform the shortest CPU microoperation, and the unit of measure is usually in microseconds (msec) or nanoseconds (nsec).

For example, the PDP-8, a mini or medium-size computer built in the early 1960s, had a cycle time of 1.6 μsec. Interestingly, only 15 years after introduction of the PDP-8, the design of this computer, which measured 9.5 cubic feet and weighed 250 pounds, could be placed on a single silicon chip. The CRAY-1, a supercomputer developed in the 1970s, has a cycle time of 12.5 nsec. A smaller cycle time (here, 12.5 nsec as opposed to 1.6 μsec) means a faster computer since the most basic computer operation is faster. The CRAY-1, in fact, has a clock cycle 128 times faster than the PDP-8:

$$\frac{1.6 \ \mu\text{sec}}{12.5 \ \text{nsec}} = \frac{1.6 \ \mu\text{sec}}{0.0125 \ \mu\text{sec}} = 128. \quad \square$$

A related measure of speed is **clock rate** or **frequency** which is simply the reciprocal of cycle time. Cycle time gives time for one cycle, while clock frequency gives the number of cycles per unit of time. Often the latter measurement is in **megahertz (MHz),** or million cycles per second.

EXAMPLE 1

The CRAY-1 with cycle time of 12.5 nsec has what clock frequency expressed in MHz?

Solution

$$\text{clock frequency} = \frac{1}{12.5 \ \text{nsec/cycle}} = 0.08 \ \text{cycles/nsec}$$

$$= \frac{0.08 \ \text{cycles}}{\text{nsec}} \times \frac{10^9 \ \text{nsec}}{\text{sec}} \times \frac{1 \ \text{Mcycle}}{10^6 \ \text{cycles}}$$

$$= 0.08 \cdot 10^3 \ \text{Mcycles/sec} = 80 \ \text{MHz.} \quad \blacksquare$$

Still another way of measuring speed is to give the number of floating-point operations (operations on real numbers stored in exponential form) that can be executed in a second, called by the amusing acronym **flops**.

EXAMPLE 2

To simulate aerodynamic flow, weather, or other activities in continuous space on the computer, a three-dimensional grid is established with several measurements stored for each grid point. Repeatedly, for a designated time interval, new values are computed at the grid points by performing perhaps 500 operations per data item using the old values. Some applications require as many as 10^{13} arithmetic operations to complete the problem. On a 50-Mflops machine, how long will it take to complete such an application?

Solution

$$50 \text{ Mflops} = 50 \times 10^6 \text{ ops/sec.}$$

Therefore,

$$\frac{10^{13} \text{ ops}}{50 \times 10^6 \text{ ops/sec}} = 2 \times 10^5 \text{ sec.}$$

200,000 sec seems rather abstruse. What is this time in days?

$$2 \times 10^5 \text{ sec} \times \frac{1 \text{ min}}{60 \text{ sec}} \times \frac{1 \text{ hr}}{60 \text{ min}} \times \frac{1 \text{ day}}{24 \text{ hr}} = 2.315 \text{ days!} \quad \blacksquare$$

When data is being transferred, say from a computer's main memory to an external storage medium such as magnetic tape or to a terminal, the **data-transfer rate** or **bandwidth** is often expressed in **baud** which usually means "bits per second." If you are using a computer terminal that is tied to the main computer via a telephone line where you actually telephone the computer, you are probably connected on a 300- or 1200-baud line. If your terminal is "hard-wired" to the computer so that you have a direct line, the data-transfer rate might be 9600 baud.

HISTORICAL NOTE

The term "baud" is named after Baudot who in 1882 developed a memory wheel to be used on teletypes for the telegraph. With a circle of 32 0's and 1's, 32 characters could be encoded. Starting at any point on the wheel, the next five bits represented a character. Since each character's representation overlapped with those of four other characters, the entire encoding scheme could be held in a small circle of bits. ☐   A technique for generating such schemes is discussed in Chapter 8.

EXAMPLE 3

HISTORICAL NOTE

Development of the ENIAC, the first general-purpose, electronic digital computer, began during World War II in response to the need to compute trajectories of shells. Completed after the war under the direction of Presper Eckert and John Mauchly, the ENIAC could compute in 20 sec a trajectory that would have taken 2 days for a person to calculate. The story goes, however, that turning on this machine with its 18,000 vacuum tubes would cause all the lights in West Philadelphia to dim. Input and output was on punched cards at an input rate of 125 cards per minute. What was the maximum data transfer rate? (Each card contained as many as 80 characters. Assume each character is encoded in 8 bits.)

Solution

$$125 \text{ cards/min} = \frac{125 \text{ cards}}{\text{min}} \times \frac{80 \text{ char}}{\text{card}} \times \frac{8 \text{ bits}}{\text{char}}$$

$$= \frac{80,000 \text{ bits}}{\text{min}} \times \frac{\text{min}}{60 \text{ sec}} = 1333 \text{ baud} \quad \square \quad \blacksquare$$

Magnetic tape was introduced in the early 1950s as an external storage medium for the UNIVAC, the first commercially produced computer. At first some people would not believe there was information recorded on the tape since they could not see holes as they did on cards. At one plant, a fellow, not knowing about computer tape, received a delivery of tape he thought was no good since it had no adhesive back!   □

Data is usually stored on a tape on 9 parallel **tracks** running the length of the tape as depicted in Figure 3.3. Thus, across the tape there are 8 bits or a **byte** of information plus an extra bit, called a **parity bit**, to help check that the information is correctly recorded. Several measurements associated with tape are as follows:

1. **storage density** expressed in bytes/inch or bits/inch (**bpi**), the number of bits in one track of length 1 inch
2. **tape speed** expressed in inches/sec
3. **data-transfer rate** expressed in bytes/sec

Given any two quantities, you can find the third.

Figure 3.3

Magnetic tape with 9 tracks

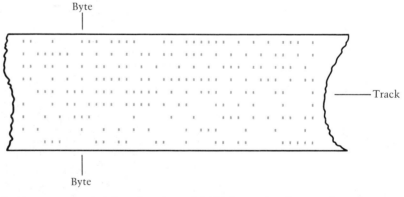

Byte

Byte

Track

EXAMPLE 4

Suppose a tape with a density of 1600 bytes/inch transfers data to and from the computer at a rate of 30,000 bytes/sec. What is the speed of the tape in inches/sec when this data is being transferred?

Solution

We are given measurements in bytes/inch and bytes/sec, and we want an answer in units of inches/sec. Thus, we have

$$\frac{30{,}000 \text{ bytes/sec}}{1600 \text{ bytes/inch}} = 18.75 \text{ inches/sec.} \quad \blacksquare$$

Information is stored on tape in chunks called **records**. For instance, in a tape file containing inventory for a company, each record might contain all the information about a particular item such as inventory number, description, cost, and quantity on hand.

On tape if records are accessed one at a time, there is a gap between records called the **interrecord gap** or **IRG** as pictured in Figure 3.4. This gap generally ranges from 0.6 to 0.75 inch. To avoid so much wasted space, records are often **blocked** or grouped together as in Figure 3.5 so that a fixed number of records are stored together with an **interblock gap (IBG)** between blocks of records.

Figure 3.4

Unblocked records on a tape

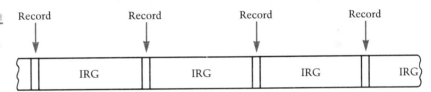

Figure 3.5

Blocked records on a tape

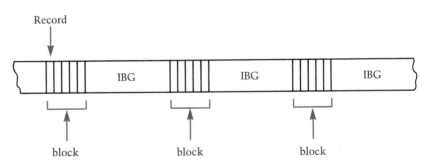

EXAMPLE 5

Suppose a standard 2400-ft tape is recorded at 1600 bpi with 80-byte records unblocked so that there is an IRG of 0.75 inch between each record. How many megabytes of information can be stored on the tape?

Solution

Without all the gaps we could store the following number of records:

$$\frac{2400 \text{ ft}}{\text{tape}} \times \frac{12 \text{ in}}{\text{ft}} \times \frac{1600 \text{ bytes}}{\text{inch}} \times \frac{\text{rec}}{80 \text{ bytes}} = \frac{576{,}000 \text{ rec}}{\text{tape}}.$$

How many records could have been stored in each 0.75-inch gap?

$$0.75 \text{ inch} \times \frac{1600 \text{ bytes}}{\text{inch}} \times \frac{1 \text{ rec}}{80 \text{ bytes}} = 15 \text{ rec.}$$

Thus, 15 out of every 16 record positions are wasted as shown in Figure 3.6. The actual number of bytes that can be stored on the tape is

$$\frac{576,000 \text{ rec}}{16} \times \frac{80 \text{ bytes}}{\text{rec}} = 2,880,000 \text{ bytes} = 2.88 \text{ Mbytes.} \quad \blacksquare$$

Figure 3.6

Tape of Example 5

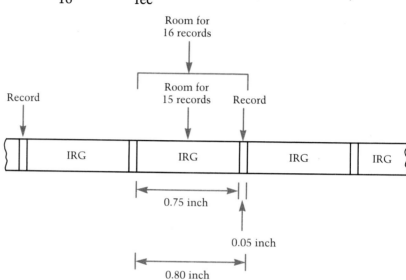

Data on disk storage is faster to access than on magnetic tape. Mainframe computer disks, which look much like phonograph records, are arranged into packs of perhaps 12 disks each. On a disk as in Figure 3.7, data is stored in concentric circles called **tracks**. A **cylinder**, as indicated in Figure 3.8, is a set of tracks that are all the same radius from the center and thus are directly under or above each other. If you wish to store data on disk, you probably will need to calculate the amount of space for your data set.

Figure 3.7

Surface of a disk

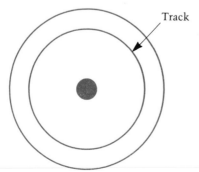

Figure 3.8

Cylinder in a disk pack

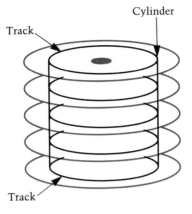

Cylinder

Track

Track

EXAMPLE 6

Let's say you are writing a COBOL program for a college registrar's office that will store a data set consisting of student records. Each record contains 324 characters of information. Since it takes a byte to encode one character, there are 324 bytes in a record. Suppose we also know that there are at most 6000 students in the school. If a disk contains 47,476 bytes per track, how many tracks, expressed as an integer, must you request to store the data set? If there are 15 tracks per cylinder, how many cylinders, expressed as an integer, are needed? *Note:* To simplify the problem we will ignore any extra bytes that are needed for "house-keeping" details about the data set.

Solution

First, how many records can be stored on a track?

$$\frac{47{,}476 \text{ bytes/track}}{324 \text{ bytes/rec}} = 146.53 \text{ rec/track.}$$

But a fraction of a record cannot be stored on the track, so 146 records can be stored as a block on the track. For the entire data set we have

$$\frac{6000 \text{ rec/data set}}{146 \text{ rec/track}} = 41.096 \text{ track/data set.}$$

Thus, we must have at least 42 tracks:

$$42 \text{ tracks} \times \frac{\text{cyl}}{15 \text{ tracks}} = 2.8 \text{ cylinders}$$

gives a final answer of 3 cylinders. ■

EXERCISES 3.4

1. A Cray X-MP supercomputer has a 9.5-nsec clock cycle. Give its clock frequency in megahertz.

2. **a.** Suppose computer A has a bigger cycle time than computer B. Which is faster?
   b. Suppose computer C has a larger clock frequency than computer D. Which is faster?

3. A Macintosh microcomputer has a 7.8336-MHz clock frequency. Find its clock cycle.

4. An Apple II microcomputer has a cycle time of 0.9775171 μsec.
   a. What is its frequency?
   b. How many times faster is the clock from the Macintosh?

5. The Whirlwind computer project was started in 1945 as an aircraft flight simulator/trainer for the U.S. Navy. With a 1-MHz clock rate, what was its cycle time?

6. The ENIAC's frequency ranged from 60 to 125 KHz. What was its clock rate range?

7. The STAR-100 supercomputer, built in the early 1970s, was designed to produce 10^8 32-bit floating-point results per second. Express this quantity in megaflops.   □

8. Suppose a 2400-ft tape reel can move at a speed of 75 inches/sec and store a maximum of 560 Kbytes of information.
   a. What is its storage density?
   b. What is the data-transfer rate?

9. For a tape with density of 1600 bpi, how many inches are used by one record of 120 bytes?

10. Repeat Exercise 9 for a tape density of 6250 bpi.

11. **a.** How much room in feet on the tape on Example 5 was wasted in gaps?
    b. What percent was wasted?

12. Suppose a 3600-ft tape is recorded at 1600 bpi with 100-byte records and an IRG of 0.75 inch.
    a. How many Mbytes of information can be stored on the tape?
    b. How many feet are wasted in gaps?

13. Suppose on the same tape of Exercise 12, 80-byte records are grouped into blocks of 8 records each. There is no gap between records within a block, but there is a gap of 0.75 inch between blocks.
    a. How many Mbytes can be stored on the tape?
    b. How many feet are wasted in gaps?

14. Repeat Exercise 13 with 100-byte records and blocks of 12 records each.

15. Suppose a 3600-ft tape is recorded at 1600 bpi with 160-byte records, grouped into blocks of 10 records each, and an interblock gap of 0.6 inch. How many Mbytes of information can be stored on the tape?

16. With a data-transfer rate of 1170 Kbytes/sec for a tape, how long will it take to read a block of 32,768 bytes?

17. If a block of size 16,384 bytes on a tape can be read in 54.6 msec, what is the data-transfer rate in Kbytes/sec?

18. The tape used with the UNIVAC was thin electroplated magnetic film with a tape speed of 100 inches/sec and a density of 120 bpi.
   a. What was its data-transfer rate?
   b. A 720-character block could be stored in what amount of space? ☐

19. Suppose 100-byte records from a data set of 11,250 records are stored in a disk pack that can hold 13,030 bytes/track and has 19 tracks/cylinder.
   a. How many records can be stored in a block on a track?
   b. How many tracks are needed?
   c. How many cylinders are needed?

20. Consider Exercise 19 assuming we want to place three blocks on each track with each block using at most 4253 bytes.
   a. How many records are in a block?
   b. In a track?
   c. How many tracks are needed?
   d. Cylinders?

21. Suppose a computer transfers data to and from memory 16 bits at a time with one transfer every 1 μsec. What is its baud rate?

22. Suppose information is written on a 1-hr video disk with a 6-MHz clock at a rate of 1 bit per cycle. How much information can be stored on the disk?

Chapter 3 REVIEW

Terminology/Algorithms	Pages	Comments
Fundamental Counting Principle	84	"and" situation, $m \cdot n$ possibilities
n-factorial, $n!$	86	$n \cdot (n - 1) \cdots 2 \cdot 1$, $0! = 1$
Disjoint-OR counting principle	87	"or" situation, $m + n$ possibilities
Inclusion–exclusion principle	87	"or" situation, $n(E_1 \cup E_2) = n(E_1) + n(E_2) - n(E_1 \cap E_2)$
Permutation	91	Ordered arrangement
Number of permutations, $P(n, r)$	91	$\dfrac{n!}{(n - r)!}$ $0 \le r \le n$
Combination	93	Subset

Terminology/Algorithms	Pages	Comments
Number of combinations, $C(n, r)$	93	$\dfrac{n!}{r!\,(n-r)!}$ $0 \le r \le n$
Binomial Theorem	95	Expansion of $(x + y)^n$
Binomial coefficient	95	$C(n, i)$, coefficient in expansion of $(x + y)^n$
Pascal's triangle	98	Has binomial coefficients
Number of times loop executed	100	Use bounds of index
Delay	102	Loops used
Cycle time	105	Time/cycle
Clock rate, frequency	106	Number cycles/time unit
Megahertz, MHz	106	Frequency of 10^6 cycles/sec
Flops	106	Number floating-point operations/sec
Data-transfer rate	107	Amount data transferred/time unit
Baud	107	Usually number bits/sec
Track on tape	108	Runs length of tape
Storage density	108	Amount data/length unit
Bits per inch, bpi	108	Unit of density
Tape speed	108	How fast tape moves
Record	109	Information about an item
Interrecord gap, IRG	109	Space between records on tape
Block	109	Group of records stored together
Interblock gap, IBG	109	Space between blocks on tape
Track on disk	110	In concentric circles
Cylinder	110	Cylinder of tracks on disk pack

REFERENCES

Austrian, G. D. "The Machine That Carried IBM into the Electronics Business." *Annals of the History of Computing* 3(2): 186–188, April 1981.

Bonn, T. "Development of UNIVAC's Plated Thin Film Metal Recording Tape." *The Computer Museum Report* 13, Fall 1983.

Brown, G. D. *System/370 Job Control Language*. New York: Wiley, 1977.

Brualdi, R. A. *Introductory Combinatorics*. New York: North-Holland, 1977.

Cajori, F. *A History of Mathematical Notations, Higher Mathematics, Vol. 2*. LaSalle: Open Court, 1929.

Computer Museum, Boston.

Cralle, R. K. "Memory Wheels." *Tentacle* 1 (5): 23–24, December 1981.

Cray Research, "Introducing the Enhanced CRAY X-MP Series of Computer Systems." *Cray Channels* 6 (3): 2–5, 1985.

Educational Testing Service. *1984 AP Course Description in Computer Science*. Princeton, NJ: College Board Publications, 1984.

Glaser, A. *History of Binary and Other Nondecimal Numeration*. Los Angeles: Tomash, 1971.

Grant, K. "What Do You Mean, My Tape's Unreadable?" *Tentacle* 4 (1): 23–30, January 1984.

Greene, D. H., and D. E. Knuth. *Mathematics for the Analysis of Algorithms, Vol. 1*, 2nd ed. Boston: Birkhauser, 1982.

Hayes, J. P. *Computer Architecture and Organization*. New York: McGraw-Hill, 1978.

Hinkel, J. S. *Apple Machine/Assembly Language Programming*. Scottsdale, AZ: Gorsuch Scarisbrick, 1986.

Knuth, D. E. "Mathematics and Computer Science: Coping with Finiteness." *Science* 194 (4271): 1235–1242, December 1976.

Knuth, D. E. "Algorithms." *Scientific American* 236: 63–80, April 1977.

Levine, R. D. "Supercomputers." *Scientific American* 246 (1): 118–180, January 1982.

Roberts, F. S. *Applied Combinatorics*. Englewood Cliffs, NJ: Prentice-Hall, 1984.

Rosen, S. "Electronic Computers: A Historical Survey." *Computing Surveys* 1 (1): 7–36, March 1969.

Tropp, H. S. (Ed.) "FORTRAN Anecdotes." *Annals of the History of Computing* 6 (1): 61, January 1984.

Tucker, A. *Applied Combinatorics*, 2nd ed. New York: Wiley, 1984.

Wand, M. *Induction, Recursion, and Programming.* New York: North-Holland, 1980.

Wexelblat, R. L. (Ed.) *History of Programming Languages.* New York: Academic Press, 1981.

4 *RELATIONS*

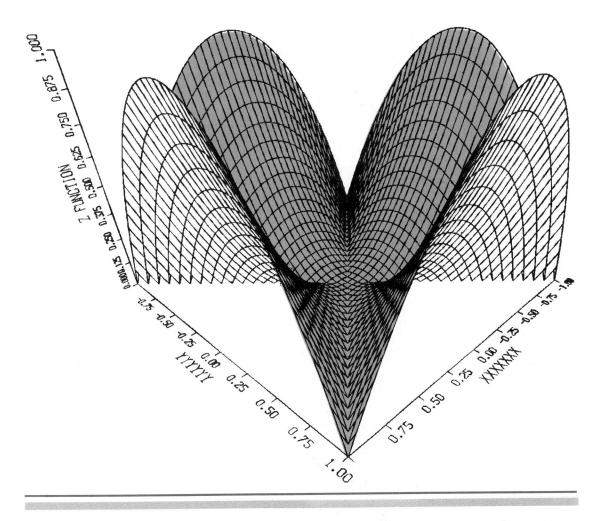

Hours of Daylight. The relationship of hours of daylight, latitude, and the day of the year is plotted using computer graphics. The two cuts indicate the amount of daylight through the year at latitudes 37.5 and −37.5.

4.1

Relation Definition

In Chapter 2 we considered several set operations such as union and intersection. Still another such operation is cross product, and a subset of a cross product of sets is called a relation. A relation in COBOL or Pascal or a relation in a relational data base is just a mathematical relation.

You have already been exposed to the idea of ordered pairs of numbers in mathematics courses: When asked to graph $y = x^2$, as in Figure 4.1, you plotted $S = \{(x, y) \mid y = x^2, x \in \mathbb{R}\}$, the set of all ordered pairs of real numbers, where the second coordinate is the square of the first. The order of the elements is important; (3, 9) is on your graph but not (9, 3). S is a subset of the cross product $\mathbb{R} \times \mathbb{R}$, which is the set of all ordered pairs of elements where the first and second elements are from $\mathbb{R}$. $\mathbb{R} \times \mathbb{R}$ is pictured on the Cartesian coordinate system by the entire plane.

Figure 4.1

Graph of $\{(x, y) \mid y = x^2, x \in \mathbb{R}\}$ in $\mathbb{R} \times \mathbb{R}$

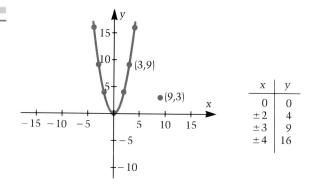

The first and second sets of the cross product do not have to be the same. For example, suppose the first set is $R = \{\text{Smith, Jones}\}$, and the second set is $T = \{M, F\}$. An ordered pair, consisting of the person's name followed by M or F, indicates name and sex. A listing of all possible pairs is the cross product of R and T:

$R \times T = \{(\text{Smith, M}), (\text{Smith, F}), (\text{Jones, M}), (\text{Jones, F})\}$.

DEFINITION

> Let A and B be sets. The **cross** or **Cartesian product** of A and B is
>
> $$A \times B = \{(a, b) \mid a \in A \text{ and } b \in B\}.$$
>
> For $A \times A$ write A^2.

EXAMPLE 1

Give the cross product of A and B when $A = \{0, 1\}$ and $B = \{1, 2, 3\}$ (see Figure 4.2).

Solution

$A \times B = \{(0, 1), (0, 2), (0, 3), (1, 1), (1, 2), (1, 3)\}$. ■

Figure 4.2

Graph of $A \times B \subseteq \mathbb{R} \times \mathbb{R}$ from Example 1

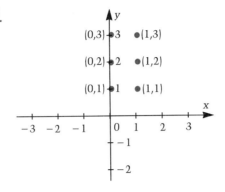

Notice that since there are 2 elements in A and 3 in B, there are $2 \times 3 = 6$ elements in $A \times B$. In general, as can be seen by the counting principle, if A has n elements and B has m, then $A \times B$ has $n \cdot m$ elements.

HISTORICAL NOTE

The Cartesian product and the Cartesian coordinate system are named after René Descartes who invented analytical geometry in the 17th century. Descartes, of fragile health, was fond of spending most of the morning in bed, resting and thinking. This habit of a lifetime was disrupted when he was invited by the Queen of Sweden to Stockholm. Her demands to work with him in the early morning are said to have contributed to his premature death within a few months from pneumonia. □

A binary relation is just a subset of the Cartesian product of two sets; it has some, but not necessarily all, the elements in the cross product. Thus, $\{(x, y) \mid y = x^2, x \in \mathbb{R}\}$ is a relation on $\mathbb{R} \times \mathbb{R} = \mathbb{R}^2$; and this set indicates how y is related to x. The set $\{(\text{Jones, F}), (\text{Smith, F})\}$, giving the specific relationship between the people and their sexes, is a subset of the cross product of R and T. In Example 1, $\{(1, 3)\}$ is a relation on $A \times B$.

DEFINITION

Let A and B be sets. A **binary relation** R on $A \times B$ is a subset of $A \times B$. If $(a, b) \in R$, we say a is **related** to b and write $a \, R \, b$.

If $n(A) = n$ and $n(B) = m$, then we already know that $n(A \times B) = n \cdot m$. Since a binary relation on $A \times B$ is simply a subset of $A \times B$, the total number of binary relations on $A \times B$ is $n(\mathcal{P}(A \times B))$, the number of elements in the power set. Consequently, with $n(A \times B) = n \cdot m$, $n(\mathcal{P}(A \times B))$ is the number of binary relations, 2^{nm}.

EXAMPLE 2

LANGUAGE EXAMPLE

A **two-dimensional array** of integers in BASIC (or FORTRAN or Pascal or. . .) with dimension DIM$(n, 2)$, where n is any positive integer, can be thought of as a table with n rows and 2 columns. A row of the array can indicate an ordered pair. This type of array may be used to represent a binary relation on $\mathbb{I}_c \times \mathbb{I}_c = \mathbb{I}_c^2$. Perhaps the first column of the array has the identification numbers of students in a class while the second has the corresponding test grades. If the array GR is declared to have dimension

    DIM GR(5, 2)

there are at most 5 students and each student has a unique test grade. Thus, the array has at most $5 \times 2 = 10$ elements. They might be illustrated by the following table:

Id Number	Grade
345	94
445	67
478	81
662	59
701	72

The relation or set of rows of the array is (345, 94), (445, 67), (478, 81), (662, 59), (701, 72)}. □ ∎

One disadvantage of arrays is that all the elements must be of the same type, such as INTEGER in Example 2. Therefore, we could not list actual student names as the elements in the first column. The record structure, which is available in some languages such as Pascal, COBOL, and PL/I, overcomes this problem.

For the example of names with grades, one record can hold one ordered pair in the relation. All the records together compose the relation which is called a **file** in computer science. You could think of each student's name and grade as written on a piece of paper. On the page there is a place for the name and a place for the grade; these locations are called the **fields** of the record. Each page corresponds to a record, while all the pages together make up the file. Generally, a computer file is stored on tape or disk, and the records are read one at a time.

EXAMPLE 3

In COBOL you must declare the name of the file and the description of the record. The name of each field in the record along with a format for each field in a PIC or PICTURE clause is given. For the file description (FD) of the above GRADE-FILE we might write

```
FD  GRADE-FILE. . .
01  GRADE-RECORD.
    05  NAME        PIC X(20).
    05  TEST-ONE    PIC 999.
```

This declaration does not read any data, but establishes the structure of the record. GRADE-RECORD has two fields: NAME which can hold 20 characters (X indicates a character field) and TEST-ONE which can hold a 3-digit number grade (9 indicates a numeric field). If we READ GRADE-FILE, an ordered pair of data, say (ALAN TURING, 93), will be placed into memory. □ ■

COBOL, the most widely used business-oriented computer language, was created in the late 1950s under the auspices of the Department of Defense. One of the driving forces on the team that wrote COBOL was Grace Murray Hopper. She received her Ph.D. in mathematics from Yale in 1934 and taught at Vassar until World War II when she enlisted in the U.S. Naval Reserve. Her computing career was launched when she was assigned to work with the first electro-mechanical, large-scale digital computer, Mark I. In the early 1950s, she wrote the first **compiler,** a program that translates a program written in a higher-level language to the language of the machine. An illustrious career in computing has also seen Rear Admiral Grace Hopper retire in 1986, the oldest officer in the U.S. military. □

Returning now to Example 3 on the student record, perhaps more information, such as social security number, address, phone number, and grades from four tests, needs to be stored. The concepts of ordered pair, cross product, and relation can be extended to meet this need.

DEFINITION

> An **ordered n-tuple** is of the form
>
> $$(x_1, x_2, x_3, \ldots, x_n)$$
>
> where order is important. The ith **coordinate** or ith **element** is x_i.

DEFINITION

> Ordered n-tuples $(x_1, x_2, x_3, \ldots, x_n)$ and $(y_1, y_2, y_3, \ldots y_n)$ are **equal** if and only if
>
> $$x_1 = y_1, \qquad x_2 = y_2, \qquad x_3 = y_3, \qquad \ldots, \text{ and } \qquad x_n = y_n.$$

DEFINITION

> Let $A_1, A_2, A_3, \ldots, A_n$ be sets. The **cross product** of these sets is
>
> $$A_1 \times A_2 \times A_3 \times \cdots \times A_n = \{(x_1, x_2, x_3, \ldots, x_n) \mid$$
> $$x_i \in A_i, i = 1, 2, \ldots, n\}.$$
>
> That is, the cross product of sets is the set of all ordered n-tuples where each coordinate comes from the corresponding set.

DEFINITION

> Let $A_1, A_2, A_3, \ldots, A_n$ be sets. An ***n*-ary relation** on $A_1 \times A_2 \times A_3 \times \cdots \times A_n$ is a subset of $A_1 \times A_2 \times A_3 \times \cdots \times A_n$.

Just as ordered pairs and binary relations on $\mathbb{R} \times \mathbb{R}$ can be pictured on the two-dimensional Cartesian coordinate system, ordered triples (3-tuples) and 3-ary relations on $\mathbb{R}^3$ can be graphed on the three-dimensional Cartesian coordinate system. Though graphing in 3-space will not be covered here, Figure 5.4 in Section 5.2 does illustrate a 3-D plot. Unfortunately, this type of graphing of binary and 3-ary relations cannot be extended to n-ary relations.

EXAMPLE 4

Let $R = \{\text{Smith, Jones}\}$, $T = \{\text{M, F}\}$, and $V = \{\text{Mathematics, Computer Science, Psychology}\}$. The number of elements in $R \times T \times V$ is $2 \times 2 \times 3 = 12$. The cross product is

$R \times T \times V = \{$(Smith, M, Math), (Smith, M, CS), (Smith, M, Psy),
(Smith, F, Math), (Smith, F, CS), (Smith, F, Psy),
(Jones, M, Math), (Jones, M, CS), (Jones, M, Psy),
(Jones, F, Math), (Jones, F, CS), (Jones, F, Psy)$\}$.

One relation on this set is

{(Smith, F, Math), (Smith, F, CS), (Jones, F, CS)}. ▪

Suppose that student information is being stored for a school, not just for a class. The registrar would need to keep the student's name, social security number (SSN), home address, phone number, semesters enrolled, courses, and grades, among other things. The housing office would need the first five items along with the school address, billing information, and meal plan. The financial aid office would require the first five items along with a detailed financial report. If all this data were stored on separate files, a change in home address would result in updating three different files. Moreover, the housing office must consult three different files to verify that the student is carrying enough hours to live on campus and has a financial aid grant to pay the bills. The solution to this complex interaction of data is to store the information in a **data base** instead of separate files. Much of the information is placed in the data base only once with a method of interrelation.

In a relational data base, relations are established with various **attributes** corresponding to fields. For example, the STUDENT-ADDRESS relation may have the structure

(SSN, name, address, phone)

with attributes SSN, name, address and phone. This relation, the set of all 4-tuples for students at the school, is a subset of I9 × NM × AD × PN, where I9, NM, AD, and PN are the set of all possible social security numbers, names, addresses, and phone numbers, respectively. For each academic year there is a relation like FINANCIAL90 for those receiving aid in 1990–91. The structure of this relation might be

(SSN, financial aid grant, family income).

We see that FINANCIAL90 is a relation on the Cartesian product I9 × $\mathbb{R}$ × $\mathbb{R}$. For each semester there is a relation like FALL90 or SPRING91 having the form

(SSN, course).

The relation HOUSING has the structure

(SSN, school address, meal plan).

All the set operations that have been introduced can be used to manipulate the data base. Suppose you wanted to know for the spring of 1991 the names of the students that were retaking a course they had had in the fall of 1990. You would take the intersection of the FALL90 and SPRING91 relations to form a new relation having the desired information. Finding the set of those students who received financial aid in 1991–92 but not in 1990–91 is accomplished by taking the difference of FINANCIAL91 and FINANCIAL90. To get a list of the courses students took in the 1990–91 academic year, take the union of the two relations, FALL90 and SPRING91.

EXERCISES 4.1

1. Form $A \times B$ for $A = \{a, b, c\}$ and $B = \{x, y, z\}$.

2. Form $A \times B$ for $A = \{chip1, chip2, chip3\}$ and $B = \{bad, good\}$.

3. Consider $A = \{0, 1\}$, $B = \{x, y, z\}$, and $C = \{y, n\}$.
   a. Find the number of elements in $A \times B \times C$.
   b. Form $A \times B \times C$.   c. Find $(B \times C) \cap (C \times B)$.

4. Suppose $I_c = \{-32{,}768, \ldots, -3, -2, -1, 0, 1, 2, 3, \ldots, 32{,}767\}$.
   a. How many elements are in I_c?
   b. In $I_c \times I_c$?
   c. How many relations are on $I_c \times I_c$?
   d. How many different two-dimensional arrays are there with 4 rows, 2 columns, and elements from I_c?

5. Let $A = \{x \mid 1 \le x \le 10, x \in I\}$ and $B = \{x \mid 3 \le x \le 9, x \in I\}$. Refer to Section 1.1 for definitions of sets I and P.
   a. Find the number of elements in $A \times B$.
   b. List elements in the relation $C = \{(a, b) \mid a \in P, b \in \{3, 4\}\}$ on $A \times B$.
   c. Give the relation D on $A \times B$, where $D = \{(a, b) \mid a \in P, b = a^2\}$.
   d. Give the relation $E = \{(a, b) \mid a + b = 6\}$ on $A \times B$.
   e. Intersect relations C and D.
   f. Form the union of C and D.
   g. Find the intersection C and E.
   h. Find $C \cup E$.
   i. Intersect C, D, and E.
   j. Form the cross product of D and E.
   k. Form $D - C$.
   l. Find $C - D$.
   m. If $C \cup E$ were the universal set, find $(C \cap E)'$.

6. When is $A \times B = B \times A$?

7. What is $\phi \times \mathbb{R}$?

8. Consider the relation $R = \{(1, 1), (1, 3), (2, 2), (3, 1), (3, 3), (3, 2), (2, 3)\}$ on $A \times A$, where $A = \{1, 2, 3\}$. Indicate true or false for each of the following assertions. Give a counterexample for each false statement.
   a. For every $x \in A$, $x \, R \, x$.
   b. $x \, \cancel{R} \, x$ for every $x \in A$.
   c. For $x, y \in A$, if $(x, y) \in R$, then $(y, x) \in R$.
   d. For $x, y \in A$, if $x \, R \, y$ and $y \, R \, x$, then $x = y$.

e. For x, y, $z \in A$, if x is related to y and y is related to z, then x is related to z.

9. Do Exercise 8 again for the relation $R = \{(x, y) \mid x$ divides into y with a remainder of 0$\}$ on $\mathbb{I}^+ \times \mathbb{I}^+$, where $\mathbb{I}^+ = \{1, 2, 3, \ldots\}$.

10. Graph the relation in Exercise 8 on the Cartesian coordinate system.

11. Graph the relation in Exercise 9 on the Cartesian coordinate system.

LANGUAGE
EXERCISE

12. Suppose in a Pascal program the variables X, Y, and Z are declared to be of a type that can take on values of any integer from 1 to 4:

    VAR X, Y, Z : 1..4;

    Writing as ordered triples (3-tuples), for what values of (X, Y, Z) will the following expression be TRUE?

    (X < Y) AND (Y < Z) ☐

LANGUAGE
EXERCISE

13. In Pascal we must declare the structure of the file and declare variables to be of type file or record before using them. Thus, Example 3 written in Pascal would be

    ```
    TYPE
       GRADETYPE =
          RECORD
             NAME      : PACKED ARRAY[1..20] of CHAR;
             TESTONE : 0..100
          END;
    VAR
       GRADEFILE   : FILE OF GRADETYPE;
       GRADERECORD : GRADETYPE;
    ```

    a. Suppose only letters of the alphabet, blanks, commas, and periods were placed into the 20 characters of NAME. How many possible values would there be for NAME? (Express the answer as a base raised to a power.)

    b. How many values are possible for TESTONE if it can have any integer from 0 to 100?

    c. How many different possible values for this record are there?

    d. Using a similar format to the above declaration, write the TYPE declaration for FINANCIALTYPE, a RECORD of the form (SSN, financial aid grant, family income) with SSN as 9 characters, the aid grant from 50 to 2000, and family income from 0 to 50,000. Then write the VAR declaration for FINANCIALFILE and FINANCIALRECORD. ☐

4.2

Partial Order

We have already seen how the concept of relation is used in some data base systems, but applications of relations extend to many other areas of computer science. Certain types of relations are important in such areas as language features, sorting and searching, scheduling of computer resources, storage methods, special functions, and data structures.

In Exercise 8 and 9 in Section 4.1 you were asked to determine if two relations satisfy each of five properties. These properties are so important that they are given special names.

DEFINITION

Let R be a binary relation on set $A \times A$. R is

a. **Reflexive** if for every $x \in A$, $x \, R \, x$.
b. **Irreflexive** if for all $x \in A$, $x \, \not{R} \, x$.
c. **Symmetric** if for every $x, y \in A$, $x \, R \, y$ implies $y \, R \, x$.
d. **Antisymmetric** if for every $x, y \in A$,
   $x \, R \, y$ and $y \, R \, x$ imply $x = y$.
e. **Transitive** if for every $x, y, z \in A$,
   $x \, R \, y$ and $y \, R \, z$ imply $x \, R \, z$.

EXAMPLE 1

Consider $U = \{a, b, c\}$ with $\mathcal{P}(U) = \{\phi, \{a\}, \{b\}, \{c\}, \{a, b\}, \{a, c\}, \{b, c\}, U\}$. Define a binary relation δ on $\mathcal{P}(U)$ such that

$$A \, \delta \, B \qquad \text{if and only if } A \subseteq B$$

Which of the above properties does δ satisfy?

Solution

It will be easier to decide about a property if we express each one in terms of $\subseteq$.

a. Reflexive:
   Is it true that $A \, \delta \, A$ for every $A \in \mathcal{P}(U)$?
   Is it true that $A \subseteq A$ for every $A \in \mathcal{P}(U)$?
   Yes, every set is a subset of itself.
b. Irreflexive:
   No, since δ is reflexive, it cannot be irreflexive.
c. Symmetric:
   Does $A \, \delta \, B \Rightarrow B \, \delta \, A$ for all $A, B \in \mathcal{P}(U)$?
   Does $A \subseteq B \Rightarrow B \subseteq A$ for all $A, B \in \mathcal{P}(U)$?
   No, a counterexample is $A = \{a, b\}$ and $B = \{a, b, c\}$.
   $\{a, b\} \subseteq \{a, b, c\}$, but $\{a, b, c\} \not\subseteq \{a, b\}$.
d. Antisymmetric:
   If $A \, \delta \, B$ and $B \, \delta \, A$, does $A = B$ for all $A, B \in \mathcal{P}(U)$?
   If $A \subseteq B$ and $B \subseteq A$, does $A = B$ for all $A, B \in \mathcal{P}(U)$?
   Yes, by the definition of equality of sets.
e. Transitive:
   If $A \, \delta \, B$ and $B \, \delta \, C$, is $A \, \delta \, C$ for $A, B, C \in \mathcal{P}(U)$?
   If $A \subseteq B$ and $B \subseteq C$, is $A \subseteq C$ for $A, B, C \in \mathcal{P}(U)$?
   Yes, if $x \in A$, then $x \in B$ by definition of $A \subseteq B$. With $x \in B$, x must also belong to C since $B \subseteq C$. We started out with $x \in A$ and showed that $x \in C$. Thus, $A \subseteq C$. ■

A relation satisfying the reflexive, antisymmetric, and transitive properties is said to be a partial order. Thus, $\subseteq$ is a partial order on $\mathcal{P}(U)$, and we say $\mathcal{P}(U)$ with $\subseteq$, written $(\mathcal{P}(U), \subseteq)$, is a partially ordered set.

Pictures often aid in understanding concepts. A picture of this relation $\subseteq$, as shown in Figure 4.3, can be drawn by carefully placing all subsets so that an edge or line segment is drawn from set A pointing to set B if $A \subseteq B$, $A \neq B$, and there is no other set C with $A \subseteq C \subseteq B$. This type of picture is called a **graph;** not the type of graph plotted on the Cartesian coordinate system, but one that is discussed is greater detailed in Chapter 8.

Figure 4.3

Graph of the partially ordered set $\mathcal{P}(U)$ with $\subseteq$, where $U = \{a, b, c\}$

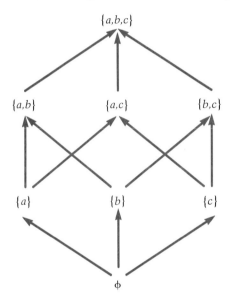

$$\{a,b,c\}$$

$$\{a,b\} \qquad \{a,c\} \qquad \{b,c\}$$

$$\{a\} \qquad \{b\} \qquad \{c\}$$

$$\phi$$

A hierarchy diagram with line segments drawn from calling programs to subprograms where each program appears once in the diagram also illustrates a partial ordering. Program A is related to program B if there is a path from A to B in the diagram.

Another application of partial ordering is in scheduling of several tasks where the completion of one or more of the tasks is necessary before others can begin. These tasks might be, for example, parts of a large research project or a system of programs to be executed in a multiprocessor computing environment. The relation $\leq$ on the set of all tasks, $\{t_i \mid i = 1, 2, \ldots, n\}$, is defined as $t_i \leq t_j$ if $i = j$ or if

HISTORICAL NOTE

$i \neq j$ and task t_i must end before task t_j can begin. The relation is clearly reflexive and transitive. Moreover, since $t_i \leq t_j$ and $t_j \leq t_i$ is possible only when $i = j$, $\leq$ is antisymmetric. One technique for dealing with this scheduling problem was developed by the U.S. Navy and used to complete the Polaris missile program in five years instead of the estimated seven. □

If a system of programs was scheduled to be executed with only one processor instead of many, then a **linear ordering** would need to be imposed upon the tasks. A decision must be made as to which task will be executed first, which second, and so forth. The process of creating a linear ordering from a partial ordering which is not linear is called a topological sort in data structures.

The graph of a linearly ordered set is a line—hence, the name "linear." Sometimes this graph is called a **chain.**

EXAMPLE 2

Let $A = \{3, 5, 15, 45\}$. Define a binary relation | on A as $a \mid b$ if and only if a divides b. By **a divides b** we mean that when we divide b by a the remainder is 0. This relation is a partial order because it is

1. Reflexive: Any number divides itself.

2. Antisymmetric: If a divides b and b divides a, then $a = b$.

3. Transitive: If a divides b and b divides c, then a divides c.

Figure 4.4

Graph of set $\{3, 5, 15, 45\}$, partially ordered by the relation |

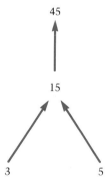

The order is not linear because 3 and 5 are not related: $3 \nmid 5$ and $5 \nmid 3$. ($\nmid$ means "does not divide.") The graph of the relation is given in Figure 4.4. Notice that an arrow is drawn from a to b if $a \mid b$, $a \neq b$, and there is no other number c such that $a \mid c$ and $c \mid b$. ■

EXAMPLE 3

Let $S = \{3, 15, 45\}$ with the same partial order $|$. Here, $|$ is a linear order on S because for any two elements in S, one divides the other. The graph is a **chain** (see Figure 4.5). ■

Figure 4.5

Graph of set $\{3, 15, 45\}$, linearly ordered by the relation $|$

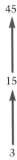

45

15

3

DEFINITION

Let R be a partial order of set S. R is a **linear** or **total order** on S if for any $a, b \in S$, $a\,R\,b$ or $b\,R\,a$. S with R, written (S, R), is a **linearly ordered set.**

DEFINITION

If S is partially ordered by R, a **topological sort** is the process of finding a linear order relation T on S such that $R \subseteq T$; that is, for $a, b \in S$ if $a\,R\,b$, then $a\,T\,b$.

EXAMPLE 4

Perform a topological sort of the partially ordered set of Example 2, $(A, |)$ where $A = \{3, 5, 15, 45\}$.

Solution

For a topological sort we wish to define a linear order relation, T, on the set A, preserving any previous relationships; that is, if $a \mid b$, then we want to have $a\,T\,b$. To form a linear ordering, however, elements that were previously unrelated, now must be related.

In Example 2, excluding the reflexive relationships like $3 \mid 3$, we find five others: $3 \mid 15$, $5 \mid 15$, $15 \mid 45$, $3 \mid 45$, and $5 \mid 45$. 3 and 5, however, are not related. For the linear ordering either $3\,T\,5$ or $5\,T\,3$ is permissible, so either of the chains in Figure 4.6 would be a proper topological sort of the partially ordered set $(S, |)$ ■

Figure 4.6

Topological sorts of the set {3, 5, 15, 45} partially ordered by the relation |

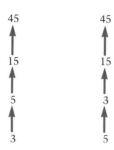

When a linear ordering exists, we can talk about successors and predecessors to elements. In Example 3, since 3 | 15, we say the predecessor of 15 is 3 and the successor of 3 is 15. 3 has no predecessor, just as 45 has no successor. The ability to find these neighboring elements is built into some languages. In Pascal suppose a variable TODAY is of type DAYS. If TODAY has a meaningful value, PRED (TODAY) is yesterday and SUCC (TODAY) is tomorrow. □

LANGUAGE EXAMPLE

A linear ordering is essential in alphabetizing a list of words. Because everything in the computer, including character data, is stored as a string of 0's and 1's, a linear order ≤ can be imposed where A ≤ B ≤ C ≤ ⋯ ≤ Z. Hence, through a sorting technique, you can put the list in ascending order.

EXERCISES 4.2

State the (r) reflexive, (i) irreflexive, (s) symmetric, (a) antisymmetric, and (t) transitive properties in terms of the given relations on each universal set U for Exercises 1–14. Also, state whether each property is true or false, and give a counterexample for each false statement.

1. $U = \mathbb{R}$ with α
   $x \, \alpha \, y$ if and only if $x \leq y$.

2. $U = \mathbb{N}$ with $\equiv_5$
   $x \equiv_5 y$ if and only if x and y have the same nonnegative remainder when each is divided by 5.

3. $U = \mathbb{N}$ with σ
   $x \, \sigma \, y$ if and only if x is even.

4. $U = \mathbb{N}$ with σ
   $x \, \sigma \, y$ if and only if y is even.

5. $U =$ set of all arrays with α
   $x \, \alpha \, y$ if and only if x and y have the same dimension.

6. $U = \mathbb{Q} - \{-1, 1\}$ with δ
   $x \, \delta \, y$ if and only if $x \cdot y = 1$.

7. $U = \mathbb{R} \times \mathbb{R}$ with β
   $(a, b) \, \beta \, (c, d)$ if and only if $a \le c$.
   Note: Suppose a record has two fields. We are saying the records are related based solely on a comparison of values in the first field. For example, the first field might be an account number at a department store, and the second field might be a charge to that account. Sorting by the account number uses this relation.

8. $U = \mathbb{I} \times \mathbb{I}$ with α
   $(a, b) \, \alpha \, (c, d)$ if and only if $a = c$ and $b \le d$.
   Note: Here, when the first coordinates are the same, we can relate the two ordered pairs; but if the first coordinates are not the same, there is no relationship.

9. $U = \mathbb{I} \times \mathbb{I}$ with δ
   $(a, b) \, \delta \, (c, d)$ if and only if $a < c$ or $(a = c$ and $b \le d)$.
   Note: One application of this relation is a file of records that are sorted primarily by the first field and secondarily by field two. If the first field values are equal, a comparison of the second field values determines the order. This relation is also similar to the ordering of words in a dictionary. The first letters are compared, but if they are equal, a comparison is made of the second letters, and so forth.

10. $U = $ set of people in Delaware with α
    $x \, \alpha \, y$ if and only if x is married to y.
    (Assume no bigamy.)

11. $U = $ set of English words with β
    $x \, \beta \, y$ if and only if the length of x is less than or equal to the length of y.

12. $U = \{2, 3, 4, 5, \ldots, 100\}$ with δ
    $x \, \delta \, y$ if and only if x is prime and is a factor of y.

13. $U = $ set of elements of array $A, A[1], A[2], \ldots, A[n]$, with α
    $A[i] \, \alpha \, A[j]$ if and only if $i \le j$.

14. $U = $ set of elements in a stack S with β
    $x \, \beta \, y$ if and only if x and y are the same or x was pushed onto the stack after y.

15. **a.** Which of the sets with relations in Exercises 1–14 are partially ordered sets?
    b. Which are linearly ordered sets?

16. Let $U = \{1, 2, 5, 20, 30, 60\}$. Define a relation δ on U: $x \, \delta \, y$ if and only if $x \mid y$.
    a. Graph the partially ordered set.
    b. Give all four-element subsets of U that are linearly ordered.
    c. Topologically sort $(U, \mid)$, and draw all possible resulting linearly ordered sets.

17. Repeat Exercise 16 with the relation in Exercise 8 and $U = \{(2, 5), (2, 9), (3, 5), (3, 2)\}$.

18. In a particular Pascal program, say statement s is related to statement t if s is the same statement as t or must come before t. For example, consider the following statements:

```
1.  READ (X) ;
2.  WRITE (X) ;
3.  Y := 5;
```

Statement 1 must come before statement 2; but statement 3 could appear anywhere and still maintain the same sense of the program. A partial order exists whose graph is

Since we must write the statements sequentially for the program, we do a topological sort to establish a linear order on the partial order. In this example any of the following graphs would yield the same results:

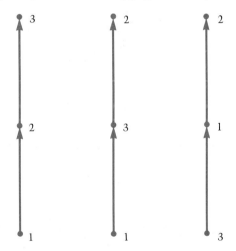

If statement 3 had been READ(Y), the linear orderings would be the same, but of course we would need to change the order of the data as well.

Consider the program below. **a.** Draw the graph of the partial ordering. **b.** How many possible arrangements (legal or illegal) of the statements are there? **c.** Considering only statements 1, 2, 3, 4, and 6, how many possible arrangements are there? **d.** How many of the linear orders in c get the same program result as the given arrangement?

```
1.  READ (A, B) ;
2.  C := 8;
3.  X := A + C;
4.  Y := 3 * X;
5.  B := B + 1;
6.  WRITE (X) ;
7.  Z := 2 * B;   □
```

19. Consider the following program segment which switches the values of X and Y:

```
1.  TEMP := X;
2.  X    := Y;
3.  Y    := TEMP;
```

a. Draw the graph of this partial ordering.
b. Is this ordering linear?
c. How many possible arrangements (legal or not) of statements are there?
d. How many arrangements of the statements are legal?   □

4.3

Equivalence Relations

Just as a partial order is a relation that meets several additional conditions, an equivalence relation satisfies the reflexive, symmetric (instead of antisymmetric) and transitive properties.

EXAMPLE 1

Consider $U = \{-3, -2, -1, 0, 1, 2, 3\}$ and the binary relation α on U, where $x \, \alpha \, y$ if and only if their absolute values are equal, $|x| = |y|$. For example, $-3 \, \alpha \, 3$ since $|-3| = |3|$. α is an equivalence relation because it is

1. Reflexive: $|x| = |x|$ for all $x \in U$.
2. Symmetric: If $|x| = |y|$, then $|y| = |x|$.
3. Transitive: If $|x| = |y|$ and $|y| = |z|$, then $|x| = |z|$.

Moreover, α splits U into disjoint subsets, where the members of a subset are related to each other and are not related to elements outside the subset.

$$U = \{-3, 3\} \cup \{-2, 2\} \cup \{-1, 1\} \cup \{0\}.$$

We say that α partitions U into four equivalence classes $\{-3, 3\}$,

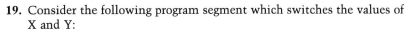

{−2, 2}, {−1, 1}, and {0} (see Figure 4.7). A notation for these equivalence classes is to place any element of the subset in brackets. Thus, [−3] = [3] = {−3, 3}. ☐

Figure 4.7

Partition of {−3, −2, −1, 0, 1, 2, 3} using the relation α: $x \; \alpha \; y \Leftrightarrow |x| = |y|$

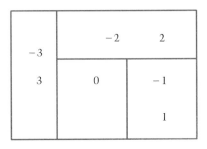

DEFINITION

A binary relation R on a set U is an **equivalence relation** if R is reflexive, symmetric, and transitive.

DEFINITION

A **partition** of a set U is a collection of nonempty disjoint subsets of U, where U is equal to the union of these subsets.

DEFINITION

Let R be an equivalence relation on U. The **equivalence class** of x, $[x]$, is the set of all elements in U related to x; that is, $[x] = \{y \mid y \; R \; x\}$.

Partitioning occurs in numerous situations. The **operating system** or overseer programs of a large computer may be one that partitions memory into variable size areas, each area holding a system or user program; or perhaps the operating system divides memory into blocks of uniform size.

Under the UNIX operating system, a **directory** provides a mapping between file names and the files themselves. A user may organize the names of all his or her files into directories, possibly partitioning further each directory into subdirectories in a hierarchical arrangement.

In file processing, a large file may be partitioned into groups, records being sorted within each group. In a data base design, it may be helpful to separate the data base into subfiles having records that are accessed at varying time intervals with records that are used frequently being grouped together. As another example, a distributed data base system has various computer sites using the same data base. Thus, data should be partitioned into subfiles appropriate for different locations, so that a location would only manipulate the data it needs.

As we will prove, anytime we have an equivalence relation, the relation defines a partition of the set.

THEOREM

An equivalence relation R on a set U partitions U into disjoint equivalence classes.

Note: By the definition of "partition" to show the collection of equivalence classes forms a partition of U we must show three things:

1. Each equivalence class is nonempty.
2. Distinct equivalence classes are disjoint.
3. The union of the equivalence classes equals U.

PROOF

1. Clearly, each equivalence class $[x]$ is nonempty since by the reflexive property the class must contain its representative element x.
2. Show that distinct equivalence classes are disjoint. Take any two equivalence classes:

$$[a] = \{z \mid z\,R\,a\} \quad \text{and} \quad [b] = \{z \mid z\,R\,b\}.$$

Show that $[a]$ and $[b]$ are equal or disjoint; i.e., show $[a] = [b]$ or $[a] \cap [b] = \phi$. To do this task, assume $[a] \cap [b] \neq \phi$ and prove $[a] = [b]$; i.e., assume $[a] \cap [b] \neq \phi$ and prove $[a] \subseteq [b]$ and $[b] \subseteq [a]$.

Since we are assuming $[a] \cap [b] \neq \phi$, $[a] \cap [b]$ contains at least one element. Take $y \in [a] \cap [b]$. By the definition of intersection, $y \in [a]$ and $y \in [b]$.

We will prove $[a] \subseteq [b]$ using the definition of subset by taking $x \in [a]$ and showing $x \in [b]$. We have the following sequence of relationships:

$x\,R\,a$	$x \in [a]$; definition of $[a]$
$y\,R\,a$	$y \in [a]$; definition of $[a]$
$a\,R\,y$	symmetric
$x\,R\,y$	transitive
$y\,R\,b$	$y \in [b]$; definition of $[b]$
$x\,R\,b$	transitive
Thus, $x \in [b]$.	definition of $[b]$

> We have shown $[a] \subseteq [b]$. By a symmetric argument (the same argument just switching the names "a" and "b"), we have $[b] \subseteq [a]$.
>
>   Thus, if $[a] \cap [b] \neq \phi$, then $[a] = [b]$; or distinct equivalence classes are disjoint.
>
> 3. To finish showing that there is a partition, we need to show that U is equal to the union of the equivalence classes. But if $u \in U$, then by the reflexive property, $u \in [u]$, one of the equivalence classes.

Every equivalence relation on a set partitions the set into subsets, and every partition establishes an equivalence relation. Consider the partition illustrated on the Venn diagram in Figure 4.8. We can define a relation by saying that two elements are related if they are in the same section. Thus, G and H are related as are D, E, and F. This relation is an equivalence relation because each element is in a subset with itself; if x and y are in the same subset, we can say y and x are in the same subset; and if x and y as well as y and z are in the same subset, then x and z are together in that set. We can let these letters be variable names and have the partition indicate locations in memory so that G and H share the same location. Then G and H are two names for the same thing. Perhaps one name is more meaningful in one part of the program, while the other is a better choice elsewhere. To establish this equivalence in the language FORTRAN, we write

EQUIVALENCE (G, H), (D, E, F).   □

In Exercise 15 you will generalize this example by proving the theorem that a partition establishes an equivalence relation on a set.

Figure 4.8

One partition of $\{A, B, C, D, E, F, G, H\}$

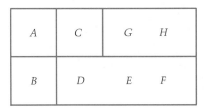

A	C	G	H
B	D	E	F

THEOREM

> Let $P = \{U_1, U_2, \ldots, U_n\}$ be a partition of the set U.
> Define a relation R on U as follows: $a\,R\,b \Leftrightarrow a$ and b are in the same U_i of P. Then R is an equivalence relation on U.

The next example indicates one way to partition the set of integers into equivalence classes.

EXAMPLE 2

Congruence modulo 3, $\equiv_3$, is an equivalence relation on the set of integers: $x \equiv_3 y$ or $x \equiv y \pmod 3$ if and only if $x - y$ is an integral multiple of 3. By "integral multiple of 3" we mean that $x - y$ is some integer times 3 or that 3 divides $x - y$.

$$6 \equiv_3 21 \quad \text{since} \quad 6 - 21 = -15 = -5 \times 3.$$

$$52 \equiv_3 28 \quad \text{since} \quad 52 - 28 = 24 = 8 \times 3.$$

$$11 \equiv_3 -1 \text{ since } 11 - -1 = 12 = 4 \times 3.$$

Notice that two elements are related if they have the same non-negative remainder, 0, 1, or 2, when divided by 3. With consideration of the remainders was how the similar relation $\equiv_5$ was defined in Exercise 2 of Section 4.2. "$x \equiv_5 y$" could have been defined there as holding if and only if "$x - y$ is an integral multiple of 5."

$\mathbb{I}$ is partitioned into three equivalence classes by $\equiv_3$ since there are three possible remainders:

$$[0] = \{\ldots, -6, -3, 0, 3, 6, \ldots\}$$

$$[1] = \{\ldots, -5, -2, 1, 4, 7, \ldots\}$$

$$[2] = \{\ldots, -4, -1, 2, 5, 8, \ldots\}$$

Consequently, $21, 6 \in [0]$; $52, 28 \in [1]$; and $11, -1 \in [2]$. ■

Congruence can be modulo any integer $n > 1$. Since there are n possible remainders $(0, 1, 2, \ldots, n - 1)$ when division is by n, there are n equivalence classes. Congruence modulo 2 might be used in a computing environment to put page numbers in the proper corners on alternating pages of a report. The correct day of the week might be calculated using congruence modulo 7.

One interesting program involving an equivalence relation is to find all the anagrams in English. **Anagrams** are different words that use the same letters, such as "below," "bowel," and "elbow." Lists of all the words in English are available on some computer systems. The "brute-force" method would be to take each word and search the dictionary word by word looking for others that contain the same letters. Such a technique would take a prohibitive amount of time. A much more elegant method by R. K. Cralle consists of assigning a prime to each letter:

a	b	c	d	e	f	g	h	i	j	k	l	m
2	3	5	7	11	13	17	19	23	29	31	37	41

n	o	p	q	r	s	t	u	v	w	x	y	z
43	47	53	59	61	67	71	73	79	83	89	97	101

Then calculate the product of the corresponding primes for each word in the dictionary. For example, the number associated with "below" is $3 \times 11 \times 37 \times 47 \times 83 = 4{,}763{,}121$. The **Fundamental Theorem of Arithmetic** tells us that each positive integer factors in exactly one way (other than different possible orders) into a product of primes. Consequently, the only factors of 4,763,121 are 3, 11, 37, 47, and 83; and only anagrams to "below," such as "bowel" and "elbow," have the same associated number. We say that two words are related if their corresponding numbers are equal, that is, if the words are anagrams of each other. This relation is an equivalence relation as you will show in Exercise 19. After pairing each word with its number, sort the list into ascending order by the number. All the anagrams for each word, the elements of its equivalence class, will be clustered with the word since they are all paired with the same number.

EXERCISES 4.3

1. Which of the relations in Exercises 1–14 of Section 4.2 are equivalence relations?

2. Consider the equivalence relation $\equiv_5$ on $\mathbb{I}$.
   a. Which of the following are related? $7, 37, 9, 2, 4, 13, 88, 66, -141, 444, -444, -65, 132, 505, 91, 9083, -72, -203, 670$
   b. How many equivalence classes are there?
   c. Give the equivalence classes using the $[n]$ notation as well as a listing of a sequence of the elements with set notation.

Give the equivalence class, $[0]$, $[1]$, $[2]$, . . . , or $[n-1]$, to which each number belongs modulo n in Exercises 3–8.

3. $14, n = 6$ 4. $14, n = 13$

5. $14, n = 27$ 6. $-6, n = 6$

7. $-7, n = 6$ 8. $-75, n = 11$

Determine if P is a partition of U in each of the Exercises 9–11. If P is not a partition, indicate why.

9. $U = \{a, b, c, d\}$, $P = \{\{c\}, \{d, b, a\}\}$
10. $U = \mathbb{I}$, $P = \{\mathbb{N}', \mathbb{I}^+\}$
11. $U = \mathbb{R}$, $P = \{\mathbb{Q}, \mathbb{R}\}$

12. Say that two conditions A and B are related if condition A always has the same value as condition B, TRUE or FALSE, for any values of their

variables. For example, condition $X - Y = 0$ is related to condition $X = Y$.
  a. Find two other conditions related to $X = Y$.
  b. Show this relation is an equivalence relation.

13. Let U be the set of all people. Define a relation R on U such that $x \, R \, y$ if x has the same birthday as y.
  a. Show R is an equivalence relation.
  b. How many equivalence classes are there?

14. Let U be the set of all programs. Say program A is related to program B if given the same input, they have the same output. Show this relation is an equivalence relation.

15. Prove the following theorem: Let $P = \{U_1, U_2, \ldots, U_n\}$ be a partition of the set U. Define a relation R on U as follows: $a \, R \, b \Leftrightarrow a$ and b are in the same U_i of P. Then R is an equivalence relation on U.
  Note: This problem is a generalization of the example in the text involving EQUIVALENCE (G, H), (D, E, F).

16. Consider $\mathbb{I} \times \mathbb{I}^+ = \{(a, b) \mid (a \in \mathbb{I}$ and $b \in \mathbb{I}^+\}$, where $\mathbb{I}^+$ is the set of positive integers. Say $(x, y) \, \alpha \, (u, v)$ if and only if $x \cdot v = y \cdot u$.
  a. Find three elements related to $(1, 2)$.
  b. How many elements are related to $(1, 2)$?
  c. Show α is an equivalence relation.
    Note: We use a relation similar to α to decide the equivalence of two fractions. We say
    $$x/y \equiv u/v \qquad \text{if and only if } x \cdot v = y \cdot u.$$

17. Consider a set of n elements. Say that two ordered arrangements of r elements are related if one is a permutation of the other.
  a. Find three ordered selections from $U = \{1, 2, 3, 4, 5, 6, 7, 8, 9\}$ related to $(5, 7, 2)$.
  b. How many ordered triples are related to $(5, 7, 2)$?
  c. How many equivalence classes in the set of 3-tuples using elements of U are there?
  d. Show this relation is an equivalence relation on the set of all r-tuples from a set of n elements for each positive integer r.
  e. How many elements are in an equivalence class of an ordered arrangement of r elements?
  f. How many equivalence classes of r-tuples are there?
  g. Use parts e and f to show that $P(n, r) = r! \, C(n, r)$.
    Note: Basically, this same proof was given in Section 3.2 without using the terms "equivalence relation" and "equivalence class."

18. a. Considering the method for finding anagrams given in this section, find the number associated with the anagrams "lager," "large," and "regal."
  b. Give ["pat"], the equivalence class of "pat."

19. Show that the relation discussed in the anagrams example at the end of this section is an equivalence relation.

Chapter 4   REVIEW

Terminology/Algorithms	Pages	Comments
Cross or Cartesian product, $A \times B$	118	$\{(a, b) \mid a \in A \text{ and } b \in B\}$
Binary relation R on $A \times B$	119	$R \subseteq A \times B$
Related, $a\ R\ b$	119	$(a, b) \in$ relation R
Ordered n-tuple	121	$(x_1, x_2, \ldots, x_n)$
ith coordinate	121	x_i in n-tuple
Cross product, $A_1 \times A_2 \times \cdots \times A_n$	122	$\{(x_1, x_2, \ldots, x_n) \mid x_i \in A_i\}$
n-ary relation R on $A_1 \times A_2 \cdots \times A_n$	122	$R \subseteq A_1 \times A_2 \times \cdots \times A_n$
Reflexive	126	$x\ R\ x$, for all x
Irreflexive	126	x never related to x
Symmetric	126	$x\ R\ y \Rightarrow y\ R\ x$
Antisymmetric	126	$x\ R\ y$ and $y\ R\ x \Rightarrow x = y$
Transitive	126	$x\ R\ y$ and $y\ R\ z \Rightarrow x\ R\ z$
Partial order relation	127	Reflexive, antisymmetric, transitive
Partially ordered set, (S, R)	127	R partial order on S
Divides, $a \mid b$	128	When dividing a into b, get a remainder of 0
Graph of (S, R)	127	Arrow from a to b if $a\ R\ b$, $a \neq b$, and there is no c with $a\ R\ c$ and $c\ R\ b$
Linear or total order, R	129	Partial order with $(a, b) \in R$ or $(b, a) \in R$
Linearly ordered set, (S, R)	129	R linear order on S
Topological sort	129	Partial ordering $\rightarrow$ linear ordering
Equivalence relation, R	133	Reflexive, symmetric, transitive

Terminology/Algorithms	Pages	Comments
Partition of U, $\{P_1, P_2, \ldots, P_n\}$	134	Sets nonempty, disjoint, union is U
Equivalence class, $[x]$	134	$\{y \mid y \; R \; x\}$
Equivalence relation $\rightarrow$ partition	136	Theorem
Partition $\rightarrow$ equivalence relation	137	Theorem
Congruence modulo n, $x \equiv_n y$ or $x \equiv y \pmod{n}$	137	$x - y$ integral multiple of n. Same remainder when x and y are divided by n

REFERENCES

Bradley, J. *File and Data Base Techniques.* New York: CBS College, 1982.

Cralle, R. K. "Brevity is the Soul of Programming." *Tentacle* 2(12): 25–27, December 1982.

Crombie, A. C. "Descartes." In *Mathematics: An Introduction to Its Spirit and Use*, 22–28. San Francisco: Freeman, 1979.

Eves, H. *An Introduction to the History of Mathematics.* New York: Holt, Rinehart & Winston, 1974.

Fleury, A. "The Discrete Structures Course: Making Its Purpose Visible," Presented at the ACM Annual Conference, 1985.

Hopcroft, J. E., and J. D. Ullman. *Introduction to Automata Theory, Languages, and Computation.* Reading, MA: Addison-Wesley, 1979.

Kline, M. *Mathematical Thought from Ancient to Modern Times.* New York: Oxford University Press, 1972.

Kroenke, D. M. *Database Processing: Fundamentals, Design, Implementation,* 2nd ed. Chicago: Science Research Associates, 1984.

Wand, M. *Induction, Recursion, and Programming.* New York: North-Holland, 1980.

Zientara, M. *The History of Computing.* Framingham, MA: CW Communications, 1981.

5 FUNCTIONS

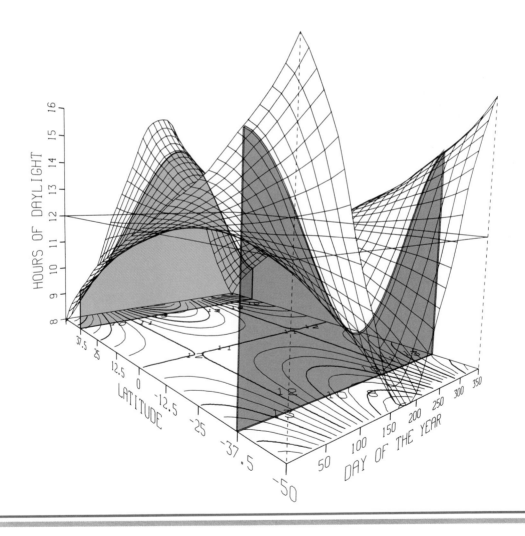

Graph of $z = f(x, y) = |x - y||x + y|$. Computer computations and graphics are used to create a graph of this function.

Function Definition

In Section 4.1 {(Smith, F), (Jones, F)}, associating a person with his or her sex, was presented as an example of a relation. In fact, this relation is also an example of a function. A particular person (and we are assuming a particular Smith and Jones here) is either male or female, so given the person, there is one associated sex. This is the essence of being a function—given an element from the first set, {Smith, Jones}, there is exactly one element from the second set, {M, F}, corresponding to it.

But suppose instead of having a relation on {Smith, Jones} × {M, F}, we have one on {Smith, Jones} × {chocolate, strawberry, pistachio} and associate the people with the flavors of ice cream they like. The relation might be {(Smith, chocolate), (Smith, strawberry), (Jones, chocolate)}. This relation is not a function since Smith really likes two flavors, chocolate and strawberry. If we wondered, "What flavor of ice cream does Smith like, chocolate, strawberry, or pistachio?," there would be two correct answers, not one. Consequently, this relation is not a function.

DEFINITION

> Let A and B be nonempty sets. A **function** or **mapping** f from A to B, denoted $f : A \to B$, is a relation on $A \times B$ such that for each element $a \in A$ there exists exactly one element $b \in B$ with $(a, b) \in f$. The set A is called the **domain,** and B is called the **codomain** of f. If $(a, b) \in f$, we say b is the **image** of a and write $b = f(a)$. The **range** is the set of all possible images, $\{f(a) \mid a \in A\}$. The **preimage** of $b \in B$ is the set of all elements from A that map to b, $\{a \mid f(a) = b, a \in A\}$.

EXAMPLE 1

For the function $f = \{(\text{Smith, F}), (\text{Jones, F})\}$ above, the domain is {Smith, Jones}; the codomain is {M, F} so that $f : \{\text{Smith, Jones}\} \to \{\text{M, F}\}$; the range is {F}, a subset of the codomain. Since $f(\text{Smith}) = F$, the image of Smith is F. But since $f(\text{Jones}) = F$, also, the preimage of F is {Smith, Jones}. Figure 5.1 illustrates this mapping. ■

Figure 5.1

Picture of the mapping δ in
Example 1

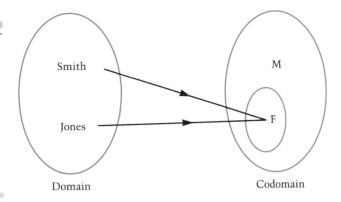

Domain   Codomain

EXAMPLE 2

Let $A = \{x \mid 1 \leq x \leq 5, x \in \mathbb{I}\}$ and
$B = \{x \mid 1 \leq x \leq 30, x \in \mathbb{I}\}$.

Define the relation f on $A \times B$ as

$f = \{(a, b) \mid b = a^2\}$ with $f : A \to B$.

Since $b = f(a)$ and $b = a^2$, we can abbreviate this relation as
$f(a) = a^2$. The relation f is a function because each integer between 1 and 5 has exactly one square which belongs to
$B : f(1) = 1; f(2) = 4; f(3) = 9; f(4) = 16;$ and $f(5) = 25$. The domain is A; the codomain, B; and the range, $\{1, 4, 9, 16, 25\}$. ■

Functions are extremely important in the study of computer
science as well as in mathematics and many other fields. In fact,
computer languages have many functions that are built into them;
and most languages have a facility for the user to create additional
ones. Once created, functions can be used over and over in a program. Functions tend to make programs more organized or structured, and, hence, easier to write, understand, and maintain. In the
early days of computers, functions and subroutines were not built
into the languages. Each person would write his/her own routines or
copy someone else's—time-consuming and error-prone tasks. □

EXAMPLE 3

The square function, SQR, is built into the language Pascal. The
following, however, is how we would define the function of Example 2 in Pascal without using SQR.

Solution

Assuming the type definitions

```
TYPE
        DOMAIN = 1..5;
      CODOMAIN = 1..30;
```

LANGUAGE EXAMPLE

we have

```
FUNCTION SQUARE (A: DOMAIN): CODOMAIN;
    BEGIN
    SQUARE := A * A
    END;
```

The more descriptive function name SQUARE was used instead of F, though F would have worked as well. The variable A is called the **parameter.** The domain element or value passed to A must be in the set {1, 2, 3, 4, 5}, while the value of the function, or a range element is an integer between 1 and 30. Defining a function does not use it, just as $f(a) = a^2$ only states the general relationship. $f(3) = 9$ indicates that the square of 3 is 9, so that 3 and 9 are related. In our Pascal example, to write the square of 3 using this function, we would have the statement

```
WRITE (SQUARE (3))
```

The value passed to SQUARE, 3, is called the **argument.** To find the image of an argument or domain element, we use the name of the function and the argument. To have the square of Y + 7 printed, where Y has been assigned a value, the statement

```
WRITE (SQUARE (Y + 7))
```

could be used. □ ■

EXAMPLE 4

The square function is not built into the language FORTRAN. The following could be used to define a function to square integer variables.

Solution

LANGUAGE EXAMPLE

```
INTEGER FUNCTION SQUARE (A)
INTEGER A
     SQUARE = A * A
RETURN
END
```

FORTRAN does not allow data types to be subsets of integers such as 1 .. 5 and 1 .. 30 in Pascal Example 3. To mimic the FORTRAN definition, the first line of the Pascal example would read

```
FUNCTION SQUARE (A: INTEGER): INTEGER;
```

otherwise, notice the strong similarity between the function definitions of both languages. □ ■

The mathematical terminology is carried over to computer science as the word FUNCTION, a reserved word in Pascal, FORTRAN, and many other languages, indicates that a function definition is to follow. Words from mathematics are used in data base theory as well; the **domain** of an attribute in a relational data base is the set of all possible values of that attribute. Thus, in the STUDENT-ADDRESS relation (SSN, name, address, phone) of Section 4.1, the domain for the SSN attribute might be the set I9 of all possible social security numbers.

Some computer languages are more functional than others. For example, the languages LISP, APL, and MACSYMA have an enormous number of built-in functions, and programs consist of new functions defined using these and other user-defined functions. John Backus, who headed the team that created the first major high-level computer language, FORTRAN, said in his 1977 ACM Turing Award Lecture that new languages need to be developed that are more functional and, hence, more mathematical in nature. ☐

HISTORICAL NOTE

A function may have the additional properties of being one-to-one and/or onto. The second coordinates of the function {(Smith, F), (Jones, F)} are both "F," representing two females. Since we do not have a pairing of one person for each sex, the function is not one-to-one. Two different domain elements have the same image.

EXAMPLE 5

The function $f(a) = a^2$, $a \in \{1, 2, 3, 4, 5\}$, of Example 2 is a one-to-one function. The preimage of any element from the range $\{1, 4, 9, 16, 25\}$ consists of exactly one element, as indicated in Figure 5.2. ■

Figure 5.2

Picture of mapping $f(a) = a^2$ in Example 5

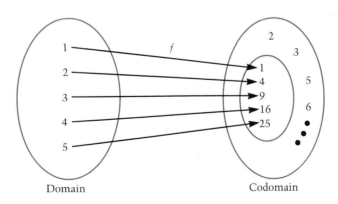

Domain   Codomain

DEFINITION

A function $f : A \rightarrow B$ is **one-to-one** if $f(a_1) = f(a_2)$ implies $a_1 = a_2$ for $a_1, a_2 \in A$.

In other words, distinct elements of domain A have distinct images. Two different ordered pairs do not exist with the same second element. In fact, with a one-to-one function we could switch all the first and second elements of the ordered pairs and still have a function whose domain would be the range of the original function. The function $g(b) = \sqrt{b}$ with $b \in \{1, 4, 9, 16, 25\}$ (Figure 5.3) is such an **inverse** of the function f in Example 5. Inverse functions are covered in greater detail in Section 5.4.

Figure 5.3

Picture of mapping $g(b) = \sqrt{b}$, $b \in \{1, 4, 9, 16, 25\}$

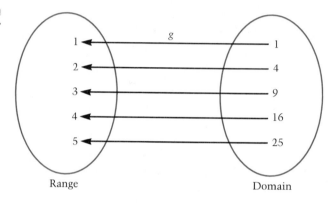

Range　　　　　　　　　　　　　　Domain

All characters are represented in the computer with strings of 0's and 1's. Any scheme that does such an encoding (ASCII and EBCDIC are the most common schemes) must be one-to-one. Each character has a unique encoded form, and each encoded form represents exactly one character. Decoding, or translating from a string of 0's and 1's to a character, is the inverse function of coding.

In early computers, the number zero was represented in two different forms as $+0$ and -0. Thus, the function that mapped a numeric computer representation to the corresponding number was not one-to-one; two different representations mapped to 0. This deficiency of not having a one-to-one function created errors when testing conditions such as

HISTORICAL NOTE

    IF　X = Y　THEN. . . .

Perhaps, the computer representation of X was $+0$ while that of Y was -0. Think how confusing it must have been to the programmer for the condition (X = Y) to be FALSE in this case. □

As another example of a one-to-one function consider a school where each student has an identification number (id-number) from 1 to 5000 and where the record of each student is stored in a file whose locations are numbered from 1 to 5000. For each id-number there is a location with the same number; for each file location there is a corresponding id-number. This type of arrangement, which is called **relative file organization,** needs a one-to-one function from the set of id-numbers to the set of file locations.

In records for a file or data base there is often a **key** or field whose value uniquely identifies the record. The student id-number, for example, uniquely identifies the student record. Social security numbers in an employee file or product codes in a file of manufactured parts are good choices for keys. There is a one-to-one function from the key to the record.

DEFINITION

A function $f : A \rightarrow B$ is **onto** if the codomain equals the range.

Though $f(a) = a^2$, $f : \{1, 2, 3, 4, 5\} \rightarrow \{1, 2, \ldots, 30\}$, of Example 5 is one-to-one, it is not onto since the codomain, $\{1, 2, \ldots, 30\}$, is not equal to the range, $\{1, 4, 9, 16, 25\}$. If the codomain had been given as $\{1, 4, 9, 16, 25\}$, f would have been onto.

Just as a function can be one-to-one but not onto, it can be onto but not one-to-one. $\{(Smith, F), (Jones, F), (Wilson, M)\}$ is a function which is not one-to-one but is onto $\{F, M\}$. If in a data base we have an onto function from a set of names to the set $\{M, F\}$, we know we will get a nonempty response to a query requesting a list of names of males.

EXERCISES 5.1

Do the following in Exercises 1–28. ***a.*** *For each of the relations evaluate any requested values.* ***b.*** *Decide whether each of the relations is a function or not. If not, state why. If it is a function, give the domain, codomain, and range.* ***c.*** *Is the function one-to-one? Onto? Refer to Section 1.1 for set definitions.*

1. $f(x) = x^2$, $f : \{x \mid -5 \leq x \leq 5, x \in \mathbb{I}\} \rightarrow \mathbb{I}$.
   Find the images of -2, 2, $2 + 1$, c, and $c + 1$.
   Find the preimages of 16, 36, and b^2.

2. $g(m) = [r]$, where $m \equiv r \pmod{5}$, and $g : \mathbb{I} \rightarrow \{[0], [1], [2], [3], [4]\}$.
   Find $g(87)$ and the preimage of $[0]$.

3. $h(x) = x + 1$, $h : \mathbb{N} \rightarrow \mathbb{I}^+$.
   Find $h(3)$, $h(2)$, $h(3 \cdot 2)$, $h(3 \cdot x)$, and preimage of 9.

4. $m(a) =$ wife of a, $m : \{x \mid x$ is a married man$\} \rightarrow \{y \mid y$ is a married woman$\}$. (Assume no bigamy.)

5. $l(w) =$ length of w, $l :$ set of English words $\rightarrow \mathbb{N}$. Find l(bit).

6. $\{(n, m) \mid n \mid m\}$, $d : U \rightarrow U$ with $U = \{1, 2, 3, 20, 30, 60\}$. What maps to 20?

7. $d(n) = m$, where $n \mid m$, $d : \mathbb{P} \rightarrow \mathbb{P}$, the set of primes. Find the image of 7. Find the preimage of 7.

8. The relation of Example 2, Section 4.1.

9. $\{(3, 15), (5, 15), (15, 45), (45, 45)\}$ on $U \times U$ with $U = \{3, 5, 15, 45\}$. Find the preimage of 15.

10. $\text{abs}(x) = |x|$, the absolute value of x, abs : $\mathbb{I} \to \mathbb{N}$. Find the preimage of 15. Find the preimage of n when $n \in \mathbb{N}$. Write $\text{abs}(a)$ and $\text{abs}(2 \cdot a)$.

11. $r = \{(a, b) \mid a = b^2, a \in \mathbb{R}\}$, $r : \mathbb{R} \to \mathbb{R}$.

12. op : $\{s \mid s$ is a string of 7 bits$\} \to \mathbb{B}$, where

$$\text{op}(s) = \begin{cases} 0 & \text{if } s \text{ has an odd number of 1's} \\ 1 & \text{if } s \text{ has an even number of 1's.} \end{cases}$$

Evaluate $\text{op}(1001111)$ and $\text{op}(0001010)$.

Note: This is a function for calculating **odd parity** (op) for a string of bits. When transmitting a string of bits from one computer to another, parity can be used as a rough check of the accuracy of transmission. The parity bit is attached to the string of 7 bits. The resulting 8 bits should always have an odd number of 1's.

13. $f : S \to D \times P$, where S is the set of students at a school; D is the set of dorm room numbers; P is the set of all phone numbers for telephones in dorm rooms; and $f(\text{student}) = (\text{dorm room number, phone number})$.

14. $p(n) = 2^n$, where $p : \mathbb{N} \to \mathbb{N}$. Evaluate $p(5)$, $p(m)$, and $p(m + 3)$.

15. $n(S) = $ the number of elements in set S, where $n : \mathscr{P}(U) \to \{m \mid 0 \le m \le \text{number of elements in } U, m \in \mathbb{N}\}$
Find $n(\{a, b, a, c, a, d\})$.

16. $r(x) = 1/x$, $r : \mathbb{R} \to \mathbb{R}$. Evaluate $r(4)$ and $r(1/4)$.

17. $f(x, y) = x + y$, $f : \mathbb{N}^2 \to \mathbb{N}$
Evaluate $f(6, 10)$ and $f(6, b)$. Find the preimage of 5.

18. $\text{wage}(\text{hours, rate}) = \text{hours} \times \text{rate}$, where wage : $\mathbb{N} \times \mathbb{Q}^+ \to \mathbb{Q}$. $\mathbb{Q}^+$ is the **set of positive rational numbers.** Evaluate $\text{wage}(50, 8.0)$.

19. $\text{wage}(\text{hours, rate}) = \begin{cases} \text{hours} \times \text{rate} & \text{if hours} \le 40; \\ 40 \times \text{rate} + 1.5 & \text{if rate} > 40. \\ \quad \times (\text{hours} - 40) \times \text{rate} & \end{cases}$

wage : $\mathbb{N} \times \mathbb{Q}^+ \to \mathbb{Q}^+ \cup \{0\}$. Evaluate $\text{wage}(50, 8.0)$.
Note: This function can be used to calculate wages taking into account overtime.

20. $\text{factorial}(n) = n!$, factorial : $\mathbb{N} \to \mathbb{N}$. Calculate $\text{factorial}(5)$, $\text{factorial}(m)$, and $\text{factorial}(m + 1)$.

21. $\text{perm}(n, r) = \begin{cases} P(n, r) & 0 \le r \le n \\ 0 & \text{otherwise} \end{cases}$
perm : $\mathbb{N} \times \mathbb{N} \to \mathbb{N}$. Find $\text{perm}(5, 0)$; preimage of 1.

22. $\text{comb}(n, r) = \begin{cases} C(n, r) & 0 \le r \le n \\ 0 & \text{otherwise} \end{cases}$
comb : $\mathbb{N} \times \mathbb{N} \to \mathbb{N}$. Find $\text{comb}(5, 0)$; preimage of 1.

23. $b(v) = \begin{cases} 0 & \text{if } 0 \le v \le 2 \\ 1 & \text{if } 2 < v \le 4 \end{cases}$
$b : \{v \mid 0 \le v \le 4, v \in \mathbb{R}\} \to \mathbb{B}$. Find $b(2.7)$ and the image of 0.3.
Note: v might stand for a voltage between 0 and 4, and b might be the function that interprets that voltage as a bit.

24. $t(s) = \begin{cases} \text{TRUE} & \text{if a condition is true} \\ \text{FALSE} & \text{if a condition is not true} \end{cases}$

t : set of conditions $\rightarrow$ {TRUE, FALSE}. Find $t(3 = 5)$.
Note: A function that has a codomain of {TRUE, FALSE} or {T, F} is a **predicate** function.

25. $\text{ODD}(n) = \begin{cases} \text{T} & \text{if } n \text{ is odd} \\ \text{F} & \text{otherwise} \end{cases}$

ODD: $\mathbb{I} \rightarrow$ {T, F}. Find the preimage of T.

26. $f(a, b) = (a - 1, b - 1)$, $f : \mathbb{R}^2 \rightarrow \mathbb{R}^2$
Find $f(7, -40)$ and $f(x^2, y)$.

27. $F(r) = (t, v)$, $F : R \rightarrow T \times V$, the relation of Example 4, Section 4.1:
{(Smith, F, Math), (Smith, F, CS), (Jones, F, CS)}

28. $f(x_1, x_2, x_3, x_4, x_5, x_6) = \begin{cases} 0 & \text{if any } x_i \text{ is } 0 \\ 1 & \text{otherwise} \end{cases}$

$f : \mathbb{B}^6 \rightarrow \mathbb{B}$. Evaluate $f(1, 0, 0, 1, 1, 0)$, $f(0, 0, 0, 0, 0, 0)$, $f(1, 1, 1, 1, 1, 1)$, $f(1, 1, 1, 0, 1, 1)$, the preimage of 1, and the preimage of 0.
Note: A function that maps from $\mathbb{B}^n$ to $\mathbb{B}$, where n is a positive integer, is called a **combinational** or **switching function** and is a mathematical model of a combinational circuit. The function of Exercise 12 also can be interpreted as a combinational function.

Find the number of functions f: U $\rightarrow$ V *for sets* U *and* V *satisfying the conditions in Exercises 29–32. (Use the Fundamental Counting Principle.)*

29. $n(U) = n$, $n(V) = 2$ **30.** $n(U) = n(V) = 50$

31. $n(U) = 35$, $n(V) = 50$ **32.** $n(U) = 50$, $n(V) = 35$

33. What is the general rule for the number of functions $f : U \rightarrow V$ where $n(U) = s$ and $n(V) = t$?

Find the number of one-to-one functions f : U $\rightarrow$ V *for each of the situations in Exercises 34–37. (Use the Fundamental Counting Principle.)*

34. $n(U) = n$, $n(V) = 2$ **35.** $n(U) = n(V) = 50$

36. $n(U) = 35$, $n(V) = 50$ **37.** $n(U) = 50$, $n(V) = 35$

38. What is the general rule for the number of one-to-one functions $f : U \rightarrow V$ for finite U and V? Consider two situations for $n(U) = s$ and $n(V) = t$: (1) $s \le t$ and (2) $s > t$.

Find the number of onto functions f : U $\rightarrow$ V *for each of the situations in Exercises 39–44. (Use the Fundamental Counting Principle.)*

39. $n(U) = n$, $n(V) = 2$

40. $n(U) = n(V) = 50$

41. $n(U) = 35$, $n(V) = 50$

42. $n(U) = 3$, $n(V) = 2$

43. $n(U) = 50$, $n(V) = 2$

44. $n(U) = s > 1$, $n(V) = 2$

45. Complete the definition of the Pascal function to add two parameters, X and Y:

____a____ F(X, Y : INTEGER) : ___b___ ;

  BEGIN

    ___c___

  END; ☐

46. Write a definition of a FORTRAN function to add the integer parameters, X and Y. ☐

47. a. Write a function in mathematical terms to calculate absolute error.
  b. Give the domain and range.
  c. Write a function in Pascal to calculate absolute error.
  d. Write a FORTRAN function to calculate absolute error. ☐
  e. Write a function in mathematical terms to calculate the relative error as a number.

48. Let f be a combinational function from $\mathbb{B}^8$ to $\mathbb{B}$ where

$$f(x_1, x_2, x_3, x_4, x_5, x_6, x_7, x_8) = \begin{cases} 1 & \text{if exactly 3 } x's = 1 \\ 0 & \text{otherwise.} \end{cases}$$

Find the number of elements
  a. In the domain,
  b. In the preimage of 1, and
  c. In the preimage of 0.

49. a. How many elements are in $\mathbb{B}^n$?
  b. How many combinational functions are there from $\mathbb{B}^n$ to $\mathbb{B}$?
  c. How many are there if $n = 4$?
  d. Suppose $n = 4$. How many combinational functions are there that have a value of 0 when $x_1 = 0$?
  e. For $n = 4$, how many combinational functions are there that have a value of 0 when $x_1 = 0$ and a value of 1 when $x_1 = 1$?

5.2

Some Mathematical Functions

There are several mathematical functions that occur frequently in applications and, hence, are important in problem solving with the computer.

Most compilers do not have the square function, $f(x) = x^2$, built-in, but it is easy to define as $f(x) = x * x$ or $f(x) = x ** 2$. Many compilers, however, do have a standard square root function, $SQRT(X) = \sqrt{X} = X^{1/2}$. Often the argument must be REAL, so that SQRT(9.0) returns 3.0, while using SQRT(9) gives an error. SQRT(−9.0) is also in error since there is no real number whose square is −9.0.

The absolute value function, $ABS(X) = |X|$, may be built-in, but examination of its mathematical definition allows us to create the function in a language in which it is lacking. As indicated in Section 2.6, use of this definition is also essential for some proofs.

> **DEFINITION** ▨▨▨▨▨▨▨▨▨▨▨▨▨▨▨▨▨▨▨▨▨▨▨▨▨▨▨▨
>
> The **absolute value function** is defined as
>
> $$ABS(X) = |X| = \begin{cases} X & \text{if } X \geq 0 \\ -X & \text{if } X < 0 \end{cases}$$

EXAMPLE 1

Though the absolute value is a part of Pascal and FORTRAN77, with this definition as the basis of our algorithm we could write the function ourselves.

Solution

In Pascal we have

```
FUNCTION ABSOLUTE (X: REAL) : REAL;
     BEGIN
     IF X >= 0 THEN
            ABSOLUTE := X
     ELSE
            ABSOLUTE := (-X)
     END;
```

Note that "greater than or equal" is written as ">=" in Pascal and as ".GE." in FORTRAN. The FORTRAN function definition varies only in syntax from that of Pascal.

```
REAL FUNCTION ABSOLUTE (X)
REAL  X
     IF  X .GE. 0  THEN
            ABSOLUTE = X
     ELSE
            ABSOLUTE = (−X)
     ENDIF
RETURN
END   □
```

Knowing a precise mathematical definition was essential to our success in writing the function. Some vague idea about dropping the sign doesn't work when we are calculating the absolute value of a variable. ■

The mathematical function projection is useful in relational data bases as well as other computer science applications. For the STUDENT-ADDRESS relation with structure

(SSN, name, address, phone)

the projection onto the first coordinate is

$\pi_1(a, b, c, d) = a.$

Thus, π_1(888-99-7777, Dale Smith, 666 Main St., 555-1111) is 888-99-7777. A projection of this 4-tuple onto the fourth coordinate is 555-1111.

DEFINITION

Let $A_1, A_2, \ldots, A_n$ be sets. The **projection onto the ith coordinate** is a function

$$\pi_i : A_1 \times A_2 \times \cdots \times A_i \times \cdots \times A_n \to A_i$$

such that $\pi_i(a_1, a_2, \ldots, a_i, \ldots, a_n) = a_i$.

Projections can be useful in analyzing data as well. Figure 5.4 is a 3-D graphics plot of data from a scientific study involving several parameters. Measurements for three variables, DELTA A, DELTA B, and DELTA LOAD, were taken and plotted as triples. The projection onto the DELTA A-DELTA B plane illustrates the relationship in the data between these two parameters. For instance, the projection in Figure 5.4 shows that for values of DELTA B near 0, DELTA A values cluster near 0. Also illustrated in this figure is the fact that shadows for a graphics display are achieved using projections.

Figure 5.4

3-D plot of data for three variables with a projection onto the DELTA A–DELTA B plane. Reprinted from "Three Dimensional Graphics for Scientific Data Display and Analysis" by Stanley L. Grotch, University of California, Lawrence Livermore National Laboratory. Work performed under the auspices of the Department of Energy

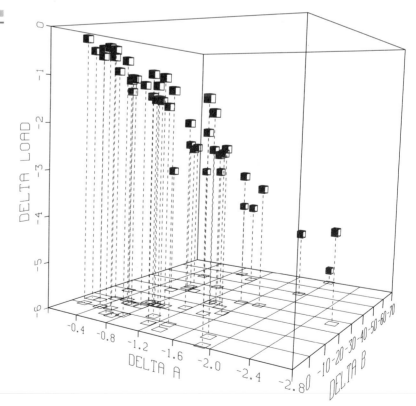

A number of statistical functions are useful in mathematics and computer science and provide information for the making of decisions. Statistical packages such as SAS (Statistical Analysis System) and SPSS (Statistical Package for the Social Sciences) have many built-in statistical procedures that are easy to use but powerful. We will discuss some elementary descriptive statistics.

The frequency of a number in a collection tells you how often the number occurs. Consider the following collection of data or data set: 11, 5, 7, 8, 13, 7, 9. The frequency of 7, freq(7), is 2 since 7 occurs twice in the list. As another example, freq(15) = 0.

DEFINITION

Let U be a set of numbers, and let L be a collection of data with values from U. The **frequency function** $freq_L : U \to \mathbb{N}$ is defined as

$$freq_L(u) = \text{the number of occurrences of } u \text{ in } L.$$

An analysis of the frequency of different types of statements in programs written in languages such as FORTRAN and Pascal shows that the assignment statement occurs with the highest frequency.

Often frequency is expressed not only in absolute or actual numbers but in relative terms as a percent of the total number. Percent is a statistic that is useful in many areas including calculation of your grade. If at the end of the semester you have earned 573 out of a possible 600 points, you have obtained

$$573/600 = 0.955 = 95.5\%$$

of the maximum possible points.

DEFINITION

Let m and n be numbers with $m \geq 0$ and $n > 0$. The **percent** that m is of n is

$$percent(m, n) = (m/n) \times 100\%.$$

Table 5.1 is from a study that examined application programming in 43 industries. The table gives the absolute and relative frequencies of each type of industry. Of the $n = 43$ companies the frequency of financial institutions was 7; freq(financial) = 7 = m. Thus, the relative frequency of financial industries or the percent of those was

$$(7/43) \times 100\% = 16.3\%.$$

TABLE 5.1

Frequency Distributions by Industry Type[a]

Financial	7	(16.3%)
Merchandising	8	(18.6%)
Transportation and utilities	6	(14.0%)
Food processing	6	(14.0%)
Other manufacturing	14	(32.6%)
Government	2	(4.7%)
	43	(100.0%)

[a]Reprinted from "A Study of Application Program Development Techniques" by Tor Guimaraes, *Communications of the Association for Computing Machinery*, 495, May 1985. Copyright 1985, Association for Computing Machinery, Inc., reprinted by permission.

Sometimes only the relative frequencies or percentages are given, as in column *A* of Table 5.2. Those numbers give the percentage of the 43 companies surveyed using various methods of documentation during development of a system of application programs. Notice that 86% of the 43 require or encourage an English narrative of what the system does during system development.

TABLE 5.2

Usage of Document Types, Methods, and Tools[a]

Document type	A	B	C		D
Systems flowchart	63	29	4.5	(52)	4.8
Comment codes	82	72	4.8	(130)	4.3
Pseudocode	9	8	1.9	(14)	2.6
English narrative	86	69	5.0	(124)	5.0
HIPO	4	5	3.9	(9)	4.0
Input/output layout	100	63	4.3	(113)	4.8
Structured programming	54	17	5.0	(31)	5.0
Program flowcharts	83	37	1.8	(67)	1.3
Program listing	—	100	5.0	(180)	5.0
Warnier diagrams	5	11	4.3	(20)	2.9
Decision tables	9	38	4.4	(68)	4.0

Note. A, Percentage of companies enforcing or encouraging the use of the particular item during system development ($n = 43$). B, Percentage of sampled application programs having the particular item as part of its documentation (5 case-study companies, $n = 180$). C, Average usefulness rating. For each sample program documented with the particular item, the maintenance programmer rated the usefulness of the item for maintenance purposes (max = 5, definitely useful; min = 1, definitely not useful). The sample size for each item is given in parentheses. D, Average general usefulness rating according to all maintenance programmers interviewed (5 case-study companies, $n = 79$).

[a]Reprinted from "A Study of Application Program Development Techniques" by Tor Guimaraes, *Communications of the Association for Computing Machinery*, p. 497, May 1985. Copyright 1985, Association for Computing Machinery, Inc., reprinted by permission.

The 43 in Table 5.1 is the sum of the numbers above it. Calculation of the sum of a collection of numbers is usually part of one of the earlier programming assignments in a first computer science course. Summation continues to be important later in both science and business applications. The following algorithm adds together a list of numbers as they are being read:

1. Initialize SUM to 0.
2. Read a number.
3. Repeat while there is data:
   a. Add the number to SUM.
   b. Read a number.

The "read" in Step 2 is called a "priming read" and avoids the problem of trying to read nonexistent data at the end-of-file.

If we wanted to calculate the average or arithmetic mean of the data, we must count as well as add:

1. Initialize SUM and COUNT to 0.
2. Read a number.
3. Repeat while there is data:
   a. Add the number to SUM.
   b. Increment COUNT.
   c. Read a number.
4. If COUNT = 0 then
      MEAN ← 0
   else
      MEAN ← SUM/COUNT.

Notice that care was taken to make sure that division by zero did not occur in SUM/COUNT. Also, nothing is done to evaluate mean until after the sum has been calculated completely. For data 11, 5, 7, 8, 13, 7, 9, the sum is 60 while the mean or average is 60/7 = 8.5714.

DEFINITION

Let U be a set of numbers, and let L be a collection of numbers, $x_1, x_2, \ldots, x_n$, from U. The **arithmetic mean** of L is

$$\overline{x} = \text{mean}(L) = \frac{x_1 + x_2 + \cdots + x_n}{n}$$

where mean is a mapping from the set of all collections of elements from U to $\mathbb{R}$.

Referring again to Table 5.2, we see in column D an analysis of responses as to the general usefulness of each type of documentation from 79 programmers who maintain systems. Each maintenance programmer rated the item on a scale of 1 to 5 with 5 being "definitely useful." The average response for each item was calculated. Thus, adding together all 79 ratings (1–5) and dividing by 79 yielded an average usefulness rating of 4.3 for comment codes.

The median is often used as a measure of central tendency or the center of the data as well. This measure provides additional information upon which to base a decision. To calculate the median, sort the data into ascending order and pick the middle number. The collection, 11, 5, 7, 8, 13, 7, 9, sorted is

5, 7, 7, 8, 9, 11, 13

with median 8, the fourth entry in this list of 7 items. Note that $4 = (7 + 1)/2$. Suppose 13 is not in the list so that after sorting we have

5, 7, 7, 8, 9, 11.

Consequently, there are two middle items, 7 and 8. The median is the mean or average of these values, 7.5. Items 7 and 8 are the third and fourth entries out of the 6 elements in the list. Note that $3 = 6/2$ and $4 = (6/2) + 1$.

DEFINITION

Let U be a set of numbers, and let L be a collection of values from U. The **median** is a function from the set of all such collections to $\mathbb{R}$ such that

median(L) = the middle value of sorted L.

If the number of entries n in L is odd, the middle value is the $(n + 1)/2$ entry. If n is even, the middle value is the mean of the $n/2$ and the $(n/2) + 1$ entries.

In Table 5.3 the mean and median values for various items are listed. Notice that the mean EDP (electric data processing) budget is \$5.2 million, while the median is much less, \$2.8 million. From these figures we can deduce that there are a few companies that have budgets significantly more than the median, pushing the mean higher.

TABLE 5.3
Attributes of the Corporate Sample[a]

	Mean	Median	Standard deviation	Range
Gross revenues	$550M	$220M	$750M	$2.98B
EDP budget				
In absolute terms	$5.2M	$2.8M	$7.3M	$34.8M
As percentage of revenues	1.2%	1.0%	1.9%	3.3%
Yearly expenditures for systems development				
In absolute terms	$845K	$335K	$1M	$4.5M
As percentage of EDP budget	19.0%	19.0%	10.0%	42.0%
Yearly expenditures for systems maintenance				
In absolute terms	$584K	$240K	$970K	$5.8M
As percentage of EDP budget	13.0%	11.0%	8.0%	31.0%
Yearly expenditures for application software packages				
In absolute terms	$50K	$30K	$85K	$500K
As percentage of systems development expenditures	1.3%	1.0%	1.5%	31.0%
Number of programs	2.2K	1.3K	2.3K	9.8K
Number of application development standards	4	4	—	10
Using chargeback (system development)	35% (additional 10% within next year)			
Using DBMS	58% (additional 29% within next year)			
Using query language	67% (additional 7% within next year)			

Note. K, thousand; M, million; B, billion. Sample size = 43.

[a]Reprinted from "A Study of Application Program Development Techniques" by Tor Guimaraes, *Communications of the Association for Computing Machinery*, 495, May, 1985. Copyright 1985, Association for Computing Machinery, Inc., reprinted by permission.

EXERCISES 5.2

Calculate the values in Exercises 1–19.

1. $\sqrt{0}$ **2.** $\sqrt{91}$ **3.** $\sqrt{-16}$ **4.** $\sqrt{(-4)^2}$

5. $|-3/4|$ **6.** $|0|$ **7.** $\sqrt{2^6}$ **8.** $(\sqrt{2})^6$

9. $(\sqrt{7})^2$ **10.** $\sqrt{18}/\sqrt{2}$ **11.** $\sqrt{4 \cdot 9}$ **12.** $\sqrt{4} \cdot \sqrt{9}$

13. $\sqrt{9 + 16}$ **14.** $\sqrt{9} + \sqrt{16}$ **15.** $\sqrt{16 + 64}$ **16.** $\sqrt{16} + \sqrt{64}$

17. $\pi_1(a, b, c, d, e)$ **18.** $\pi_4(a, b, c, d, e)$ **19.** $\pi_6(a, b, c, d, e)$

Simplify Exercises 20–29.

20. $|-y|$ **21.** $\sqrt{x^2}$ **22.** $(\sqrt{x})^2$,
      for $x \geq 0$ for $x \geq 0$

23. $\sqrt{x} \cdot \sqrt{x}$,   **24.** $\sqrt{x}/\sqrt{x}$,   **25.** $\sqrt{x \cdot y}/\sqrt{y}$,
      for $x \geq 0$ for $x > 0$ for $x \geq 0, y > 0$

26. $\sqrt{x^2 + y^2}$ (See Exercises 13, 14, 15, 16)

27. $x/\sqrt{x}$
for $x > 0$

28. $\sqrt{x^2 \cdot y^2}$,
for $x, y \geq 0$

29. $\sqrt{x^2} \cdot \sqrt{y^2}$,
for $x, y \geq 0$

30. Consider the following data: 24, 3, 25, 3, −17, 0, −6, 25, 25, 14, 7, 25, −11, 3.
  a. Calculate the frequencies of −6, 3, 25, and 26.
  b. Calculate the mean and median.

31. Consider the following data: −22, −3, 46, 9, 46, 51, −22, −73, −6.
  a. Find the frequencies of −3, 0, and 51.
  b. Find the mean and median.

32. Write an algorithm for evaluating $freq_L(u)$ while reading in the data of L.

33. **a.** Give the domain and range of $f(x) = |x|$, where x is real.
  b. Is it one-to-one? If not, why?

34. Repeat Exercise 33 for $f(x) = \sqrt{x}$, where $x \geq 0$.

35. For the Pascal (or FORTRAN) function definition of ABSOLUTE in this section, give the
  a. Domain and
  b. Range.
  c. Is it onto? ☐

36. **a.** Give the range of π_3 if the domain is $\mathbb{R} \times \mathbb{N} \times \mathbb{I}$.
  b. Is π_3 one-to-one? If not, why?
  c. Is it onto? If not, why?

37. Write a Pascal (or FORTRAN) function for SUM following the algorithm in this section. ☐

For each function in Exercises 38–41 state whether it is one-to-one or not. If not, give a counterexample.

38. $freq_L$ with $U = \mathbb{I}$

39. mean with $U = \mathbb{Q}$

40. ABS with domain $= \{n \,|\, n < 0, n \in \mathbb{I}\}$

41. median with $U = \mathbb{I}$

42. **a.** In Table 5.1 what is the meaning of 18.6% in row 2, and how was it obtained?
  b. Find the median and
  c. Mean of the 6 frequencies.

43. **a.** In Table 5.2 the average ratings of English narrative, structured programming, and program listings are 5.0. What does this indicate about how individuals rated these documentation types?
  b. 63% of the 43 companies want system flowcharts at the development stage (see column A); how many companies does this represent?
  c. How many companies want program flowcharts during systems development?
  d. In columns B and C statistics on 180 application programs from 5 companies are noted. What connection is there between the 29 for systems flowchart in column B and the 52 in parentheses in column C?

Note: Read the descriptions of B and C in the table.

44. Give the **a.** domain and   **b.** range of the percent function: percent(m, n) = (m/n) × 100.
   c. Write a Pascal definition that accepts two real inputs for M and N and returns the percent that M is of N. The function should return −1 if M or N is negative or if N is 0. ☐

45. The following appeared in the sample questions of the *1984 AP Course Description in Computer Science**:
   Write a function in Pascal to compute the average of a list of *n* numbers. Assume that the declarations

   ```
   const  n = 100; {number of items}
   type   ItemList = array[1..n] of real;
   ```

   are present and begin your function definition with the statement

   ```
   function Average (Items: ItemList): real;
   ```
   ☐

46. In an article by Litecky and Rittenberg (1981) a survey of computer auditors (CPAs) who evaluate computer systems and provide recommendations for improving control was discussed. The following is a table summarizing questionnaire distribution and response rate. Fill in the numbers that have been left blank.

Questionnaire Distribution and Response Rate

	Number	Percent
Number of questionnaires delivered		195
Number of questionnaires returned		146
Response rate		**a** %
Number of CPA firms participating in study		9
Average number of responses per firm		**b**
Job category of respondents	Number	Percent
Audit managers or partners	**c**	13.7 %
Management Advisory Services managers or partners	90	**d** %
Computer audit specialists	**e**	24.7 %
	146	100.0 %

47. **a.** In Table 5.3 how would you explain the differences in the mean and median of gross revenues for the 43 companies?
   b. On the average, what percent of the budget is used for systems development?
   c. For maintenance?   **d.** For packages?

48. The following chart is presented in *The Software Revolution* (R. T. Fertig, 1985).
   a. How much did manufacturing spend on micros in 1983?
   b. How much was projected for manufacturing to spend in 1986?
   c. Compare how much the home market will spend on micros in each of the two years.

---

*AP questions selected from *AP Course Description in Computer Science.* College entrance Examination Board, 1984. Reprinted by permission of Educational Testing Service, the copyright owner of the sample questions.

U.S. micro market by industry sector. (Reprinted by permission of the publisher from Robert T. Fertig, *The Software Revolution: Trends, Players, Market Dynamics in Personal Computer Software.* Copyright 1985 by Elsevier Science Publishing Co., Inc.)

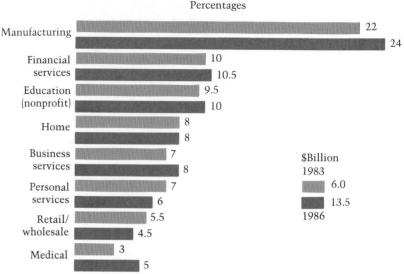

Percentages

Manufacturing 22 / 24

Financial services 10 / 10.5

Education (nonprofit) 9.5 / 10

Home 8 / 8

Business services 7 / 8

Personal services 7 / 6

Retail/wholesale 5.5 / 4.5

Medical 3 / 5

$Billion
1983
6.0
13.5
1986

*Others include agriculture, real estate, construction, legal and miscellaneous

49. **a.** Plot the following points on the Cartesian coordinate system: (3, 5), (−2, −3), (0, 4), (3, 2), (1, −3).
   b. Evaluate $\pi_1(x, y)$ for each point (x, y) of Part a.
   c. Project the points of Part a to the *x*-axis by plotting $(x, 0)$ for each point (x, y).

50. The following is a pie chart from *The Software Revolution*. Fill in the omitted numbers.

Microcomputer software purchases by distribution channel (1984–1990 cumulative projections in billions of dollars); total market value = $94.2 billion. (Reprinted by permission of the publisher from Robert T. Fertig, *The Software Revolution: Trends, Players, Market Dynamics in Personal Computer Software.* Copyright 1985 by Elsevier Science Publishing Co., Inc.)

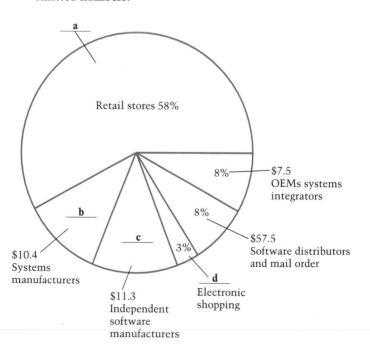

a

Retail stores 58%

8% — $7.5 OEMs systems integrators

8%

$57.5 Software distributors and mail order

3%

b

c

d

$10.4 Systems manufacturers

$11.3 Independent software manufacturers

Electronic shopping

51. In an article by Biermann *et al.* (1985) a study was presented of how well people learned to use a voice natural language processor. Fill in the numbers for the following statement found in the article: "Of the 1365 transactions actually scored, . . . , 1053, or _____ percent were successful, and _____ or 22.9 percent were unsuccessful."

52. Hiltz and Turoff (1985) analyzed the overload of information that can be experienced with electronic mail, computerized conferencing, and bulletin-board systems. Part of a table on reported overload experience is given below. Notice the numbers of responders in each system are given below the table.
    a. Fill in the blanks.
    b. How many from the EIES1 system answered "never"?

User's Reported Overload Experience
Responses are to the question, "Thinking back over your experiences with the system, how frequently have you felt overloaded with information?"

| System | Responses (by percent) | | | | | Mean[a] | Modal Use |
	Always	Almost Always	Sometimes	Almost Never	Never		
EIES1	4	18	55	16	6	—	20–49 hr 35%
EIES 2	6	23	42	20	9	3.0	10–49 hr 62%
COM	3	16	43	32	5	3.2	<10 hr 57%
PARTI	7	17	48	23	—	3.0	20–49 hr 48%
INTMAIL	2	5	25	45	23	—	5–9 hr 43%
PLANET	3	6	44	24	22	3.6	

Sources: EIES1—follow-up questionnaire 3 to 4 months after the beginning of system use; 110 scientific and professional users; EIES2—4-month follow-up survey of new users (1983), responses from 140 managers and professionals; COM—4-month follow-up survey of new users (1983), 37 responses; PARTICIPATE (on the Source)—sample of 100 new users (1983–1984); INTMAIL—follow-up at 4 months of internal corporate users of a commercial mail system, 60 responses from managerial users; PLANET—percentages and means computed from raw data, 115 scientific and professional users. Copyright 1985, Association for Computing Machinery, Inc., reprinted by permission.

Note on use categories: Cumulative time on line at end of 4 months was arranged into four categories— less than 4 hr, 5–19 hr, 20–49 hr, and 50 or more hours. The modal category and percentage in that category are shown as rough guide to the typical activity level of respondents.

[a]The mean is determined by responses from 1 for "always" to 5 for "never."

53. Weiser (1982) studied the use of a technique for debugging called slice. Program slicing is tracing backwards from a particular variable in a statement to find the source of an error. The demographics of the participants in the study revealed from the number of software courses taught a minimum of 0, a maximum of 31, a median of 1, and a mean of 5.4. How can you explain the large difference in the median and mean?

54. SPSS is a system of programs used for statistical analysis of data. For example, a data set of values for two variables might be read and analyzed with a cross tabulation as presented below. Perhaps VAR1 stands for sex (0 ≡ male and 1 ≡ female) and VAR2 stands for whether a person likes a particular TV show or not (0 ≡ doesn't like and 1 ≡ likes). The meaning of the four numbers in the four inside boxes are given in the top left corner.

a. How many men like the program?

b. How many women do not like it?

c. What percent of the women like the program?

d. What percent of those who like it are women?

e. How were the three percentages in the 0–0 block calculated?

f. What is the meaning of each of those percents?

Count *Row percentage* *Column percentage* *Total percentage*	*Var2*		
	0	*1*	*Row total*
Var1 0	22 47.8 28.2 14.3	24 52.2 31.6 15.6	46 29.9
1	56 51.9 71.8 36.4	52 48.1 68.4 33.8	108 70.1
Column total	78 50.6	76 49.4	154 100.0

5.3

Built-in Functions

LANGUAGE EXAMPLE

Two other functions useful in statistical and other studies are the maximum and minimum of a list of numbers. Even if these are built-in functions, their forms may still vary from one language to another. In some languages, the maximum or minimum function, often named MAX or MIN, returns the largest or smallest of two arguments, respectively. Thus, MAX (2, 8) returns 8. Many languages, such as FORTRAN and MACSYMA, are more flexible, allowing any finite number of arguments such that a function call similar to MAX (2, 8, 3, 27, 9) returns a value of 27. In a language like Pascal that does not provide these functions, we can define our own:

```
FUNCTION MAX (X, Y: REAL) : REAL;
     BEGIN
          MAX := X;
          IF Y > MAX THEN
               MAX := Y
     END;   □
```

Even without the more general MAX function, we can design an algorithm to find the maximum as data is read:

> 1. Read a value for X
> 2. MAX ← X
> 3. Do until EOF:
>   a. If X > MAX, then MAX ← X.
>   b. Read X

Initialization of MAX to 0 would give us erroneous results if all the data values were negative. If the compiler has some facility for assigning the smallest possible real value that can be stored in that particular computer, MAX could be given that value in Step 2.

A relational data base language will probably have MAX, MIN, and AVG built-in to find the largest, smallest, and mean values, respectively, of a numeric attribute or column in a relation. SUM and COUNT, mentioned in the last section, are usually included. SUM totals the numbers in an attribute, and COUNT computes how many tuples or rows are in a relation.

Truncation or chopping off some of the digits of a number, as discussed in Section 1.1, occurs frequently in the operation of a computer. The TRUNC function in Pascal and the INT function in FORTRAN both truncate the number from the decimal to the right, leaving the integer part. Therefore, TRUNC(5.8) or INT(5.8) returns 5 in the appropriate language. Generally, functions which truncate to a certain number of decimal places must be written by the user.

EXAMPLE 1

Let's say you want to truncate numbers to two decimal places. So, 6.8293 would truncate to 6.82. If we only have a function to truncate to the integer part, we must move the decimal to the right of 2, truncate, and then move the decimal back. Since multiplication by 10^2 or 10^{-2} moves the decimal two places to the right or left, respectively, we have the following algorithm to truncate to two decimal places:

1. Multiply by 100:	682.93
2. Truncate to the integer part:	682
3. Divide by 100:	6.82 ■

The process in Example 1 can be very handy when we need to express answers in dollars and cents. A generalization of that algorithm follows.

**Algorithm for Truncating Real Number X
to n Decimal Places**

> 1. Multiply X by 10^n.
> 2. Truncate the result of Step 1 to the integer part.
> 3. Divide the result of Step 2 by 10^n.

**LANGUAGE
EXAMPLE**

If you want your answer rounded to the nearest integer instead of truncated, in Pascal use the ROUND function. As discussed in Section 1.2, ROUND(6.8293) is 7 since the number after the decimal, 8, is 5 or greater.   □   To round to n decimal places, follow the same algorithm as for truncation to n decimal places, rounding at Step 2 instead of truncating.

Many languages that do have a truncate function do not have a built-in round function. We can round a positive (negative) number to the nearest integer easily, however, by adding 0.5 (-0.5) before truncation. For example, $6.8293 + 0.5 = 7.3293$ is truncated to 7. But in a different example, $6.24 + 0.5 = 6.74$ is truncated to 6 since $0.2 + 0.5 = 0.7$, not enough to increase 6. Similarly, $-6.8293 + -0.5 = -7.3293$ truncates to -7, and $-6.24 + -0.5 = -6.74$ truncates to -6.

We may want to round to n decimal places. Perhaps we want to round our dollars and cents answer instead of truncating it. The trick is to add 5 to the decimal place just beyond where truncation is to occur. In this case, add 0.005 or 0.5×10^{-2} before truncating to two decimal places.

Algorithm for Rounding Number X to n Decimal Places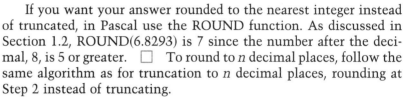

> 1. If X is positive, then add $0.5 \cdot 10^{-n}$ to X.
> 2. If X is negative, then add $-0.5 \cdot 10^{-n}$ to X.
> 3. Truncate X to n decimal places.
>   (Step 3, of course, can be expanded using the truncation algorithm discussed earlier.)

Two functions similar to TRUNC and ROUND which are built into some languages are ceiling and floor, rather amusing but descriptive names. The floor of a number X is the integer closest to X that is less than or equal to X; and the ceiling is the closest integer greater than or equal to X. Thus, the floor of 5.78, $\lfloor 5.78 \rfloor$, is 5; and the ceiling, $\lceil 5.78 \rceil$, is 6. Floor is not truncation, however, since $\lfloor -5.78 \rfloor = -6$, while TRUNC$(-5.78) = -5$. Similarly, $\lceil -5.78 \rceil = -5$, since $-5.78 < -5$.

DEFINITION

> The **floor** or **greatest integer function**, $f(x) = \lfloor x \rfloor$, is the largest integer less than or equal to x. The **ceiling function**, $g(x) = \lceil x \rceil$, is the smallest integer greater than or equal to x.

The floor and ceiling functions are useful in many computations in computer science. Do not get bogged down in worrying about the computer science in this paragraph, but study the examples for how they illustrate the utility of the floor and ceiling functions. The **pigeonhole principle** in combinatorics states that if you have more pigeons (m) than holes (k) and every pigeon occupies a hole, then at least one hole must contain more than one pigeon. More precisely, at least one hole contains more than $\lfloor (m - 1)/k \rfloor$ pigeons. Applying pigeons to computers, suppose $k = 4$ processors are working in parallel on the same program so that different instructions can be executed simultaneously by different processors if the logic of the program allows. By the pigeonhole principle, if there are $m = 6$ independent instructions, then at least one processor must execute more than $\lfloor (6 - 1)/4 \rfloor = 1$ instruction. The floor function is just what we need to obtain this integer lower bound. Now, suppose each instruction takes time t to execute. The total time it takes for k processors, working in parallel, to execute m independent instructions is

$$T = \lceil m/k \rceil t.$$

If there are 4 processors and 6 independent instructions, with each instruction taking 7 nsec to execute, then the total time is

$$T = \lceil 6/4 \rceil \times 7 = 2 \times 7 \text{ nsec} = 14 \text{ nsec}.$$

Each processor executes a different instruction at the same time, but there are two instructions left over that then are executed simultaneously. Therefore, the whole process takes time $2t$. Dividing the number of processors k into the number of instructions m might not give us the correct answer. We need the smallest integer that is greater than or equal to m/k which application of the ceiling function gives us.

LANGUAGE EXAMPLE

While most languages have functions that manipulate numeric data, LISP, a *list* processing language, is particularly good at manipulating symbols. Consequently, LISP is used extensively in **artificial**

intelligence, the study of having the computer exhibit human-like intelligence. MACSYMA, a programming system which performs symbolic and numeric mathematical manipulations, is also written in LISP. The emphasis in LISP and MACSYMA is not assignment statements but application of functions to expressions. ☐

The fundamental unit in LISP is an **atom**, a string of letters and/or digits. A **list** is a set of parentheses enclosing any number of atoms and lists; a **symbolic expression** or **s-expression** is an atom or list. Thus, the word MATHEMATICS is an atom while (ESSENCE OF MATHEMATICS) is a list; both are s-expressions. Three of the basic functions in LISP are CAR, CDR, and CONS. The CAR of a list returns the first element:

(CAR ' (ESSENCE OF MATHEMATICS)) = ESSENCE.

The quote indicates not to look for a value for each element in the list. CDR yields the rest of the list:

(CDR ' (ESSENCE OF MATHEMATICS))   =
 (OF MATHEMATICS) .

While CAR and CDR pull apart, CONS concatenates or glues an item to a list:

(CONS 'THE' (ESSENCE OF MATHEMATICS))
     = (THE ESSENCE OF MATHEMATICS) .

Using these and other LISP functions we can have the computer do such things as generate sentences, find synonyms from a dictionary, write a synopsis of an article, play games, be an expert problem solver in calculus, diagnose diseases, or help our understanding of how people hear and see. ☐

EXERCISES 5.3

Calculate the values in Exercises 1–19.

1. TRUNC(17.299)   **2.** $\lfloor 17.299 \rfloor$ **3.** TRUNC(-17.299)

4. $\lfloor -17.299 \rfloor$ **5.** ROUND(17.299)   **6.** $\lceil 17.299 \rceil$

7. ROUND(-17.299)   **8.** $\lceil -17.299 \rceil$ **9.** TRUNC(17.5299)

10. $\lfloor 17.5299 \rfloor$ **11.** TRUNC(-17.5299)   **12.** $\lfloor -17.5299 \rfloor$

13. ROUND(17.5299)   **14.** $\lceil 17.5299 \rceil$ **15.** ROUND(-17.5299)

16. $\lceil -17.5299 \rceil$

17. MAX($-222, -3456, -113, -245$)

18. MIN($-222, -3456, -113, -245$)

19. MAX(234/235, 141/170, 1), where all arithmetic yields integer values.

20. Using the algorithm at the first of this section, give the values of MAX and X at each step of the calculation of the maximum for the following data: 13, 4, 27, 18.

21. a. Write a Pascal or FORTRAN function to calculate MIN with two arguments.

   b. Give an algorithm to calculate MIN with any number of arguments where the values will be read.   ☐

22. a. If MIN has two real arguments, what is the domain and range?

   b. Six real arguments?   **c.** Four integer arguments?

23. Give algorithms for calculating the

   a. Floor and

   b. Ceiling functions assuming the TRUNC function is available to you.

24. Give the domain and range of the floor, ceiling, truncate, and round functions.

25. a. Following the algorithm of this section, show the steps to truncate 32.456522 to 3 decimal places.

   b. Repeat for 5 decimal places.

26. a. Following the algorithm of this section, show the steps to round 32.456522 to 3 decimal places.

   b. Repeat for 5 decimal places.

27. If the parameters of the MAX function are declared to be of type HUNDRED, a previously defined type in Pascal, give the declaration for the function type.   ☐

28. State whether each of the following functions is one-to-one. If it is not, give a counterexample:

   a. MAX   **b.** TRUNC   **c.** ROUND
   d. Floor   **e.** CDR   **f.** CAR

29. Give the domain and range of CDR.

In LISP evaluate Exercises 30–32.

30. (CDR '(IT IS NOT WHO CAN BUT WHO WILL))

31. (CAR '(IT IS NOT WHO CAN BUT WHO WILL))

32. (CONS 'NOT, '(IF I CAN HELP IT))   ☐

33. The following question is one of the sample questions in the *1984 AP Course Description in Computer Science**:
Suppose that the variable *d* represents the number of dollars in a bank account after interest has just been credited to that account (e.g., $d = 123.456$). Which of the following Pascal expressions would round that amount to the nearest cent (e.g., to 123.46)?

   (A) *round* (100 ∗ *d*)/100

   (B) *round* (*d*/100) ∗ 100

   (C) *round* (100 ∗ *d*)

   (D) *round* (*d*/100)

   (E) *round* (*d*)   ☐

---

* AP questions selected from *AP Course Description in Computer Science*. College Entrance Examination Board, 1984. Reprinted by permission of Educational Testing Service, the copyright owner of the sample questions.

34. Consider real numbers X and Y. Verify that the algorithm for MAX(X, Y) in this section works for the two cases $X \geq Y$ and $X < Y$.

35. Let X, Y, and Z be real numbers.
   a. How many different arrangements are there of these numbers in the inequality _____ $\leq$ _____ $\leq$ _____ . For instance, $X \leq Y \leq Z$ is one such arrangement.
   b. Verify that the algorithm for MAX(X, Y, Z) in this section holds for all the cases of part a.

5.4

Composition of Functions

LANGUAGE
EXAMPLE

Suppose in a Pascal program you wish to take the absolute value of X and then take the square root of the result, assigning the final value to Z. You could accomplish these two steps in one statement:

Z := SQRT(ABS(X)).

Applying one function and then another on the result is called composition of functions. In mathematical terms, if $f(x) = \sqrt{x}$ and $g(x) = |x|$, we want $f(g(x)) = \sqrt{|x|}$, which has a special notation, $f \circ g(x)$. The function ABS maps from $\mathbb{R}_c$ to the set of nonnegative real numbers on the computer. The domain of SQRT is the set of nonnegative real numbers, so SQRT can use the range value of ABS as a domain element. ☐

> **DEFINITION**
>
> Let $g: A \rightarrow B$ and $f: B \rightarrow C$ be functions. The **composition** of f and g is a function $f \circ g: A \rightarrow C$ such that $f \circ g(x) = f(g(x))$ for all $x \in A$. $f \circ g$ is called the **composite function**.

EXAMPLE 1

Consider the functions $g = \{(1, a), (2, a), (3, b)\}$ and $f = \{(a, x), (b, y), (c, y), (d, z)\}$ with $g: \{1, 2, 3\} \rightarrow \{a, b\}$ and $f: \{a, b, c, d\} \rightarrow \{x, y, z\}$. The range of g is $\{a, b\}$, but certainly one codomain is $\{a, b, c, d\}$. We can picture the mappings as in Figure 5.5. To evaluate $f \circ g(1)$ use the definition of composition.

$$f \circ g(1) = f(g(1)) = f(a) = x$$

Repeating the process for other elements of the domain of g, we can write the function $f \circ g$ in set notation,

$$f \circ g = \{(1, x), (2, x), (3, y)\}. \quad ∎$$

Figure 5.5

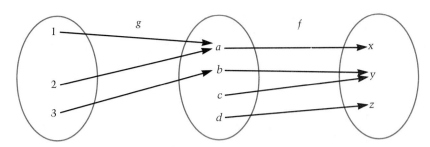

Mappings $f = \{(a, x), (b, y),$ $(c, y), (d, z)\}$ and $g = \{(1, a),$ $(2, a), (3, b)\}$

EXAMPLE 2

In LISP apply functions to pick out the second element in a list.

Solution

First, we apply CDR to obtain a list with the leftmost item eliminated. Then CAR is applied to obtain the first element in the remaining list, which is the desired element.

> LANGUAGE
> EXAMPLE

(CAR (CDR '(THE ESSENCE OF MATHEMATICS)))

returns ESSENCE. In LISP there is a special notation, CADR, for this composite function so that if X is a variable whose value is a list,

(CADR X) = (CAR (CDR X))

As with the composition above and with mathematical expressions in general, we work from the inside parentheses out. ☐ ■

Sometimes the domains and ranges of functions f and g are such that both $f \circ g$ and $g \circ f$ can be evaluated, even though the composite functions probably are not equal.

EXAMPLE 3

Let $f(x) = 3x^2 + 5$ and $g(x) = 2x + 1$. Simplify $f \circ g$ and $g \circ f$ as much as possible.

Solution

$$f \circ g(x) = f(g(x)) = f(2x + 1) = 3(2x + 1)^2 + 5$$
$$= 3(4x^2 + 4x + 1) + 5 = 12x^2 + 12x + 8.$$

To reiterate, work from the inside out with parentheses. Substitute $2x + 1$ for $g(x)$ to obtain $f(2x + 1)$. We are now taking f of $2x + 1$, *not* f of x. x is a dummy variable, like a parameter in a program. We replace every occurrence of x in the definition of f with $2x + 1$, which is like an argument in a program. We invoke the function f with the argument $g(x) = 2x + 1$.

Composing in the other direction

$$g \circ f(x) = g(f(x)) = g(3x^2 + 5)$$
$$= 2(3x^2 + 5) + 1 = 6x^2 + 11.$$

Here, $f(x)$ or $3x^2 + 5$ replaces the x in $g(x)$. Notice that $f \circ g(x) \neq g \circ f(x)$. ■

EXAMPLE 4

Let $f(x) = \sqrt{x}$ and $g(x) = x^2$ for $x \geq 0$ (Figure 5.6). Evaluate $f \circ g$ and $g \circ f$.

Solution

$f \circ g(x) = f(g(x)) = f(x^2) = \sqrt{x^2} = |x| = x$ since $x \geq 0$.
$g \circ f(x) = g(f(x)) = g(\sqrt{x}) = (\sqrt{x})^2 = x$. ■

Figure 5.6

Mapping of $f(x) = \sqrt{x}$ and $g(x) = x^2$

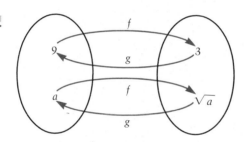

In Example 4, $f \circ g(x) = g \circ f(x) = x$. When this situation occurs, we say that f and g are inverses of each other and write $g = f^{-1}$ and $f = g^{-1}$. The -1 is not an exponent so that $f^{-1}(x) = x^2, x \geq 0$, does not mean $1/f(x)$ or $1/\sqrt{x}$. With functions that are inverses of each other, one function will "undo" what the other has done. $f(9) = \sqrt{9} = 3$, while $g(3) = 3^2 = 9$ (see Figure 5.6). (9, 3) is an ordered pair in f, and (3, 9) is in g. The image of 9 under f is 3, and the preimage of 9 under g is {3}. Because coordinates of a function appear switched in its inverse, the domain and range are also exchanged. Thus, the domain of f is the range of f^{-1}, and f's range is f^{-1}'s domain.

As noted in Section 5.1, for a function f to have an inverse f must be one-to-one. Each element in the range of f must be associated with exactly one element in the domain, so that when we switch the first and second elements of the ordered pairs to obtain f^{-1}, we have a function. The way we force $g(x) = x^2$ in Example 4 to be one-to-one and thus to have an inverse is by restricting the domain to be $\{x \mid x \geq 0\}$.

DEFINITION

> Let $f: A \rightarrow B$ be a function with range B. A function g: $B \rightarrow A$ is the **inverse function** of f, written $f^{-1} = g$, if $f \circ g(x) = x$ for all $x \in B$ and $g \circ f(x) = x$ for all $x \in A$.

LANGUAGE
EXAMPLE

There are many situations in the study of computer science where inverse functions are involved. When a character is read in the computer, it is encoded into a string of 0's and 1's using some encoding scheme such as ASCII and EBCDIC. When the corresponding character is printed, the string of 0's and 1's is decoded. In Pascal the functions ORD and CHR do much the same processes. If we are working on a computer that uses the ASCII coding scheme, ORD('A') = 65, the decimal number corresponding to the binary encoding of 'A'. (The conversion between the decimal and binary number systems is discussed in Chapter 10.) CHR(65), on the other hand, decodes 65 to its corresponding character, 'A' (see Figure 5.7). Note that

Figure 5.7

Mappings CHR and ORD

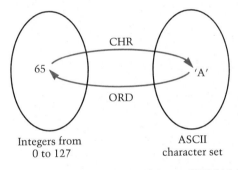

Integers from 0 to 127

ASCII character set

$$\text{ORD (CHR (65))} = \text{ORD ('A')} = 65$$
$$\text{and CHR (ORD ('A'))} = \text{CHR (65)} = \text{'A'}.$$

The domain of ORD, which is exactly the range of CHR, is a set of 128 characters, some of which are listed in Appendix A. Moreover, the set $\{0, 1, 2, \ldots, 127\}$ is both the range of ORD and domain of CHR.

We have already discussed the functions PRED and SUCC that give the predecessor and successor, respectively, in a linearly ordered set. Certainly, these are inverses of each other. For example, in WEEKDAYS

$$\text{SUCC (PRED (WED))} = \text{SUCC (TUE)} = \text{WED}$$
$$\text{and PRED (SUCC (WED))} = \text{PRED (THU)} = \text{WED} \quad \square$$

The exponential and natural logarithmic functions, which are discussed in Chapter 11, are also inverses of one another. These functions, which have so many applications in the sciences, are useful in studying the efficiency of different algorithms in computer science.

EXERCISES 5.4

For Exercises 1–5 consider the following functions:

$f = \{(2, 7), (0, 4)\}$ $g = \{(4, 2), (7, 6), (9, 8)\}$
$h(x) = 2^x$ with domain $= \{0, 1, 2, 3, 4, 5\}$
$k(x) = \sqrt{x}$ with domain $= \{0, 1, 4, 9, 16\}$
$s(x) = x^2$ with domain $= \{0, 1, 2, 3\}$
$t(x) = \min(x, 3)$ with domain $= \{0, 1, 2, \ldots, 9\}$

1. If possible evaluate   a. $f \circ g$ and   b. $g \circ f$ as sets of ordered pairs. If not possible for either, say why.

2. Repeat Exercise 1 for $h \circ k$ and $k \circ h$.

3. Repeat Exercise 1 for $s \circ t$ and $t \circ s$.

4. Which of the functions above, f, g, h, k, s, t, are one-to-one?

5. Consider each function above (f, g, h, k, s, t) that has an inverse.
   a. Give the inverse as a set of ordered pairs.
   b. Find the domain and range of the function.
   c. Give the domain and range of its inverse.

Find and simplify $f \circ g(x)$ and $g \circ f(x)$ in Exercises 6–10.

6. $f(x) = |x|$, $g(s) = 9x + 1$

7. $f(x) = (4x + 2)/3$, $g(x) = 6x - 11$

8. $f(x) = \min(x, 2)$, $g(x) = \max(x, 2)$

9. $f(x) = x + 0.5$, $g(x) = \text{TRUNC}(x)$

10. $f(x) = x^2 + 2x + 5$, $g(x) = 3x - 1$

Which functions in Exercises 11–17 are inverses of each other?

11. $f(x) = 2x + 8$, $g(x) = 1/2\, x - 4$

12. $f(x) = \lfloor x \rfloor$, $g(x) = \lceil x \rceil$

13. $f(x) = 1/x$, $g(x) = 1/x$

14. $f(x) = x^3$, $g(x) = x^{1/3}$

15. $f(x) = $ wife of x, $g(x) = $ husband of x
    $f: M \rightarrow W$, $g: W \rightarrow M$,
    $M = $ set of married men, $W = $ set of married women. Assume no bigamy.

16. $f(x) = |x|$, $g(x) = -x$

17. $f(x) = -x$, $g(x) = -x$

18. Use the percent and absolute error functions to write a function to evaluate relative error as a percent.

19. a. Define a function that computes the absolute error when a real number is truncated.
    b. Repeat for relative error.

20. Which of the functions in Exercises 1–28 of Section 5.1 have inverses?

For Exercises 21–23 let X be an arbitrary list with at least three elements.

LANGUAGE EXERCISES

21. Write a function in LISP to return a list with the two leftmost elements missing.

22. Write a function in LISP to return the third element of the list.

23. Write a function in LISP to return a list with the second element missing. □

24. Find $f \circ g \circ h(x) = f(g(h(x)))$ for $f(x) = \sqrt{x}$, $g(x) = |x|$, and $h(x) = 3x + 7$.

25. Repeat Exercise 24 for $f(x) = 1/x$, $g(x) = x^2$, and $h(x) = 5x$.

5.5

Mod Function

The equivalence relation congruence modulo 3 presented in Example 2 of Section 4.3 is not a function because, for example, 11 is not congruent modulo 3 to just one number—$11 \equiv 2 \pmod 3$, $11 \equiv -1 \pmod 3$, etc. We can, however, define a function, mod_3, from this relation by restricting its value to be 0, 1, or 2, the nonnegative remainder when dividing by 3 or, in other words, the equivalence class with representative from 0, 1, or 2. Thus, $\text{mod}_3 (11) = 2$. Of course, a mod_n function can be defined for any positive integer n. For instance, $\text{mod}_7 (53) = 4$ and $\text{mod}_{30} (89) = 29$.

DEFINITION

For $n \in \mathbb{I}^+$, the **mod function**, $\text{mod}_n (m)$, is the nonnegative remainder, 0, 1, 2, ..., or $n - 1$, when m is divided by n. Or, equivalently, $\text{mod}_n (m) = k$, where $m \equiv k \pmod n$ and $0 \le k < n$.

We can use the mod_2 function to define EVEN and ODD functions that will return TRUE or FALSE depending on the value of the parameter. One application is in evaluating the median of a sorted list of numbers, where we need to know if there are an odd or an even number of elements in the list. The function could also be helpful in printing a document using the computer when even page numbers need to be on the left and odd on the right.

An algorithm for the EVEN function of parameter N follows:

```
If mod₂(N) is 0 then
     return TRUE
else
     return FALSE.
```

The mod function is also used in public-key cryptosystems for encoding and decoding secret messages. Two very large primes, p and q, are chosen and a number for encoding, e, is derived from p and q using number theory. Another number, d, used for decoding, is determined from e, p, and q. (See Exercise 31 for additional detail concerning proper choice of e and d.) Each character of the message is mapped to an integer, resulting in a string of digits, M. Encoding M consists of applying the function mod_n to M^e:

$$C = \text{mod}_n\, (M^e).$$

The number C is transmitted and then decoded by again using mod_n:

$$M = \text{mod}_n\, (C^d).$$

The numbers n and e are made public. Without knowing p and q, however, d and the way to decode cannot be derived. Coding and decoding can easily be accomplished with the aid of a computer. But a spy, trying to crack your code by factoring n, would be out of luck. Even with the best algorithm available on a supercomputer, for an n of several hundred digits it is estimated that factoring could take several billion years!

Another important application of the mod function is in generating **random numbers.** The computer can be used to form a sequence of what appears to be numbers picked at random. Of course, since some algorithm must be performed for their generation, these numbers are not actually random. Consequently, they are sometimes called **pseudo-random numbers.** One significant application of random numbers is in the area of simulation where the computer is used to imitate and study such occurrences as nuclear reactions or the workings of a large corporation. In checking the accuracy of your program, it is often best to generate random test data to ensure thorough debugging. Surveying a sampling of people or picking winners in a lottery, also, should be at random.

HISTORICAL NOTE

John von Neumann is attributed with originating the idea of storing programs as well as data in the memory of the computer. He also introduced the first algorithm for generating random numbers with the computer. The brilliant von Neumann, born in Budapest, Hungary, in 1903, received his Ph.D. in mathematics at age 22. He contributed significantly to a variety of areas including the mathematical foundation of quantum theory, logic, theory of games, economics, nuclear weapons, meteorology, as well as theory and applications in early computer science. Many stories tell of his phenomenal memory, reasoning ability, and computational speed. It is said that he could memorize a column of the telephone book at a glance, and that he had mastered calculus by age 8. Halmos wrote in *Legend of John von Neumann*,

> When his electronic computer was ready for its first preliminary test, someone suggested a relatively simple problem involving powers of 2. (It was something of this kind: what is the smallest power of 2 with the property that its decimal digit fourth from the right is 7? This is a completely trivial problem for a present-day computer: it takes only a fraction of a second of machine time.) The machine and Johnny started at the same time, and Johnny finished first. □

 **HISTORICAL NOTE**

Many besides von Neumann have contributed to the theory of random numbers. One of the best techniques for generating random numbers, called the linear congruential method, was presented by D. J. Lehmer in 1949. □

EXAMPLE 1

One such linear congruential random number function is

$$r(X) = \text{mod}_{10,000}(229X + 1).$$

X is given an initial or **seed** value, say of 8367. After the "random number," $r(X)$, is calculated, to generate another value, substitute $r(X)$ for X and repeat the evaluation of the function. Thus, for this function with a seed of 8367 the following sequence of pseudo-random numbers can be generated:

$$229 \times 8367 + 1 = 1{,}916{,}044 \equiv 6044 (\text{mod } 10{,}000)$$

$$229 \times 6044 + 1 = 1{,}384{,}077 \equiv 4077 (\text{mod } 10{,}000)$$

$$229 \times 4077 + 1 = 933{,}634 \equiv 3634 (\text{mod } 10{,}000)$$

$$229 \times 3634 + 1 = 832{,}187 \equiv 2187 (\text{mod } 10{,}000)$$

$$229 \times 2187 + 1 = 500{,}824 \equiv 824 (\text{mod } 10{,}000)$$

$$229 \times 824 + 1 = 188{,}697 \equiv 8697 (\text{mod } 10{,}000).$$

Notice that division by 10,000 moves the decimal place four places to the left so that the remainder is the number to the right of the decimal. If a decimal number between 0 and 1 is desired, since the generated numbers are between 0 and 9999, divide $r(X)$ by $m = 10,000$, giving the sequence 0.6044, 0.4077, 0.3634, 0.2187, 0.0824, 0.8697. ■

LINEAR CONGRUENTIAL METHOD

Initially, four values are chosen:

   1. modulus, m, with $0 < m$
   2. multiplier, a, with $0 \le a < m$
   3. increment, c, with $0 \le c < m$
   4. seed or initial value of X, $0 \le X < m$

A "random number" is generated as indicated by the following assignment statement:

$$X \leftarrow \text{mod}_m(aX + c).$$

A sequence of values for X can be formed, each new generation of X using the previous value of X in the expression on the right. If a number between 0 and 1 is desired, divide X by m.

Through much computer testing of results, some guidelines for the choice of m, a, c, and the seed have evolved. With a good selection of the values of these identifiers, the maximum number of random numbers can be generated before the sequence of random numbers starts repeating itself. Since there are m remainders, there are at most m numbers to be calculated before repetition.

LANGUAGE EXAMPLE

Many large computers have a routine available for random number generation that is optimal for that particular machine. Some languages such as BASIC have a built-in random number function, though the method of generation varies from one type of computer to another. Usually, $X = \text{RND}(1)$ will return the next random number in the sequence with $0 < X < 1$. If you desire an integer number, say between 0 and 24, inclusive, first multiply by 25 since $0 < 25 \cdot X < 25$. Then truncate the positive result with the INT function (the floor function in BASIC) to obtain an element in the set $\{0, 1, 2, \ldots, 24\}$. If you really want integers between 1 and 25, inclusive, add 1 to the result because $1 < 25 \cdot X + 1 < 26$. Thus, in BASIC

```
X = INT(25 * RND(1)) + 1
```

assigns to X a "random" integer in the desired range, $\{1, 2, 3, \ldots, 25\}$. Multiplication by 25 widens the range, while addition of 1 shifts or translates it.   □

An application of the mod function similar to random number generation is **hashing functions.** Suppose you want to store a file on a disk where records can be accessed directly. If there is a numeric key, like SSN, that uniquely identifies the record, you can use this key to compute on which track of the disk pack the record should be placed or found. Suppose there are 97 tracks, numbered 0 to 96, set aside for the file. Evaluated by the hashing function

$$h(\text{key}) = \text{mod}_{97}(\text{key})$$

the record with key 358442241 would be placed on track $\text{mod}_{97}(358442241) = 81$. If that track becomes filled, additional records will be placed in an overflow area. The value of m in mod_m is picked to be the largest prime less than or equal to the number of tracks for the file. This prime can be obtained from a table of primes such as the one in Appendix B. A hashing function can also be used in accessing items in a table or entries in a data base.

It should be noted that the random number and hashing functions discussed are not the only ones available. Much work has been done and continues to be done on these topics; and though some of the best methods involve using mod, not all of the techniques employ this function.

EXERCISES 5.5

Evaluate Exercises 1–6.

1. $\text{mod}_7(349)$ 2. $\text{mod}_{100}(4621)$ 3. $\text{mod}_3(-23)$

4. $\text{mod}_{542}(11{,}382)$ 5. $\text{mod}_{100}(-4621)$ 6. $\text{mod}_{13}(-19{,}568)$

7. If possible evaluate   **a.** $f \circ g$ and   **b.** $g \circ f$ as sets of ordered pairs. If not possible for one or both, say why.

   $f(x) = x^2$, $f: \{0, 1, 2, 3, 4, 5\} \to \mathbb{I}$;
   $g(m) = \text{mod}_3(m)$, $g: \mathbb{I} \to \{0, 1, 2\}$

8. Repeat Exercise 7 with the following functions:

   $f(x) = \text{mod}_5(x)$ and $g(x) = \text{mod}_2(x)$ $0 \le x \le 10$ $x \in \mathbb{I}$

9. In Example 1, compute the next three random numbers after 8697.

10. In Example 1, give the first three numbers in the sequence if the results are (mod 10^5).

11. **a.** What is the maximum number of random numbers that can be generated using a linear congruential method with (mod 10^7)?
    b. With (mod 2^{20})?

12. **a.** Give the first five integers generated by the random number function
$$f(x) = \text{mod}_{1024}(467\,X)$$
with seed 1021.
Note for using the calculator: Store 1024 in memory since it will be used several times. (1) Compute $467 \times 1021 \div 1024$ to obtain 465.63184. The integer quotient is 465. To figure out the remainder, (2) calculate $465.63184 - 465 = 0.63184$. (3) Then 0.63184×1024 (in memory) yields 646.99999 which rounds to 647, the remainder and next "random number."
    b. Give the range of f.
    c. From the sequence of integers in part a, compute an appropriate sequence of four-digit real numbers from 0 to 1.

13. a. Give the first five integers generated by the function $f(X) = \text{mod}_{1000}(250X)$ with seed 767.

b. Why is this a bad random number generator?

14. Repeat Exercise 13 with $f(X) = \text{mod}_{10}(3X)$ with seed 3.

LANGUAGE EXERCISES

For Exercises 15–19 write a BASIC statement that will assign to X a "random" integer in each of the indicated sets.

15. a. $\{0, 1, 2, \ldots, 35\}$ **b.** $\{3, 4, 5, \ldots, 38\}$
c. $\{-3, -2, -1, 0, 1, \ldots, 32\}$

16. a. $\{0, 1, 2, \ldots, 99\}$ **b.** $\{100, 101, 102, \ldots, 199\}$
c. $\{-100, -99, -98, \ldots, -1\}$

17. $\{1, 2, 3, \ldots, 80\}$ **18.** $\{5, 6, 7, \ldots, 62\}$

19. $\{-8, -7, -6, \ldots, 15\}$ □

20. Suppose in a program you read the day of week on which the last day of the previous month falls. From that input, how can you use mod_7 to find the day of the week for any date in this month?

21. Write an algorithm to evaluate the median of a list L. The first step should be to sort the list, but omit the details of how the sorting is accomplished.

22. Write a definition for the EVEN function in Pascal.
Note: In Pascal, $\text{mod}_2(N)$ is written as N MOD 2. Since EVEN returns a value of TRUE or FALSE, the type of the function is BOOLEAN.   □

LANGUAGE EXERCISE

23. Give the domain and range of the EVEN function.

24. a. For the hashing function $h(\text{key}) = \text{mod}_{97}(\text{key})$, find the track on which to place the record with key = 649822424.

b. Give the domain and range for h where the key is a social security number.

25. Suppose there are 300 tracks available for the records. What should be the value of m for the hashing function $h(\text{key}) = \text{mod}_m(\text{key})$.

26. a. If there are 800 records and you want the file to be 80% filled, how many available locations should there be in the file? *Note:* See if your answer is reasonable. Do you want more or less locations than records?

b. If 11 records can be stored on a track how many tracks are needed?

c. What is the prime number m for the hashing function $h(\text{key}) = \text{mod}_m(\text{key})$?

27. Repeat Exercise 26 for 32,000 records and 10 records per track.

28. Sometimes in a program you need to isolate one of the digits of a number.

a. What value of m would you use so that $\text{mod}_m(X)$ returns the unit's digit? For example, with $X = 3549$ we want to have 9 returned.

b. To isolate the hundred's digit, in this case 5, you would perform integer division by what number before applying the mod function?

LANGUAGE
EXERCISE

Note: In COBOL suppose the variable X can only hold nonnegative integer values. The statement

        DIVIDE X BY 10 GIVING X REMAINDER R.

divides the value of X by 10, placing the integer quotient in X and the remainder in R. Thus, the remainder is $\text{mod}_{10}(X)$. Performed repeatedly, this statement isolates each digit in turn into R.   □

29. Halmos wrote of an early computer test which involved finding "the smallest power of 2 with the property that its decimal digit fourth from the right is 7." How can you use integer division and the mod function to isolate the fourth digit from the right in X?

30. The following is a sample question from the *1984 AP Course Description in Computer Science**:

A retail merchant wishes to keep an inventory of items identified by six-digit identification numbers. The quantity of each item ranges from 0 to 1,000 and must be updated each time that item is bought or sold. New identification numbers are created, and old ones discontinued, on a regular basis. How might one best maintain the inventory if one wishes to economize on both storage and processing time?

(A) Use a single list A, letting $A[n]$ be the quantity of item n.

(B) Use a single list A, letting $A[n]$ be the quantity of item $2*n-1$ plus 1,000 times the quantity of item $2*n$.

(C) Use two lists A and I, placing identification numbers in I in the order in which they are encountered and letting $A[n]$ be the quantity of item $I[n]$.

(D) Use two lists A and I, placing identification numbers in I in increasing numeric order and letting $A[n]$ be the quantity of item $I[n]$.

(E) Use two lists A and I, placing identification numbers in I in the first available location beyond that indicated by a hashing function and letting $A[n]$ be the quantity of item $I[n]$.   □

31. In a public-key cryptosystem large primes p and q are chosen. For the sake of illustration, let us pick very small primes $p = 3$ and $q = 7$. Then, $n = 3 \times 7 = 21$ is made public. Calculate $m = (p-1)(q-1) = 2 \times 6 = 12$. We pick d such that the greatest common divisor of m and d is 1. One choice of d is 5.

a. Find the smallest positive integer e such that $\text{mod}_m(d \cdot e) = 1$. This number is also published.

b. If the letter "B" is mapped to 2, encode $M = 2$ using $C = \text{mod}_n(M^e)$.

c. Decode C using $M = \text{mod}_n(C^d)$.

Note: The following guidelines for choice of m, a, and c have been developed through computer testing of random number generators:

1. If working on a decimal machine, choose m to be a large power of 10 for easy computation of mod_m. Most computers, however, are binary

*AP questions selected from *AP Course Description in Computer Science.* College Entrance Examination Board, 1984. Reprinted by permission of Educational Testing Service, the copyright owner of the sample questions.

LANGUAGE
EXERCISE

machines so that m should be a large power of 2, like 2^{32}. Division by 2^{32} moves the binary point 32 places to the left in a binary number. (The binary number system will be discussed further in Chapter 10.)

2. If on a binary computer, choose a such that $mod_8(a) = 5$ and $0.01m < a < 0.99m$.

3. No integer greater than 1 should divide both c and m.

32. Give three choices for a that meet the suggested criteria above with $m = 2^{20}$.

33. **a.** If $m = 2^{32}$ and $0 \le c < 10$, list the choices for c based on the above guideline that no integer greater than 1 should divide both c and m.
    b. List choices for c, $0 \le c < 10$, if $m = 10^{19}$.

Chapter 5   REVIEW

Terminology/Algorithms	Pages	Comments
Function or mapping f on $A \times B$	144	For element in domain there corresponds exactly 1 element in range
Domain of f	144	A
Codomain of f	144	B
Image of a	144	$f(a)$
Preimage of b	144	Set of elements from A that map to b
Range of f	144	Set of all images
One-to-one function	147	For each element in range there corresponds exactly 1 element in domain
Onto function	149	Codomain = range
Absolute value function $ABS(x) = \lvert x \rvert$	153	$\begin{cases} x & \text{if} & x \ge 0 \\ -x & \text{if} & x < 0 \end{cases}$
Projection, π_i	154	$\pi_i(a_1, a_2, \ldots, a_i, \ldots, a_n) = a_i$
Frequency, $freq_L$	155	$freq_L(u) =$ number of occurrences of u in L
Percent	155	$Percent(m, n) = (m/n)100\%$
SUM	157	Algorithm

Terminology/Algorithms	Pages	Comments
MEAN	157	Algorithm
Arithmetic mean	157	$(x_1 + x_2 + \cdots + x_n)/n$
Median	158	Middle value of sorted collection
MAX function	164	Maximum value
Find maximum	165	Algorithm
TRUNC or INT	165	Truncate
Truncate to n decimal places	166	Algorithm
ROUND	166	Round
Round to n decimal places	166	Algorithm
Floor, $\lfloor x \rfloor$	167	Greatest integer $\leq x$
Ceiling, $\lceil x \rceil$	167	Smallest integer $\geq x$
CAR	168	Returns first element of list
CDR	168	Returns rest of list
CONS	168	Returns concatenation of item and list
Composition, composite function, $f \circ g$	170	$f \circ g(x) = f(g(x))$
Inverse function, f^{-1}	173	$f \circ f^{-1}(x) = x = f^{-1} \circ f(x)$
Mod function, mod_n	175	$\mathrm{mod}_n(m)$ = nonnegative remainder when m divided by n
Linear congruential method	178	Algorithm to generate random numbers
Hashing function	178	$h(\text{key}) = \mathrm{mod}_n(\text{key})$

REFERENCES

Backus, J. "Can Programming Be Liberated from the von Neumann Style?" *Communications of the Association for Computing Machinery* 21(8): 613, August 1978.

Basinger, D. "Cryptology—Where Number Theory and Encryption Meet." *Tentacle* 4 (8): 2–3, August 1984.

Bierman, A. W., R. D. Rodman, D. C. Rubin, and J. F. Heidlage. "Natural Language with Discrete Speech as a Mode for Human-to-Machine Communication," *Communications of the Association for Computing Machinery* 28 (6): 628–636, June 1985.

Blazewicz, J., and J. R. Nawracki. "Dynamic Storage Allocation with Limited Compaction." *Discrete Applied Mathematics* 10: 241–253, 1985.

Burkhart, H. "On Jargon, A Revolution in Secret Codes." *UMAP,* 121–124, August 1981.

Durst, M. "Math and Statistics News." *Tentacle* 2 (4): 4–5, April 1982.

Educational Testing Service. *1984 AP Course Description in Computer Science.* Princeton, NJ: College Board Publications, 1984.

Fertig, R. T. *The Software Revolution: Trends, Players, Market Dynamics in Personal Computer Software,* New York: Elsevier Science, 1985.

Grotch, S. L., "Three-Dimensional Graphics for Scientific Data Display and Analysis." Lawrence Livermore National Laboratory, University of California, UCRL-90099, Livermore, 1983.

Guimaraes, T. "A Study of Application Program Development Techniques." *Communications of the Association for Computing Machinery* 28 (5): 494–499, May 1985.

Halmos, P. R. "The Legend of John von Neumann." *American Mathematical Monthly* 80: 382–394, 1973.

Hayes, J. P. *Computer Architecture and Organization,* New York: McGraw-Hill, 1978.

Hiltz, S. R., and M. Turoff. "Structuring Computer-Mediated Communication Systems to Avoid Information Overload." *Communications of the Association for Computing Machinery* 28 (7): 680–689, July 1985.

Karp, R. M., "Two Combinatorial Problems." In *Combinatorial Algorithms* (R. Rustin, Ed.), pp. 17–29. New York: Algorithmics Press, 1972.

Knuth, D. E. *The Art of Computer Programming,* Vol. 2, *Seminumerical Algorithms,* 2nd ed. Reading, MA: Addison-Wesley, 1981.

Litecky, C. R., and L. E. Rittenberg. "The External Auditor's Review of Computer Controls." *Communications of the Association for Computing Machinery* 24 (5): 288–295, May 1981.

Roberts, F. S. *Applied Combinatorics.* Englewood Cliffs, NJ: Prentice-Hall, 1984.

Wand, M. *Induction, Recursion, and Programming.* New York: North-Holland, 1980.

Weiser, M. "Programmers Use Slices When Debugging." *Communications of the Association for Computing Machinery* 25 (7): 446–452, July 1982.

Welburn, T. *Advanced Structured COBOL, Batch, On-Line, and Data-Base Concepts.* Palo Alto, CA: Mayfield, 1983.

Wexelblat, R. L., Ed. *History of Programming Languages.* New York: Academic Press, 1981.

6 *SUBSCRIPTS*

Ice Crystal. This picture is from a 3D film shown at the 1984 World's Fair in Tokyo. The film used computer graphics to simulate the molecular structure of water as it freezes.

6.1

Vectors or One-Dimensional Arrays

Most computer languages have the array as a **data structure,** a formal skeleton that can hold data. In fact the entire memory of a computer can be thought of as an array. Arrays allow us to collect a great deal of similar data together under one name instead of thinking of perhaps hundreds of individual identifiers. For example, let RATE be an array that holds the hourly rate of pay for a company's employees. To refer to a particular employee, RATE with a number in parentheses or brackets can be used. Thus, RATE(4) or RATE[4], depending on the language, refers to the rate of the fourth employee in the array. Another advantage of arrays is the ability to use a variable like I in parentheses instead of a constant like 4. Then this variable, called an **index,** can be programmed to change values in a loop, allowing us to perform the same operations on all array elements. Usually the programmer must declare the dimension of the array before processing it. In this section we consider one-dimensional arrays, where the **dimension** is the maximum number of items or elements that can be held in the array.

EXAMPLE 1

Let us declare the dimension and read exactly five values, 6.25, 8.00, 7.50, 7.50, and 6.25, into RATE using the language Pascal.

Solution

LANGUAGE EXAMPLE

```
CONST
      NUMBEREMP := 5;
VAR
      I:        1..NUMBEREMP;
      RATE:     ARRAY[1..NUMBEREMP] OF REAL;
BEGIN
FOR  I := 1 TO NUMBEREMP DO
      READ (RATE[I]);
. . . .
```

After execution the array elements have the indicated values:

```
RATE[1] ← 6.25
RATE[2] ← 8.00
RATE[3] ← 7.50
RATE[4] ← 7.50
RATE[5] ← 6.25
```

Execution of the statement WRITE (RATE[4]) would accomplish the printing of 7.50. □ ■

The index I is used to reference individual array elements. Whether 5 or 5000 elements, the program would be the same except for the value of the upper bound on the number of elements in the array, the constant NUMBEREMP.

This array is called a one-dimensional array because it has just one index. In mathematics such a structure is sometimes called a vector. Usually a single letter like **r** indicates the name of the vector, and the index appears as a subscript. Thus, the RATE array written mathematically would be $\mathbf{r} = (r_1, r_2, r_3, r_4, r_5)$ with r_i, like RATE(I), standing for the ith element. The difference in notation between vectors and one-dimensional arrays arise from difficulties in typing subscripts with a computer terminal keyboard. The number of elements in **r** corresponds to the dimension of the filled RATE array. Notice also that the vector $(r_1, r_2, r_3, r_4, r_5)$ is just an ordered 5-tuple from $\mathbb{R}^5$ or $\mathbb{Q}^5$. In fact, we have already been using vectors or ordered n-tuples without calling them by the former name.

DEFINITION

> A real **vector v** is an ordered n-tuple
>
> $$\mathbf{v} = (v_1, v_2, \ldots, v_n)$$
>
> where $v_1, v_2, \ldots, v_n \in \mathbb{R}$.

Recall from Section 4.1 that an ordered tuple of two or three elements can be interpreted geometrically as a point in two- or three-dimensional space, respectively. Another interpretation of a vector is as a directed line segment, or arrow, that has the same direction and length as the arrow from the origin to the point indicated by the ordered tuple. The vector $(3, 4)$ in **standard position** with initial point at the origin of the Cartesian coordinate system is illustrated in Figure 6.1. Actually, the vector can be pictured anywhere in the plane as long as the arrow maintains its length and direction.

Figure 6.1

The vector (3, 4)

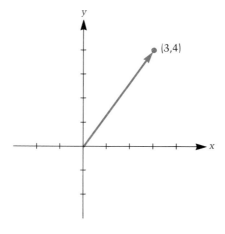

Two ordered *n*-tuples (or vectors or one-dimensional arrays) are equal if and only if they are of the same size and corresponding elements are equal.

EXAMPLE 2

LANGUAGE
EXAMPLE

Solution

Suppose OLDRATE holds the hourly rate of pay 6 months ago for the same employees from Example 1. Write a program segment in Pascal to let the (Boolean) variable SAME have the value TRUE if OLDRATE and RATE are equal and FALSE if at least one person has gotten a raise or cut in rate of pay in the 6 months.

```
SAME   := TRUE;
I      := 1;
WHILE (I <= NUMBEREMP) AND SAME DO
     BEGIN
          IF  OLDRATE[I]<> RATE[I]  THEN
               SAME := FALSE;
          I := I + 1
     END;
```

As I increments with values from 1 to NUMBEREMP, corresponding elements of arrays OLDRATE and RATE are tested one at a time. As soon as the corresponding rates do not agree (OLDRATE[I] <> RATE[I]), SAME's value is changed and execution of the loop is terminated. If each pair is equal, the arrays are identical and SAME remains TRUE.   □ ■

DEFINITION

> Vectors $(x_1, x_2, \ldots, x_n)$ and $(y_1, y_2, \ldots, y_n)$ are **equal** if and only if $x_i = y_i$ for $i = 1, 2, \ldots, n$.

Various operations can be performed on vectors just as operations can combine numbers or sets. In fact, some computers, such as the CRAY-2 supercomputer pictured in Figure 6.2, operate most efficiently upon vectors of numbers instead of upon individual data items. Addition and subtraction of vectors is accomplished by adding corresponding elements. Suppose the five employees of Examples 1 and 2 are paid bimonthly with vectors **f** and **g** containing the hours each employee works during the first and second weeks of the pay period, respectively. Suppose

$$\mathbf{f} = (40, 40, 0, 45, 43) \quad \text{and} \quad \mathbf{g} = (40, 38, 32, 40, 46).$$

Thus, employee 4 works 45 hr the first week and 40 the second for a total of 85 hr during the pay period. This addition must be performed for each employee:

Figure 6.2

The CRAY-2™ supercomputer, designed by Seymour Cray, has four processors working in parallel along with another processor as an overseer. It usually operates fastest on data that is arranged in one-dimensional arrays which are multiples of 64 elements each. The CRAY-2 has 65 million words, a 4.1-nsec clock cycle, and costs about $15 million. The dense circuitry is cooled by direct contact with Fluorinert, a fluid that looks much like water. However, this inert liquid does not conduct electricity and is so rich in oxygen that it can be used as a blood substitute. The CRAY-2 supercomputer was delivered to Lawrence Livermore National Laboratory in 1985. Cray Engineer Jeff Evanko is pictured standing inside the hollow center of the CRAY-2.
Photo courtesy James E. Stoots, Jr., Lawrence Livermore National Laboratory, run by the University of California for the Department of Energy. Printed with permission of Cray Research, Inc.

$$\mathbf{h} = \mathbf{f} + \mathbf{g} = (40, 40, 0, 45, 43) + (40, 38, 32, 40, 46)$$
$$= (80, 78, 32, 85, 89).$$

Here, the vectors are written horizontally, as **row vectors.** It may be easier for you to read them as **column vectors:**

$$\mathbf{h} = \begin{bmatrix} 40 \\ 40 \\ 0 \\ 45 \\ 43 \end{bmatrix} + \begin{bmatrix} 40 \\ 38 \\ 32 \\ 40 \\ 46 \end{bmatrix} = \begin{bmatrix} 80 \\ 78 \\ 32 \\ 85 \\ 89 \end{bmatrix}.$$

DEFINITION

Let $\mathbf{x} = (x_1, x_2, \ldots, x_n)$ and $\mathbf{y} = (y_1, y_2, \ldots, y_n)$ be vectors of n elements each. The **sum** of $\mathbf{x}$ and $\mathbf{y}$ is the vector

$$\mathbf{x} + \mathbf{y} = (x_1 + y_1, x_2 + y_2, \ldots, x_n + y_n).$$

Suppose everyone in our company is given a 12% raise. Since each employee is actually getting 112% of the previous salary, we multiply the array by the scalar or constant 1.12. The new rate vector **v** is

$$\mathbf{v} = 1.12 \cdot \mathbf{r} = 1.12 \cdot (6.25,\ 8.00,\ 7.50,\ 7.50,\ 6.25)$$
$$= (1.12 \cdot 6.25,\ 1.12 \cdot 8.00,\ 1.12 \cdot 7.50,\ 1.12 \cdot 7.50,\ 1.12 \cdot 6.25)$$
$$= (7.00,\ 8.96,\ 8.40,\ 8.40,\ 7.00).$$

DEFINITION

A **scalar** is a real number.

DEFINITION

Let $\mathbf{x} = (x_1, x_2, \ldots, x_n)$ be a vector. The **product** of a scalar a and the vector $\mathbf{x}$ is the vector

$$a\mathbf{x} = a(x_1, x_2, \ldots, x_n)$$
$$= (ax_1, ax_2, \ldots, ax_n).$$

A geometric interpretation of sum and difference of two- and three-element vectors is needed in computer graphics. The sum of vectors (3, 1) and (2, 3) is illustrated in Figure 6.3 as a diagonal of the parallelogram formed by the vectors. The difference, which is pictured in two positions in Figure 6.4, is evaluated using addition of vectors and multiplication by the scalar -1:

$$(3,\ 1) - (2,\ 3) = (3,\ 1) + -1 \cdot (2,\ 3)$$
$$= (3,\ 1) + (-2,\ -3) = (1,\ -2).$$

Figure 6.3

Vector sum
(3, 1) + (2, 3) = (5, 4)

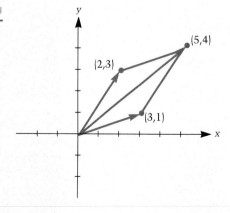

Figure 6.4

Vector difference
$(3, 1) - (2, 3) = (1, -2)$
shown in two positions

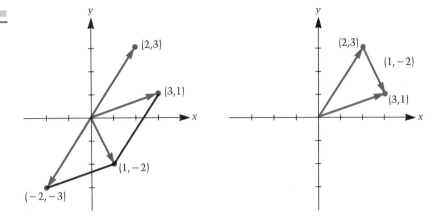

We now return to the five employees to discuss another operation on vectors. Ignoring overtime pay, each person in calculating his or her pay check, multiplies hours worked by the rate of pay. The fourth person, after the raise, gets

85 hr $\times$ \$8.40/hr = \$714.00.

Performing this operation for each person, we get an array containing bimonthly salaries:

$$\mathbf{s} = (80 \times 7.00, 78 \times 8.96, 32 \times 8.40, 85 \times 8.40, 89 \times 7.00)$$
$$= (560.00, 698.88, 268.80, 714.00, 623.00).$$

Thus, the company must pay

$$560.00 + 698.88 + 268.80 + 714.00 + 623.00 = \$2864.68$$

in wages for that 2-week period. The process of multiplying corresponding elements and then summing the results is called the dot or scalar product. Notice the final result, 2864.68, is a number, not a vector.

DEFINITION

Let $\mathbf{x} = (x_1, x_2, \ldots, x_n)$ and $\mathbf{y} = (y_1, y_2, \ldots, y_n)$ be vectors of n elements each. The **dot product** (or **scalar product** or **inner product**) of $\mathbf{x}$ and $\mathbf{y}$ is

$$\mathbf{x} \cdot \mathbf{y} = x_1 \cdot y_1 + x_2 \cdot y_2 + \cdots + x_n \cdot y_n.$$

Often in writing the dot product, the first term is written as a row vector while the second is written as a column vector. Multiplication of corresponding elements below follows the arrows, across and down. This notation will be useful when we cover matrix multiplication in Chapter 7.

$$\mathbf{h} \cdot \mathbf{v} = \overrightarrow{(80,\ 78,\ 32,\ 85,\ 89)} \cdot \begin{bmatrix} 7.00 \\ 8.96 \\ 8.40 \\ 8.40 \\ 7.00 \end{bmatrix} \Big\downarrow = 2864.68$$

The dot product is useful geometrically as well. Taking the dot product of the vector $(3, 4)$ with itself, we have

$$(3,\ 4) \cdot (3,\ 4) = 3 \cdot 3 + 4 \cdot 4 = 3^2 + 4^2 = 9 + 16 = 25$$

which you may recognize as the square of the distance from the origin $(0, 0)$ to the point $(3, 4)$. Thus, $\sqrt{25} = 5$ is the length of the vector $(3, 4)$. This length, which is written $|(3, 4)| = 5$, is sometimes called the norm of $(3, 4)$. If we need to find the distance between two points, we find the norm of the difference of the corresponding vectors. Thus, the distance between points $(3, 1)$ and $(2, 3)$ is the norm of $(3,\ 1) - (2,\ 3) = (1,\ -2)$

$$|(-1,\ 2)| = \sqrt{(-1,\ 2) \cdot (-1,\ 2)} = \sqrt{1 + 4} = \sqrt{5}.$$

You may be accustomed to seeing the general formula for the distance d between two points, (x_1, y_1) and (x_2, y_2), written in a different way.

$$d = \sqrt{(x_1 - x_2)^2 + (y_1 - y_2)^2}$$

which is the norm of the vector difference:

$$d = |(x_1,\ y_1) - (x_2,\ y_2)|.$$

The same idea can be extended to vectors in 3-space.

DEFINITION

The **norm** or **length** of a vector $\mathbf{v}$ is

$$|\mathbf{v}| = \sqrt{\mathbf{v} \cdot \mathbf{v}}.$$

EXERCISES 6.1

Given the scalars a = 7 *and* b = 3 *and the vectors* **u** = *(3, −4, 8, 0)*, **v** = *(−9, 4, 21, 2),* **y** = *(8, 8, 1, −2), and* **x** = *(7, 17, 6), where possible, compute the values of Exercises 1–20.*

1. $a\mathbf{u}$	2. $b\mathbf{v}$	3. $a\mathbf{u} + b\mathbf{v}$
4. $\mathbf{u} + \mathbf{v}$	5. $\mathbf{v} + \mathbf{u}$	6. $(\mathbf{u} + \mathbf{v}) + \mathbf{y}$
7. $\mathbf{u} + (\mathbf{v} + \mathbf{y})$	8. $\mathbf{u} + \mathbf{x}$	9. $(a + b)\mathbf{y}$

10. $a\mathbf{y} + b\mathbf{y}$ **11.** $0\mathbf{x}$ **12.** $\mathbf{u}\cdot\mathbf{v}$

13. $\mathbf{y}\cdot(2\mathbf{v})$ **14.** $2(\mathbf{y}\cdot\mathbf{v})$ **15.** $(2\mathbf{y})\cdot\mathbf{v}$

16. $\mathbf{x}\cdot\mathbf{y}$ **17.** $\mathbf{v} - \mathbf{y}$ **18.** $a(\mathbf{u} + \mathbf{y})$

19. $a\mathbf{u} + a\mathbf{y}$ **20.** $(0, 0, 0)\cdot\mathbf{x}$

Compute, if possible, the dot products in Exercises 21–23.

21. $(5, 7)\cdot\begin{bmatrix} -1 \\ 4 \end{bmatrix}$ **22.** $(6, 2, 3)\cdot\begin{bmatrix} 1 \\ 1 \end{bmatrix}$ **23.** $(7, -7, 1)\cdot\begin{bmatrix} 3 \\ 3 \\ 1 \end{bmatrix}$

24. Suppose the following items are for sale one week in a grocery store at the indicated prices: tooth brush, \$2.20; comb, \$3.42; gum, \$0.60; tissue, \$0.94.
   a. Write these prices in a vector $\mathbf{v}$.
   b. Suppose there is a 25%-off sale. What scalar is multiplied by $\mathbf{v}$ to give the sale prices?
   c. Perform this multiplication.
   d. Suppose 83 tooth brushes, 17 combs, 145 packs of gum, and 108 boxes of tissue are sold during the sale. Write a dot product of vectors to calculate the amount of money brought in from the sale, and perform this dot product.
   e. Suppose the next week the store sells 20 tooth brushes, 3 combs, 76 packs of gum, and 37 boxes of tissue. Write the vector sum to indicate the number of each item sold during the 2-week period, and perform this addition.

Determine the values of the unknowns to make the vectors equal in Exercises 25–27.

25. $(3, 5, 7) = (a, b, 7)$

26. $(-6, 2, 1) = (-6, 2, 1, a)$

27. $2(6, 1, a) = b(3, c, 4)$

LANGUAGE EXERCISE

28. Write Pascal program segments to implement the following array operations based on the vector operations for the employee examples of this section.
   a. Add FIRSTWEEK and SECONDWEEK arrays (vectors $\mathbf{f}$ and $\mathbf{g}$) together giving the HOURS array (vector $\mathbf{h}$) for the 5 employees in a company.
   b. Multiply 1.12 times the RATE array (vector $\mathbf{r}$), assigning these values to RATE.
   c. Calculate the dot product of HOURS and RATE.
   d. Print each person's hours, rate, and salary (vector $\mathbf{s}$). □

29. Let a and b be scalars and $\mathbf{u} = (u_1, u_2, u_3)$ and $\mathbf{v} = (v_1, v_2, v_3)$ be vectors containing 3 elements. Verify $\mathbf{u} + \mathbf{v} = \mathbf{v} + \mathbf{u}$ using the definitions. *Note:* Do not substitute particular numbers to prove. Look at Exercises 4 and 5 for hints from a specific example.

30. Repeat Exercise 29 verifying $a(\mathbf{u} + \mathbf{v}) = a\mathbf{u} + a\mathbf{v}$. *Note:* Look at Exercises 18 and 19 for a specific example.

31. Repeat Exercise 29 verifying $(a + b)\mathbf{u} = a\mathbf{u} + b\mathbf{u}$. *Note:* Look at Exercises 9 and 10 for a specific example.

32. Suppose in a course with five homework assignments, four tests, and an exam, you make 85, 70, 90, 100, and 55 on the homework; 80, 62, 76, and 86 on the tests; and 73 on the exam. For simplicity, write your grades as a vector of length 3 with cumulative homework score, cumulative test score, and exam grade. Each grade is out of 100 possible points, but is weighted. Homework carries a weight of 1, tests a weight of 5, and the exam a weight of 8.
   a. Write a vector for the weight.
   b. Write your final weighted sum as a dot product.
   c. Write the total possible weighted sum as a dot product.
   d. Calculate your average.

LANGUAGE EXERCISE

33. Suppose an exam has 100 true-false questions. Write a program sequence in Pascal (or FORTRAN) to calculate the grade of a student. Let CORANS be an array containing the correct answers, and let STUANS be an array containing the student's answers. □

Consider the vectors $\mathbf{u} = (-2, 4)$, $\mathbf{v} = (-1, -3)$, *and* $\mathbf{z} = (2, 0)$. *Sketch the vectors in Exercises 34–43.*

34. $\mathbf{u}$ **35.** $\mathbf{v}$ **36.** $\mathbf{z}$ **37.** $\mathbf{u} + \mathbf{v}$

38. $\mathbf{v} - \mathbf{u}$ **39.** $\mathbf{u} - \mathbf{v}$ **40.** $-2\mathbf{v}$ **41.** $2\mathbf{v}$

42. $\mathbf{z} + \mathbf{u}$ **43.** $1/3(\mathbf{z} + \mathbf{u})$

Let $\mathbf{u} = (-2, 4)$, $\mathbf{v} = (-1, -3)$. *Calculate the values of Exercises 44–47.*

44. $|\mathbf{u}|$ **45.** $|\mathbf{v}|$ **46.** $|\mathbf{u} + \mathbf{v}|$

47. The distance between the endpoints of the vectors $\mathbf{u}$ and $\mathbf{v}$ in standard position.

6.2

Sigma and Pi

We saw in the last section that calculation of the total payroll involved taking the sum of the elements in the salary array or vector

$$\mathbf{s} = (560, 698.88, 268.80, 714, 623).$$

The total is

$$s_1 + s_2 + s_3 + s_4 + s_5$$
$$= 560 + 698.88 + 268.80 + 715 + 623 = 2864.68.$$

LANGUAGE EXAMPLE

There is no problem writing the addition of the vector elements with only five employees, but 5000 employees would be an entirely different matter. The logic of our program to compute the total payroll, however, is the same whether we have 5 or 5000 employees:

```
SUM := 0;
FOR I := 1 TO NUMBEREMP DO
    SUM := SUM + SALARY[I];
```

Similarly, in mathematics, we have a succinct abbreviation for the sum of the elements of the vector $\mathbf{s} = (s_1, s_2, s_3, s_4, s_5)$, namely

$$\Sigma_{i=1}^{5} s_i \quad \text{or} \quad \sum_{i=1}^{5} s_i = s_1 + s_2 + s_3 + s_4 + s_5$$

The Greek capital letter Σ, **sigma,** indicates lots of plus signs. In the program segment, I initially takes on the value 1 and increments by 1 each time through the loop until it exceeds NUMBEREMP. As I changes, different array elements, SALARY[I], are accessed. Similarly, in $\Sigma_{i=1}^{5} s_i$, i starts with a value of 1 and increments to a maximum of 5. With each new value of i, a new s_i is accumulated. This handy notation is used in many computer science articles and books.

Just as in the programming segment we could have used another variable, say K or NDX or SUB, for the index, any convenient letter can be used for the vector subscript. Thus

$$\sum_{i=1}^{5} s_i = \sum_{j=1}^{5} s_j = \sum_{k=1}^{5} s_k.$$

The lowercase letters i, j, and k are some of the most common choices, however. ☐

NOTATION

For the positive integer n

$$\Sigma_{i=1}^{n} v_i \quad \text{or} \quad \sum_{i=1}^{n} v_i = v_1 + v_2 + \cdots + v_n.$$

The **lower bound** of the **index,** i, is 1, and the **upper bound** of the index is n.

EXAMPLE 1

Write the dot product of vectors $\mathbf{x} = (x_1, x_2, \ldots, x_n)$ and $\mathbf{y} = (y_1, y_2, \ldots, y_n)$ using sigma notation.

Solution

Since $\mathbf{x} \cdot \mathbf{y} = x_1 y_1 + x_2 y_2 + \cdots + x_n y_n$ is a sum of products of the form $x_i y_i$, we can define the dot product as

$$\mathbf{x} \cdot \mathbf{y} = \sum_{i=1}^{n} x_i y_i. \quad ■$$

In situations where a pattern exists for the vector elements, the formula for the general term can be given instead of an abstract symbol like v_i. In the algorithm from the last chapter for evaluating the mean of a data set, an on-going COUNT of the number of processed items was kept. Basically, every time a data item was read and added to the sum of numbers, the COUNT was incremented by 1. Mathematically speaking,

$$\sum_{i=1}^{n} 1 = \underbrace{1 + 1 + \cdots + 1}_{n \text{ terms}}$$

was computed, where n was the number of data items. If there were 5 numbers added, then

$$\text{COUNT} = \sum_{i=1}^{5} 1 = 1 + 1 + 1 + 1 + 1 = 5.$$

Clearly, $\sum_{i=1}^{n} 1 = n$ since n 1's are added together.

EXAMPLE 2

Evaluate $\sum_{i=1}^{7} i$, $\sum_{j=1}^{4} (3j + 2)$, and $\sum_{k=1}^{5} k^2$.

Solution

$$\sum_{i=1}^{7} i = 1 + 2 + 3 + 4 + 5 + 6 + 7 = 28$$

$$\sum_{j=1}^{4} (3j + 2) = (3 \cdot 1 + 2) + (3 \cdot 2 + 2) + (3 \cdot 3 + 2) + (3 \cdot 4 + 2)$$

$$= 5 + 8 + 11 + 14 = 38$$

$$\sum_{k=1}^{5} k^2 = 1^2 + 2^2 + 3^2 + 4^2 + 5^2$$

$$= 1 + 4 + 9 + 16 + 25 = 55. \quad \blacksquare$$

 **HISTORICAL NOTE**

Interesting formulas can arise from these sums. One was discovered by Carl Friedrich Gauss as a child. His teacher asked the class to add together the first 100 positive integers, $\sum_{i=1}^{100} i$, undoubtedly hoping to have a few moments of uninterrupted quiet. But almost immediately Gauss blurted out the correct answer of 5050 to the astonishment of the teacher and class. His method consisted of pairing each element of the vector with a copy of the vector in reverse order.

$$1 \quad 2 \quad 3 \quad 4 \quad \ldots \quad 97 \quad 98 \quad 99 \quad 100$$
$$100 \quad 99 \quad 98 \quad 97 \quad \ldots \quad 4 \quad 3 \quad 2 \quad 1.$$

Notice that each of the 100 pairs sums to 101. The total of all these numbers is therefore $100 \cdot 101$; but since each term of the vector is added twice, the total must be divided by 2:

$$\frac{100 \cdot 101}{2} = \frac{\overset{50}{\cancel{100}} \cdot 101}{\cancel{2}} = 5050.$$

Thus, $\Sigma_{i=1}^{100} i = (100 \cdot 101)/2$. ☐   By this same reasoning we can verify the general formula:

$$\sum_{i=1}^{n} i = \frac{n(n + 1)}{2}.$$

The language MACSYMA has a very mathematical way of writing $\Sigma_{i=1}^{n} v_i$:

    SUM (V (I), I, 1, N);

In fact,

    SUM (I, I, 1, N);

correctly returns

$$\frac{N (N + 1)}{2}. \quad ☐$$

Suppose it is not the positive integers from 1 to 100, but the even positive integers, 2, 4, 6, 8, ..., 100, we wish to add. An even integer, a multiple of 2, can be written as $2i$, where i is an integer. Consequently, we want to evaluate the sum

$$\sum_{i=1}^{50} 2i.$$

The index i ranges from 1 to 50 since $2 \cdot 1 = 2$ and $2 \cdot 50 = 100$. Let us consider the problem on a smaller scale, say we are adding 2, 4, 6, 8, and 10 with

$$\sum_{i=1}^{5} 2i = 2 \cdot 1 + 2 \cdot 2 + 2 \cdot 3 + 2 \cdot 4 + 2 \cdot 5.$$

Since 2 appears in every term, this number can be factored out of the sum, leaving

$$2(1 + 2 + 3 + 4 + 5) = 2 \cdot \sum_{i=1}^{5} i$$

so that we could employ Gauss's formula to give

$$\cancel{2} \cdot \frac{5 \cdot 6}{\cancel{2}} = 30.$$

Going back to $\Sigma_{i=1}^{50} 2i$, we again factor out 2 to obtain

$$\sum_{i=1}^{50} 2i = 2\sum_{i=1}^{50} i = \cancel{2} \, \frac{50 \cdot 51}{\cancel{2}} = 50 \cdot 51 = 2550.$$

In general, for scalar a we have

$$\sum_{i=1}^{n} a v_i = a \sum_{i=1}^{n} v_i.$$

EXAMPLE 3

Evaluate $\sum_{i=1}^{400} 162$.

Solution

$$\sum_{i=1}^{400} 162 = \sum_{i=1}^{400} 162 \cdot 1 = 162 \sum_{i=1}^{400} 1 = 162 \cdot 400 = 64{,}800. \quad \blacksquare$$

Another simplification can be used when taking a sum of a sum such as $\sum_{j=1}^{4} (j + 7)$. Expanding, we have

$$\sum_{j=1}^{4} (j + 7) = (1 + 7) + (2 + 7) + (3 + 7) + (4 + 7).$$

But all the operations are addition, so the grouping of the numbers and the order in which they appear do not matter by the associative and commutative properties for addition of real numbers. Thus, rearranging we have

$$(1 + 2 + 3 + 4) + (7 + 7 + 7 + 7) = \sum_{j=1}^{4} j + \sum_{j=1}^{4} 7.$$

Thus, note that

$$\sum_{j=1}^{4} (j + 7) = \sum_{j=1}^{4} j + \sum_{j=1}^{4} 7.$$

By similar reasoning we have in general

$$\sum_{i=1}^{n} (u_i + v_i) = \sum_{i=1}^{n} u_i + \sum_{i=1}^{n} v_i.$$

Hitherto, we have written all sums with 1 as the initial value of the index, but any reasonable integer value is acceptable. For example, perhaps our index has a lower bound of 0 so that we need to evaluate $\sum_{k=0}^{n} v_k$. Here, the index is initialized to be 0. Thus, with a lower bound for the index of 2, we have

$$\sum_{i=2}^{5} i^2 = 2^2 + 3^2 + 4^2 + 5^2 = 4 + 9 + 16 + 25 = 54.$$

NOTATION

For the integers m and n with $m \leq n$,

$$\sum_{i=m}^{n} v_i = v_m + v_{m+1} + \cdots + v_n.$$

The **lower bound** is m, and the **upper bound** is n.

EXAMPLE 4

Evaluate $\sum_{k=8}^{80} 30k$ using formulas when appropriate.

Solution

Factoring out 30 we have

$$\sum_{k=8}^{80} 30k = 30 \sum_{k=8}^{80} k.$$

Using the formula for addition of consecutive positive integers is easier than the actual addition. In this case, however, we must take away the first 7 terms, since our lower bound is 8.

$$30\left(\sum_{k=1}^{80} k - \sum_{k=1}^{7} k\right) = 30\left(\frac{80 \cdot 81}{2} - \frac{7 \cdot 8}{2}\right)$$
$$= 30(40 \cdot 81 - 7 \cdot 4) = 96{,}360. \quad \blacksquare$$

EXAMPLE 5

Evaluate $s = \sum_{i=0}^{25} 2^i$.

Solution

This sum, where we are adding increasing powers of a constant, is called a **finite geometric series.** Sums of powers of 2 are particularly useful in some areas of computer science such as data structures and computational complexity. Instead of using brute force and adding together 26 terms, we will develop a general formula. First, multiply s by 2 to obtain

$$2s = 2 \sum_{i=0}^{25} 2^i = \sum_{i=0}^{25} 2 \cdot 2^i = \sum_{i=1}^{26} 2^i.$$

Now

$$2s - s = \sum_{i=1}^{26} 2^i - \sum_{i=0}^{25} 2^i = 2^{26} - 2^0 = 2^{26} - 1$$

because $\sum_{i=1}^{25} 2^i$, which is part of both terms, is canceled from the expression. Thus,

$$s(2 - 1) = 2^{26} - 1$$

so

$$s = \frac{(2^{26} - 1)}{(2 - 1)} = 2^{26} - 1.$$

By the same reasoning we have

$$\sum_{i=0}^{n-1} a^i = \frac{(a^n - 1)}{(a - 1)} \quad \text{for } a \neq 1. \quad \blacksquare$$

Summary of some basic formulas involving sigma; m and n are positive integers with $m \leq n$, and a is a real number.

$$\sum_{i=1}^{n} 1 = n \qquad\qquad \sum_{i=1}^{n} (u_i + v_i) = \sum_{i=1}^{n} u_i + \sum_{i=1}^{n} v_i$$

$$\sum_{i=1}^{n} i = \frac{n(n + 1)}{2} \qquad\qquad \sum_{i=m}^{n} v_i = \sum_{i=1}^{n} v_i - \sum_{i=1}^{m-1} v_i$$

$$\sum_{i=1}^{n} av_i = a \sum_{i=1}^{n} v_i \qquad\qquad \sum_{i=0}^{n-1} a^i = \frac{a^n - 1}{a - 1}, a \neq 1$$

The sigma notation is used frequently in books and journal articles to abbreviate sums in formulas and definitions. You may recall examples of polynomials such as

$$5x^2 + 8x - 4 \quad \text{or} \quad 7x^{12} - 3/2\, x^{10} - \sqrt{5}\, x$$

from other mathematics classes. Since we have sums of terms, the sigma notation can be employed for the definition of a polynomial. Each term of a polynomial consists of a real number coefficient times a variable that is raised to a nonnegative, integral exponent. Even the constant -4 of the first polynomial above can be written in this fashion as $-4 \cdot x^0$. The largest exponent is the degree of the polynomial, so that $5x^2 + 8x - 4$ has degree 2 while the other polynomial has degree 12. As will be seen in Section 11.2, polynomials are used to describe the speed at which certain algorithms are executed.

DEFINITION

$$P = \sum_{i=0}^{n} a_i x^i = a_0 + a_1 x + a_2 x^2 + \cdots + a_n x^n$$

with $a_i \in \mathbb{R}$ for $i = 0, 1, 2, \ldots, n$, $n \in \mathbb{N}$, and $a_n \neq 0$, is a **polynomial of degree n**. If $P = 0$, then P is the **zero polynomial** with no degree.

If repeated multiplication instead of addition is to be done, another symbol, capital **pi, Π,** is used. Thus,

$$\prod_{i=1}^{5} i = 1 \cdot 2 \cdot 3 \cdot 4 \cdot 5 = 5! = 120$$

and $\displaystyle\prod_{i=1}^{5} 2 = 2 \cdot 2 \cdot 2 \cdot 2 \cdot 2 = 2^5 = 32.$

NOTATION

$$\prod_{i=m}^{n} v_i \quad \text{or} \quad \Pi_{i=m}^{n} \, v_i = v_m \cdot v_{m+1} \cdots v_n,$$

where m and n are integers with $m \leq n$.

EXERCISES 6.2

Evaluate Exercises 1–12 by expanding the sums or products.

1. $\displaystyle\sum_{i=1}^{4} \frac{i+3}{i}$ **2.** $\displaystyle\sum_{i=4}^{7} i^2$ **3.** $\displaystyle\sum_{i=-2}^{2} i^3$

4. $\displaystyle\sum_{i=1}^{10} \lfloor i/2 \rfloor$ **5.** $\displaystyle\sum_{i=1}^{10} \lceil i/3 \rceil$ **6.** $\displaystyle\sum_{k=0}^{4} k!$

7. $\displaystyle\sum_{k=0}^{3} P(10, k)$ **8.** $\displaystyle\sum_{k=0}^{3} C(10, k)$ **9.** $\displaystyle\sum_{k=0}^{7} (-1)^k k$

10. $\displaystyle\prod_{i=3}^{6} 2i$ **11.** $\displaystyle\prod_{j=1}^{8} \lceil j/4 \rceil$ **12.** $\displaystyle\prod_{k=-2}^{2} |2k + 1|$

Evaluate Exercises 13–21 by using the properties of Σ notation.

13. $\displaystyle\sum_{i=1}^{7000} i$ **14.** $\displaystyle\sum_{i=800}^{7000} i$ **15.** $\displaystyle\sum_{j=1}^{482} 1$

16. $\displaystyle\sum_{j=1}^{94} 9$ **17.** $\displaystyle\sum_{k=6}^{60} 17$ **18.** $\displaystyle\sum_{k=1}^{20} (k + 4)$

19. $\displaystyle\sum_{k=21}^{85} (k + 4)$ **20.** $\displaystyle\sum_{k=1}^{350} (6k + 5)$ **21.** $\displaystyle\sum_{i=0}^{10} 7 \cdot 5^i$

Evaluate Exercises 22–32.

22. $\displaystyle\prod_{i=1}^{7342} 1$ **23.** $\displaystyle\prod_{i=1}^{7} 2$ **24.** $\displaystyle\prod_{i=1}^{4} 3$

25. $\displaystyle\prod_{i=1}^{6} i$ **26.** $\displaystyle\prod_{i=1}^{10} i$ **27.** $\displaystyle\prod_{i=1}^{4} 5i$

28. $\displaystyle\left(\prod_{i=1}^{4} 5\right)\left(\prod_{i=1}^{4} i\right)$ **29.** $\displaystyle 5^4 \cdot \prod_{i=1}^{4} i$ **30.** $\displaystyle 5 \cdot \prod_{i=1}^{4} i$

31. $\displaystyle\prod_{i=4}^{7} i$ **32.** $\displaystyle\left(\prod_{i=1}^{7} i\right)\Big/\left(\prod_{i=1}^{3} i\right)$

In Exercises 33–38 you will develop the properties for the Π notation. Refer to the indicated exercises for specific examples. Simplify each product for positive integers m *and* n *with* m $\leq$ n *and constant* a.

33. $\displaystyle\prod_{i=1}^{n} 1$ (See Exercise 22)   **34.** $\displaystyle\prod_{i=1}^{n} a$ (See Exercises 23 and 24)

35. $\displaystyle\prod_{i=1}^{n} i$ (See Exercises 25 and 26) **36.** $\displaystyle\prod_{i=1}^{n} (u_i v_i)$ (See Exercises 27 and 28)

37. $\displaystyle\prod_{i=1}^{n} a v_i$ (See Exercises 27 and 29)

38. $\displaystyle\prod_{i=m}^{n} v_i$ (See Exercises 31 and 32)

39. Using Exercises 35 and 36 find formulas for

  a. $\displaystyle\prod_{i=1}^{n} i^2$ **b.** $\displaystyle\prod_{i=1}^{n} i^3.$

40. Consider the vector $\mathbf{s} = (40, 38, 32, 40, 46)$. Find

  a. $\displaystyle\sum_{i=1}^{5} s_i$ **b.** $\displaystyle\sum_{i=3}^{5} s_i$ **c.** $\displaystyle\sum_{i=0}^{4} s_i$

  Note: The last sum would get an out-of-range error on the computer.

41. The vectors $\mathbf{f} = (40, 40, 0, 45, 43)$ and $\mathbf{g} = (40, 38, 32, 40, 46)$ from the last section indicate the number of hours 5 employees worked the first and second weeks, respectively, of a pay period. $\mathbf{h} = \mathbf{f} + \mathbf{g} = (80, 78, 32, 85, 89)$ is the vector for the hours worked by employees during the pay period.

  a. Give two ways to calculate the total man-hours of work in the company for the two-week period.

  b. What property involving Σ does the equality of the results of these two methods illustrate?

42. Consider the 10-tuple of 5 homework grades, 4 test grades, and 1 exam grade: $\mathbf{g} = (5, 7, 10, 8, 8, 63, 75, 92, 81, 144)$. Write in sigma notation the sum of

  a. All grades   **b.** The 5 homework grades
  c. The 4 test grades   **d.** The exam grade.

LANGUAGE EXERCISE

43. Write a Pascal program segment to evaluate $\Pi_{I=1}^{N} A(I)$ for array A. □

44. a. Evaluate $\Sigma_{j=1}^{3}\,\Sigma_{i=1}^{20}\,5ij$ by calculating $\Sigma_{i=1}^{20}\,5ij$ first then $\Sigma_{j=1}^{3}$ of the result. Note in calculating $\Sigma_{i=1}^{20}\,5ij$ that both 5 and j can be taken outside (to the left of) the Σ notation since both are constants as i changes. Simplify as you are going along.
  b. Write $\Sigma_{i=1}^{20}\,5ij$ in MACSYMA.
  c. Write the sum of a sum, $\Sigma_{j=1}^{3}\,\Sigma_{i=1}^{20}\,5ij$, as a composition of SUM functions in MACSYMA. ☐

45. a. Evaluate $\Sigma_{j=1}^{29}\,\Sigma_{i=0}^{9}\,j\,2^{i}$.
  b. Write this composition of sums in MACSYMA. ☐

46. Let a and d be constants and n a positive integer. Using the properties of sigma notation, show that

$$\sum_{i=0}^{n} (a + id) = (1/2)(n + 1)(2a + nd).$$

47. In an article by Chin and Wong, the finishing time for printing if there are no extra typesettings is shown to be

$$f_0 = \sum_{i=1}^{n} a_i + (n - 1) + r$$

and if there are k extra typesettings

$$f_k = \sum_{i=1}^{n} a_i + \sum_{i=1}^{k} (1 - P_i) + (n - k - 1) + r$$

with variables appropriately defined. (Ignore the meaning of the variables for the purpose of this problem.) The saving S_k is $f_0 - f_k$. Verify, as given in the article, that

$$f_0 - f_k = \sum_{i=1}^{k} P_i.$$

Fill in the appropriate lower and upper bounds for simplifying Exercises 48–51.

48. $\displaystyle\sum_{i=1}^{33} 2^3 \cdot 2^i = \sum 2^i$　　　　**49.** $\displaystyle\sum_{i=0}^{8} 5^i \cdot 5^i = \sum 5^{2i}$

50. $\displaystyle\sum_{i=1}^{19} 7^{i+2} = \sum 7^i$　　　　**51.** $\displaystyle\sum_{i=2}^{36} (i + 6) = \sum i$

For each of Exercises 52–60 state whether the expression is a polynomial or not. If your answer is yes, give the degree.

52. $6x + 3$　　　　**53.** $18.2\,x^{45}$　　　　**54.** $\displaystyle\sum_{i=-3}^{3} a_i x^i,\ a_1 \in \mathbb{R}$

55. $\sqrt{2}x + \pi x^{17} - 3/2$　　　　**56.** 0　　　　**57.** $2\sqrt{x} + 28\,x^{\pi}$

58. 42　　　　**59.** $x^{3/2} + 4x^{1/2}$　　　　**60.** $(x^6 + 8)/x$

Simplify Exercises 61–68 using Σ notation.

61. $\left(\sum_{i=1}^{5} i\right) + 6$ **62.** $\left(\sum_{i=1}^{n} i\right) + (n + 1)$ **63.** $\left(\sum_{j=1}^{9} j^2\right) + 100$

64. $\left(\sum_{j=1}^{n-1} j^2\right) + n^2$ **65.** $\left(\sum_{k=2}^{5} k!\right) + 1$ **66.** $\left(\sum_{k=2}^{5} k!\right) + 1 + 6!$

67. $\left(\sum_{k=3}^{n} k^3\right) + 9$ **68.** $\left(\sum_{k=1}^{n} 3k\right) + 3(n + 1) + 3(n + 2)$

69. Write the equality in Exercise 32, Section 3.2, using Σ notation.

70. Write the Binomial Theorem using Σ notation. (See Section 3.2.)

6.3

Recursion

Without calling it such, we have already used recursion. In Example 1 of Section 5.5 we considered a function to generate a sequence of random numbers. An initial value is given to X, say 8367; the next value of X is always computed from the previous value by taking $\mathrm{mod}_{10,000}(229X + 1)$. Since we now have the subscript notation at our disposal, let us redefine the function of Example 1 as follows:

$$\begin{cases} x_0 = 8367 \\ x_n = \mathrm{mod}_{10,000}(229x_{n-1} + 1) \qquad n > 0. \end{cases}$$

In the definition of the sequence, an initial condition, $x_0 = 8367$, is given as well as a generating rule or method of computing one value from the previous value(s), $x_n = \mathrm{mod}_{10,000}(229x_{n-1} + 1)$; such is a recursive function. This function can be defined in Pascal, assuming NONNEGATIVE is of type $0\,..\,$MAXINT, as follows:

LANGUAGE EXAMPLE

```
FUNCTION RAND (N: NONNEGATIVE):  INTEGER,
BEGIN
      IF  N = 0  THEN
          RAND := 8367
      ELSE
          RAND := ((229 * RAND(N - 1) + 1)MOD 10000)
   END;   □
```

If we prefer not writing the index as a subscript in the mathematical definition, we can use the following:

$$\begin{cases} x(0) = 8367 \\ x(n) = \mathrm{mod}_{10,000}(229x(n - 1) + 1) \qquad n > 0. \end{cases}$$

The first few values of this function from the set of positive integers to the set of integers between 0 and 7999, inclusive, can be written in sequence notation:

8367,   6044,   4077,   3634,   2187,   824,   8697, . . .

x_0, x_1, x_2, x_3, x_4, x_5, x_6, . . .

or $x(0)$, $x(1)$, $x(2)$, $x(3)$, $x(4)$, $x(5)$, $x(6)$, . . .

DEFINITION

A **recursive function** or **recurrence relation** or **difference equation** for a sequence $x_0, x_1, x_2, x_3, \ldots$ is a relation that defines x_n in terms of $x_0, x_1, x_2, x_3, \ldots, x_{n-1}$. The formula relating x_n to earlier values in the sequence is called the **generating rule.** The assignment of a value to one of the $x's$ is called an **initial condition.**

Formulas that we write using Σ notation can be defined recursively, also.

EXAMPLE 1

Write a recurrence relation to define the sum

$$s_n = \sum_{i=0}^{n} (3i^2), \, n \in \mathbb{N}.$$

Solution

We want to write s_n in terms of $s_0, s_1, s_2, \ldots, s_{n-1}$. It will help us to observe the expansion for a few of these s_n's.

$$s_0 = \sum_{i=0}^{0} 3i^2 = 3 \cdot 0^2$$

$$s_1 = \sum_{i=0}^{1} 3i^2 = 3 \cdot 0^2 + 3 \cdot 1^2$$

$$s_2 = \sum_{i=0}^{2} 3i^2 = 3 \cdot 0^2 + 3 \cdot 1^2 + 3 \cdot 2^2$$

$$\vdots$$

$$s_{n-1} = \sum_{i=0}^{n-1} 3i^2 = 3 \cdot 0^2 + 3 \cdot 1^2 + 3 \cdot 2^2 + \cdots + 3 \cdot (n-1)^2$$

$$s_n = \sum_{i=0}^{n} 3i^2 = 3 \cdot 0^2 + 3 \cdot 1^2 + 3 \cdot 2^2 + \cdots + 3 \cdot (n-1)^2 + 3 \cdot n^2.$$

Notice the difference between the nth and $(n-1)$th elements of the sequence is the term $3 \cdot n^2$; so, the generating rule is

$$s_n = s_{n-1} + 3n^2.$$

Also, note the initial condition should be

$$s_0 = 3 \cdot 0^2 = 0.$$

Thus, the recursive definition for s_n is as follows:

$$\begin{cases} s_0 = 0 \\ s_n = s_{n-1} + 3n^2 & n > 0. \end{cases} \blacksquare$$

Actually, many functions we have defined otherwise could be defined recursively. For instance, the factorial function $f(n) = n! = n \cdot (n-1) \cdots 3 \cdot 2 \cdot 1 = n \cdot [(n-1) \cdots 3 \cdot 2 \cdot 1] = n \cdot [(n-1)!]$ can be written as follows:

$$\begin{cases} f(0) = 1 \\ f(n) = n \cdot f(n-1) & n > 0. \end{cases}$$

Certainly, $f(0) = 1 = 0!$ Let's use this definition to show that $f(4) = 4!$

$$\begin{aligned} f(4) &= 4 \cdot f(3) \\ &= 4 \cdot (3 \cdot f(2)) \\ &= 4 \cdot 3 \, (2 \cdot f(1)) \\ &= 4 \cdot 3 \cdot 2 \cdot (1 \cdot f(0)) \\ &= 4 \cdot 3 \cdot 2 \cdot 1 \cdot 1 = 4!. \end{aligned}$$

LANGUAGE EXAMPLES

When allowed by a computer language, recursion is a very useful feature for function definition. Pascal, LISP, and APL permit recursive functions while COBOL, BASIC, and FORTRAN do not. A definition of the factorial function in Pascal follows with type NONNEGATIVE being $0 \, . \, . \, \text{MAXINT}$:

```
FUNCTION FACT  (N:  NONNEGATIVE) :  INTEGER;
    BEGIN
            IF  N  =  0  THEN
                    FACT  := 1
            ELSE
                    FACT  := N * FACT (N - 1)
    END;  □
```

Recursive functions are implemented with stacks in the computer. For example, to evaluate FACT(4), FACT(3) must be known first, so the definition of FACT as $4 \cdot \text{FACT}(3)$ is pushed onto a stack as is each successive value of FACT. When a value for FACT is encountered which stops the process, values are popped off the stack to be assigned to the various FACTs in reverse order as illustrated in Figure 6.5.

Figure 6.5

Stack for the FACT function

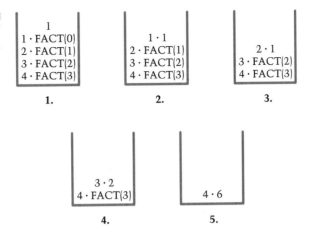

Given a sequence, we can often find a difference equation definition; conversely, given a recurrence relation, many times we can find an equivalent nonrecursive function. Derivation of a non-recursive function is useful in situations where an algorithm is easier to develop, but more difficult to implement recursively. For instance, perhaps we need to evaluate the 360th term of the sequence. With a recursive function we might get a STACK OVERFLOW error, meaning the stack simply isn't large enough to maintain 360 values. To overcome this problem, we solve, or write nonrecursively, the recurrence relation as illustrated in Example 2.

DEFINITION

> To **solve a recurrence relation** means to find a nonrecursive definition for the function.

EXAMPLE 2

Solve the following recurrence relation:

$$\begin{cases} x_0 = 7 \\ x_n = 2x_{n-1} + 3 \qquad n > 0 \end{cases}$$

Solution

One method of discovering a nonrecursive definition is to evaluate the first few terms for the sequence and by inspection to discover a pattern. Do not evaluate completely, but simplify, each expression.

$$x_0 = 7$$
$$x_1 = 2x_0 + 3 = 2 \cdot 7 + 3$$
$$x_2 = 2x_1 + 3 = 2(2 \cdot 7 + 3) + 3 = 2^2 \cdot 7 + 2 \cdot 3 + 3$$

$$x_3 = 2x_2 + 3 = 2(2^2 \cdot 7 + 2 \cdot 3 + 3) + 3$$
$$= 2^3 \cdot 7 + 2^2 \cdot 3 + 2 \cdot 3 + 3$$
$$x_4 = 2x_3 + 3 = 2(2^3 \cdot 7 + 2^2 \cdot 3 + 2 \cdot 3 + 3)$$
$$= 2^4 \cdot 7 + 2^3 \cdot 3 + 2^2 \cdot 3 + 2 \cdot 3 + 3.$$

Grouping the terms with a factor of 3 we have

$$x_4 = 2^4 \cdot 7 + (2^3 + 2^2 + 2 + 1) \cdot 3$$
$$= 2^4 \cdot 7 + (2^3 + 2^2 + 2^1 + 2^0) \cdot 3$$

or

$$x_4 = 2^4 \cdot 7 + \left(\sum_{i=0}^{3} 2^i \right) \cdot 3.$$

Generalizing to x_n, we have

$$x_n = 2^n \cdot 7 + \left(\sum_{i=0}^{n-1} 2^i \right) \cdot 3.$$

But $\sum_{i=0}^{n-1} 2^i$ is just a finite geometric series whose sum is $(2^n - 1)/(2 - 1) = (2^n - 1)$. Substituting, we obtain

$$x_n = 2^n \cdot 7 + (2^n - 1) \cdot 3 = 2^n \cdot 7 + 2^n \cdot 3 - 3$$
$$= 2^n \cdot (7 + 3) - 3 = 2^n \cdot 10 - 3.$$

Particularly since we are deriving the formula by inspection, we should be careful to check our answer for both the initial condition and the generating rule. Using $x_n = 2^n \cdot 10 - 3$, we verify that the initial condition is satisfied by letting $n = 0$:

$$x_0 = 2^0 \cdot 10 - 3 = 10 - 3 = 7.$$

To check the generating rule we must evaluate x_n and x_{n-1} and show that $2x_{n-1} + 3 = x_n$:

$$x_n = 2^n \cdot 10 - 3$$

and

$$x_{n-1} = 2^{n-1} \cdot 10 - 3$$

Thus,

$$2x_{n-1} + 3 = 2(2^{n-1} \cdot 10 - 3) + 3$$
$$= 2 \cdot 2^{n-1} \cdot 10 - 6 + 3$$
$$= 2^n \cdot 10 - 3$$
$$= x_n. \quad \blacksquare$$

Solving a Difference Equation by Inspection

1. Write the first few terms of the sequence, simplifying but not performing the arithmetic.
2. Find a pattern.
3. If possible, use formulas for sigma and pi notation to simplify the result.

HISTORICAL NOTE

Recursion can be used in the formal definition of languages, also. **Backus-Normal form (BNF)** was first used to give an exact definition of the language ALGOL in the late 1950s. ☐ Many language manuals and texts today use BNF to display their syntax. In BNF ::= is much like an assignment symbol; the term on the left of the ::= is being defined by what appears on the right. A vertical bar, |, means "or." Terminal symbols, such as digits and letters of the alphabet, are not defined but can be used on the right. Nonterminals, which can appear on the right or left, are indicated by being surrounded by pointed brackets, < >. Concatenation or the "gluing" of things together is indicated by placing the two things side-by-side. The BNF description of an ALGOL variable is as follows:

LANGUAGE EXAMPLE

```
<variable> ::= <letter>|<variable><letter>
<letter> ::= a|b|c|d|e|f|g|h|i|j|k|l|m|n|o|p|q|r|
             s|t|u|v|w|x|y|z
```

For example, "rate" is a variable. It can be built up from the BNF description as shown in Figure 6.6.

Figure 6.6

"rate" satisfies the BNF description of a variable

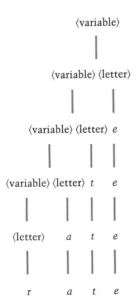

The BNF description of ALGOL is a particular example of the use of a formal grammar, and languages so described can be recognized and interpreted more easily by a compiler. A detailed study of **formal grammars** is part of **automata theory.** A grammar has four parts, a **vocabulary** or set of symbols (V), a nonempty subset of **terminal symbols** (V_T), the **start symbol** (S), and a finite set of **productions** or **rules** (P). A **word** consists of a finite-length string of terminals, and a **language** is the set of all words generated using the vocabulary and the productions. The vocabulary of ALGOL consists of the start symbol; letters; digits; some special symbols, such as a comma; reserved words, such as WHILE; and other words, such as "variable." The above definition of "variable" could be written with the following productions:

$$
\begin{aligned}
S &\rightarrow \texttt{variable} \\
\texttt{variable} &\rightarrow \texttt{letter}\,|\,\texttt{variable letter} \\
\texttt{letter} &\rightarrow \texttt{a}|\texttt{b}|\texttt{c}|\texttt{d}|\texttt{e}|\texttt{f}|\texttt{g}|\texttt{h}|\texttt{i}|\texttt{j}|\texttt{k}|\texttt{l}|\texttt{m}|\texttt{n}|\texttt{o}|\texttt{p}|\texttt{q}|\texttt{r}|\texttt{s}| \\
&\quad\ \texttt{t}|\texttt{u}|\texttt{v}|\texttt{w}|\texttt{x}|\texttt{y}|\texttt{z} \quad \Box
\end{aligned}
$$

EXAMPLE 3

For the vocabulary $V = \{0, 1, S\}$ with set of terminals $V_T = \{0, 1\}$, find the language generated by the following productions:

1. $S \rightarrow 0S$ 2. $S \rightarrow 1S$
3. $S \rightarrow 0$ 4. $S \rightarrow 1$

Also, indicate how the string 10011 is generated by this grammar.

Solution

The language contains all finite strings of 0's and 1's. The string 10011 is formed by the next sequence with the indicated productions:

$$
S \xrightarrow{2} 1S \xrightarrow{1} 1\,0S \xrightarrow{1} 10\,0S \xrightarrow{2} 100\,1S \xrightarrow{4} 10011
$$

Notice the recursive nature of the definition as S appears on the left and right of productions 1 and 2; S is defined in terms of S. Also, a way to stop the process is given by productions 3 and 4. Productions 3 and 4 are comparable to the initial conditions, while productions 1 and 2 give the generating rule. ■

As demonstrated in the next example, recursion can be used in developing the grammar for and in counting the words of a language.

EXAMPLE 4

Consider the language containing all finite nonempty strings of bits that do not contain 11.

a. Give a recurrence relation that describes how many words of length n, $n \in \mathbb{I}^+$, are in this language.

b. Give the grammar for this language.

Solution

Before tackling the problem, let us note several words that are in the language: 0, 1, 10101, 10000000, 00010. The following strings, however, are not in the language because they each contain two adjacent 1's: 110000, 01011, 11, 0100110110.

a. Let c_n be the number of words of length n in this language, and define c_n recursively. An n-bit string can begin with 0 or 1. If the n-bit string begins with 0, it is followed by an $(n-1)$-bit string which is also in the language:

$$\underbrace{0x \cdots x}_{n-1 \text{ bits}}$$

If the string begins with 1, however, the 1 cannot be immediately followed by another 1. Thus, the word must start with 10 and be followed by an $(n-2)$-bit string from the language:

$$\underbrace{0x \cdots x}_{n-1 \text{ bits}} \quad \text{or} \quad \underbrace{10x \cdots x}_{n-2 \text{ bits}}$$

Therefore, the number of n-bit strings, c_n, is the sum of the number of $(n-1)$-bit strings and the number of $(n-2)$-bit strings, or

$$c_n = c_{n-1} + c_{n-2}.$$

Since c_n is defined using two previous terms in the sequence, we need two initial conditions. There are two strings of length 1, 0 and 1, so that

$$c_1 = 2.$$

Since the words of length 2 are 00, 01, and 10, but not 11, we have

$$c_2 = 3.$$

Thus, the recursion relation is

$$\begin{cases} c_1 = 2 \\ c_2 = 3 \\ c_n = c_{n-1} + c_{n-2} \quad n > 2. \end{cases}$$

From this formula we anticipate that c_3, the number of 3-bit strings in the language should be $c_3 = c_2 + c_1 = 3 + 2 = 5$. In fact, we can derive these strings using the above reasoning. Concatenate 0 with all 2-bit words and concatenate 10 with all 1-bit words:

000, 001, 010, 100, 101.

Since $c_4 = c_3 + c_2 = 5 + 3 = 8$, we concatenate 0 with the above 3-bit strings and 10 with the 2-bit strings in the language to obtain eight 3-bit words:

0000, 0001, 0010, 0100, 0101, 1000, 1001, 1010.

b. Using a similar reasoning we can find the productions for the grammar of this language with vocabulary $\{0, 1, S\}$, set of terminal symbols $\{0, 1\}$, and start symbol S. The generating rules to define the n-bit words,

$$\underbrace{0x \cdots x}_{n-1 \text{ bits}} \quad \text{and} \quad \underbrace{10x \cdots x}_{n-2 \text{ bits}}$$

in terms of $(n - 1)$- and $(n - 2)$-bit words are as follows:

1. $S \rightarrow 0S$
2. $S \rightarrow 10S$

We now only need to consider initial conditions, where $n = 1$ or 2. The bit strings 0 and 1 are both in the language, so we have productions

3. $S \rightarrow 0$
4. $S \rightarrow 1$

To form the 2-bit words, 00, 01, 10, we have the rules

5. $S \rightarrow 00$
6. $S \rightarrow 01$
7. $S \rightarrow 10$

Those strings starting with 0, however, can be generated using productions 1, 3, and 4 as follows:

$$S \rightarrow 0S \rightarrow 00$$

and

$$S \rightarrow 0S \rightarrow 01$$

Thus, the necessary productions are as follow:

$$\begin{cases} S \rightarrow 0 \\ S \rightarrow 1 \\ S \rightarrow 10 \\ S \rightarrow 0S \\ S \rightarrow 10S \end{cases}$$

EXERCISES 6.3

Write the first five terms in the defined sequences of Exercises 1–3.

1. $\begin{cases} x_1 = 3 \\ x_n = 2x_{n-1} - 1 \end{cases}$
2. $\begin{cases} a_1 = -1 \\ a_n = -a_{n-1} \end{cases}$
3. $\begin{cases} y_1 = 8 \\ y_n = y_{n-1}/2 \end{cases}$

Find a recurrence relation with initial condition that defines the sequence in each of Exercises 4–9.

4. 1, 4, 7, 10, 13, . . .

5. 2, 4, 6, 8, 10, . . .

6. 3, 5, 7, 9, 11, . . .

7. 3, 6, 9, 12, 15, . . .

8. 5, −5, 5, −5, 5, . . .

9. 1, 4, 16, 64, 256, . . .

Write a recurrence relation with initial condition to define each sum s_n *in Exercises 10–15.*

10. $s_n = \displaystyle\sum_{i=1}^{n} 1$

11. $s_n = \displaystyle\sum_{i=1}^{n} i$

12. $s_n = \displaystyle\sum_{i=1}^{n} u_i$

13. $s_n = \displaystyle\sum_{i=1}^{n} 5i^3$

14. $s_n = \displaystyle\sum_{i=0}^{n-1} 7^i$

15. $s_n = \displaystyle\sum_{i=1}^{n} (4i^2 - 9)$

Write the recurrence relation to determine each product p(n) *in Exercises 16 and 17.*

16. $p(n) = \displaystyle\prod_{i=1}^{n} 8$

17. $p(n) = \displaystyle\prod_{i=1}^{n} 3i$

18. Suppose the principal P of \$5000 is deposited in a bank and is compounded at a yearly interest rate R of 10%. Let $A(N)$ be the amount present after N years. At the end of each year, suppose the interest is deposited in the account. Thus,

$A(0) = P = 5000$
$A(1) = P + P \cdot R = P \cdot (1 + R) = P \cdot (1 + 0.10) = 5500$
$A(2) = A(1) + A(1) \cdot R = A(1) \cdot (1 + R) = 5500 \cdot 1.1 = 6050$

LANGUAGE
EXERCISE

a. Find $A(3)$, $A(4)$, and $A(5)$.
b. Write the formula for $A(N)$ recursively.
c. Write this function in Pascal. □

19. The Fibonacci sequence is a common example of recursion. The initial conditions consist of a definition of the first two terms of the sequence. Any subsequent term is the sum of the two immediately preceding numbers in the sequence. Thus, for $x_0 = x_1 = 1$,

$$x_2 = x_1 + x_0 = 1 + 1 = 2,$$
$$x_3 = x_2 + x_1 = 2 + 1 = 3,$$
$$x_4 = x_3 + x_2 = 3 + 2 = 5.$$

The sequence is defined as

$$\begin{cases} x_0 = 1 \\ x_1 = 1 \\ x_n = x_{n-1} + x_{n-2} \quad \text{for } n > 1. \end{cases}$$

a. Find terms x_5 through x_{10}.

b. We could define a function f as $f(n) = x_n$, the nth term of the Fibonacci sequence. Write the definition for this function.

c. Write a Pascal function FIBONACCI to generate the Nth term of the sequence. □

LANGUAGE
EXERCISE

20. a. For Example 4 find c_5.

b. List all 5-bit strings in the language.

Solve each of the recurrence relations in Exercises 21–35.

21. $\begin{cases} x_0 = 5 \\ x_n = 5x_{n-1} \end{cases}$ **22.** $\begin{cases} x_0 = a \\ x_n = ax_{n-1} \end{cases}$ **23.** $\begin{cases} x_0 = a \\ x_n = bx_{n-1} \end{cases}$

24. $\begin{cases} x_0 = 5 \\ x_n = x_{n-1} + 1 \end{cases}$ **25.** $\begin{cases} x_0 = a \\ x_n = x_{n-1} + 1 \end{cases}$ **26.** $\begin{cases} x_0 = 5 \\ x_n = x_{n-1} + 2 \end{cases}$

27. $\begin{cases} x_0 = a \\ x_n = x_{n-1} + c \end{cases}$ **28.** $\begin{cases} x_0 = 2 \\ x_n = 3x_{n-1} + 7 \end{cases}$ **29.** $\begin{cases} x_0 = a \\ x_n = bx_{n-1} + c \quad b \ne 1 \end{cases}$

30. $\begin{cases} x_0 = 1 \\ x_n = x_{n-1} + n^2 \end{cases}$ **31.** $\begin{cases} x_0 = 0 \\ x_n = x_{n-1} + n - 1 \end{cases}$ **32.** $\begin{cases} x_0 = 1 \\ x_n = x_{n-1} + a^n \quad a \ne 1 \end{cases}$

33. $\begin{cases} x_0 = 0 \\ x_n = x_{n-1} + 2n \end{cases}$ **34.** $\begin{cases} x_0 = 1 \\ x_n = 2nx_{n-1} \end{cases}$ **35.** $\begin{cases} x_0 = a \\ x_n = nx_{n-1} \end{cases}$

In each of the Exercises 36–39 describe the language that is generated from the productions in part a on the vocabulary $\{0, 1, S\}$. For each show how the string in each part b is generated using the productions from part a.

36. a. $\begin{cases} S \rightarrow 0S \\ S \rightarrow 0 \end{cases}$

b. 000

37. a. $\begin{cases} S \rightarrow 01S \\ S \rightarrow 01 \end{cases}$

b. 0101

38. a. $\begin{cases} S \rightarrow 0S0 \\ S \rightarrow 1 \end{cases}$

b. 0001000

39. a. $\begin{cases} S \rightarrow 0S0 \\ S \rightarrow 1S1 \\ S \rightarrow 1 \end{cases}$

b. 0011100

Consider the vocabulary {0, 1, S}. Write productions to generate the languages in Exercises 40–42.

40. The set of all nonempty strings (strings that contain at least one character) of 1's.

41. The set of all nonempty strings of an even number of 0's.

42. The set of all palindromes, strings that read the same left to right as right to left, such as 1001001 and 011110.

43. Find the difference equation for the number of character strings of length n where the first character is a letter and the other characters are letters or digits.
Note: For most computer languages you are finding the number of legal variable names of length n.

*For each of the languages in Exercises 44–46 **a.** Give the recursive function to calculate the number of strings of length $n \in \mathbb{I}^+$. **b.** Give the productions to generate the language.*

44. The set of all bit strings that do not contain 0.

45. The set of all bit strings that do not contain 00.

46. The set of all bit strings that do not contain 001.

47. The set of all bit strings that do not contain 10.

48. How many n-bit strings are there? By referring to Example 4 give a recursive definition for the number of n-bit strings that *do* contain 11.

49. Write a recursive definition for the function p such that $p(n) = P(n, k)$, where k is fixed and the initial condition is $P(k, k) = k!$.

50. Write a recursive definition for the function $t(n) = 2^n$.

The following algorithm, named after Euclid who first developed it, finds the greatest common divisor, GCD, of two numbers. 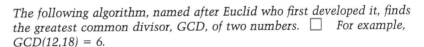 *For example, GCD(12,18) = 6.*

```
GCD (X,  Y):
    If X > Y then
        X ← GCD (X − Y,  Y)
    else if Y > X then
        X ← GCD (X,  Y − X)
    else GCD ← X.
```

Verify the GCD algorithm works for the pairs of numbers in Exercises 51–53.

51. 12, 18   **52.** 28, 7   **53.** 8, 9

54. Write GCD above as a Pascal function. □

55. Write the recursive function of Example 2 in Pascal.

56. a. Write a recursive function to evaluate the function

$$f(n) = na = \underbrace{a + a + \cdots + a}_{n \text{ summands}} \qquad n \in \mathbb{N}$$

    b. Write a recursive Pascal function, F, to evaluate N * A, for parameters N and A. ☐

57. Complete the following definition of a recursive Pascal function to find the maximum element in an array A of N elements. In this definition the maximum of the first N − 1 elements is found and compared with the Nth element of the array. Assume ARRAYTYPE is type ARRAY[1 . . N] OF REAL.

```
FUNCTION MAXA (N:   a   ; A: ARRAYTYPE):   b   ;
BEGIN
    IF N = 1 THEN
        MAXA := A[N]
    ELSE
        MAXA := MAXA (  c  ,   d  );
    IF A[N] > MAXA THEN
        MAXA :=   e
END;  ☐
```

58. Write a recursive function in Pascal to calculate the sum of the elements of the array A of N elements. ☐

59. For a sequence $u_1, u_2, u_3, \ldots$, we can define a function, mean, where mean(*n*) is the arithmetic mean of the first *n* terms of the sequence. Therefore

$$\text{mean}(n) = \frac{\sum\limits_{i=1}^{n} u_i}{n}.$$

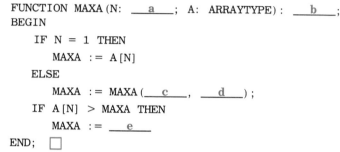
    a. Write mean(*n* − 1).
    b. Write the function mean recursively.
    c. Give the Pascal definition of MEAN, where the parameters consist of N and, an array A of N elements. ☐

6.4

Induction

With the emphasis on structured programming has come the development of an area called **program verification,** which means proving your program is correct as you are writing it. Thus, the time spent debugging can be decreased dramatically. Also, the same techniques can be useful in creating a more efficient, as well as a correct, program.

    Bugs in programs are quite a problem, illustrated by the fact that much commercial software is being released with its share. One leader in the computer field and a superior programmer reporting on a major system he wrote, noted a rate of 40 errors per 1000 lines of code undetected by the compiler and 12 bugs undetected by any testing before release.

You might be interested in the origin of the term "bug." The Mark I, completed in 1943 at Harvard, was the first electromechanical, general purpose computer; and unlike the modern electronic computer, it had moving parts. One hot summer day the machine was not working properly. After hours of checking, the problem was discovered—a moth on the contacts of one of the relays had stopped the current from flowing. The insect was carefully removed and pasted in the log book with the remark that a "bug" had been found in the computer.   □

One technique essential to program verification is **mathematical induction,** a method of proof that has been useful in every area of mathematics as well. It is used to prove statements that assert something is true for every positive integer. Consider an arbitrary loop in Pascal starting with the statement

    FOR I := 1 TO N DO

If you want to verify that the loop does something regardless of the particular integral value of N, you need mathematical induction.   □   Recall also from Section 6.2 formulas like

$$\sum_{i=1}^{n} i^2 = \frac{n(n + 1)(2n + 1)}{6}$$

hold for all positive integers n. Identities such as this one are often useful in analysis of algorithms.

We will prove this assertion using mathematical induction, but let's examine the method first. Suppose we have an infinitely long ladder (use your imagination!) where the rungs are numbered 1, 2, 3, . . . from the bottom up. And suppose we brag to friends that we can climb to any rung we want. Since there are an infinite number of rungs, we can't possibly go through all the cases; but our friends will believe us if we can show two things:

1. We can get on the ladder. (You've got to begin somewhere.)
2. Wherever we are on the ladder (say, an arbitrary rung k), we can always manage somehow to get to the next rung (rung $k + 1$).

Then, clearly, we can get anywhere we want on this superladder.

The same process holds for mathematical induction. We are trying to prove that some statement, abbreviated $P(n)$, is true for any positive integer n. In our example, $P(n)$ is the statement, "we can climb to rung n." The steps of mathematical induction are as follows:

MATHEMATICAL INDUCTION METHOD OF PROOF ▰▰▰

> 1. Prove $P(1)$. (In the above, prove you can get on the ladder at rung 1.)
> 2. Prove if $P(k)$ is true, then so is $P(k + 1)$.

In the second step, $P(k)$ is called the **induction hypothesis.** Assuming $P(k)$ for an arbitrary k, we need to show the next assertion $P(k + 1)$ is also true—on an arbitrary rung we can always climb to the next rung.

EXAMPLE 1

Use the technique of mathematical induction to prove

$$\sum_{i=1}^{n} i^2 = \frac{n(n + 1)(2n + 1)}{6}$$

for all positive integers n. *Note:* Proofs of formulas like this will almost always proceed the same way, so carefully mimic this proof in doing your homework. Remarks to lead you through the proof are in parentheses.

Solution

$P(n)$ is the equality

$$\sum_{i=1}^{n} i^2 = \frac{n(n + 1)(2n + 1)}{6}$$

or $1^2 + 2^2 + 3^2 + \cdots + (n - 1)^2 + n^2 = \frac{n(n + 1)(2n + 1)}{6}$.

(Before embarking on the proof, it might be enlightening to evaluate $P(n)$ for $n = 2$ and 3. $P(2)$ is the equality

$$\sum_{i=1}^{2} i^2 = \frac{2(2 + 1)(2 \cdot 2 + 1)}{6}$$

where 2 is substituted for n. Expanding the sum with i going from 1 to 2, we have

$$1^2 + 2^2 = \frac{2(3)(5)}{6} \quad \text{or} \quad 1 + 4 = 5.$$

In $P(3)$ the sum has an additional term for $i = 3$, namely

$$\sum_{i=1}^{3} i^2 = \frac{3(3 + 1)(2 \cdot 3 + 1)}{6}$$

or $1^2 + 2^2 + 3^2 = \frac{3(4)(7)}{6} \quad \text{or} \quad 1 + 4 + 9 = 14.)$

We will use mathematical induction to prove $P(n)$ true for all positive integers n.

1. Prove $P(1)$: Substitute 1 for n in $P(n)$ and show

$$\sum_{i=1}^{1} i^2 = \frac{1(1 + 1)(2 \cdot 1 + 1)}{6}.$$

Since the sum is only evaluated for $i = 1$, we have

$$1^2 = \frac{1(2)(3)}{6}$$

2. Prove: $P(k) \Rightarrow P(k + 1)$.

(It is always a good idea to write down precisely what you are given and what you want to prove. For $P(k)$ we just copy the assertion $P(n)$, taking out n and substituting k in its place. For $P(k + 1)$, throw away the n in $P(n)$ and put in $k + 1$.)

Assume $P(k)$:
$$\sum_{i=1}^{k} i^2 = \frac{k(k + 1)(2k + 1)}{6}$$

or $1^2 + 2^2 + 3^2 + \cdots + k^2 = \dfrac{k(k + 1)(2k + 1)}{6}.$

We must show $P(k + 1)$ is true:

$$\sum_{i=1}^{k+1} i^2 = \frac{(k + 1)(k + 1 + 1)(2(k + 1) + 1)}{6}$$

or

$$1^2 + 2^2 + 3^2 + \cdots + k^2 + (k + 1)^2 = \frac{(k + 1)(k + 2)(2k + 3)}{6}.$$

(Start with $P(k)$. After all, we are assuming the induction hypothesis is true, and, thus, we can use that statement.)

$$\sum_{i=1}^{k} i^2 = \frac{k(k + 1)(2k + 1)}{6} \qquad P(k)$$

(Notice that the left sides of $P(k)$ and $P(k + 1)$ are both sums of i^2; the former has i going to k and the latter has i going to $k + 1$. Thus, they differ only by the $(k + 1)$ term, $(k + 1)^2$. Add this term to both sides of $P(k)$, maintaining the equality.)

$$\sum_{i=1}^{k} i^2 + (k + 1)^2 = \frac{k(k + 1)(2k + 1)}{6} + (k + 1)^2.$$

(The left side is just what we want for $P(k + 1)$, so we use algebra to show the right side is also what we need.)

$$\sum_{i=1}^{k+1} i^2 = \frac{k(k+1)(2k+1)}{6} + \frac{6(k+1)^2}{6}$$

$$= \frac{k(k+1)(2k+1) + 6(k+1)^2}{6} \qquad \text{Add fractions}$$

$$= \frac{[k(2k+1) + 6(k+1)](k+1)}{6} \qquad \text{Factor out } k+1$$

$$= \frac{(2k^2 + k + 6k + 6)(k+1)}{6} \qquad \text{Multiply}$$

$$\sum_{i=1}^{k+1} i^2 = \frac{(k+1)(k+2)(2k+3)}{6}. \qquad \text{Factor}$$

Starting with $P(k)$, we have shown $P(k+1)$. Thus, $P(n)$ is true for all positive integers n. ■

It is important to show both the statement $P(1)$ and the implication $P(k) \Rightarrow P(k+1)$ are true in an induction proof. As we will observe in the next two examples, if either step fails, the statement $P(n)$ for $n \in \mathbb{I}^+$ is false.

EXAMPLE 2

Suppose we *incorrectly* conjecture that

$$\sum_{i=1}^{n} 2i^2 = n^3 + n \qquad \text{for } n \in \mathbb{I}^+.$$

Find where the inductive proof breaks down.

Solution

1. For $n = 1$ we see that $P(1)$ does hold since

$$\sum_{i=1}^{1} 2i^2 = 1^3 + 1$$

or $\qquad 2 \cdot 1^2 = 2 = 1^3 + 1$

is true.

2. Assume $P(k)$: $\displaystyle\sum_{i=1}^{k} 2i^2 = k^3 + k$

or $\qquad 2 \cdot 1^2 + 2 \cdot 2^2 + 2 \cdot 3^2 + \cdots + 2 \cdot k^2 = k^3 + k$

Prove $P(k+1)$: $\displaystyle\sum_{i=1}^{k+1} 2i^2 = (k+1)^3 + (k+1)$

or expanding

$$2 \cdot 1^2 + 2 \cdot 2^2 + 2 \cdot 3^2 + \cdots + 2 \cdot k^2 + 2(k + 1)^2$$
$$= (k + 1)^3 + (k + 1)$$
$$= k^3 + 3k^2 + 3k + 1 + k + 1$$
$$= k^3 + 3k^2 + 4k + 2.$$

Proceeding as we did in Example 1, we start with the induction hypothesis, $P(k)$, and add the $(k + 1)$ term of the sum, $2(k + 1)^2$, to both sides.

$$\sum_{i=1}^{k} 2i^2 = k^3 + k \qquad P(k)$$

$$\sum_{i=1}^{k} 2i^2 + 2(k + 1)^2 = k^3 + k + 2(k + 1)^2$$

$$\sum_{i=1}^{k+1} 2i^2 = k^3 + k + 2(k^2 + 2k + 1)$$
$$= k^3 + k + 2k^2 + 4k + 2$$
$$= k^3 + 2k^2 + 5k + 2.$$

But here is where we have a problem. Since $k^3 + 2k^2 + 5k + 2 \neq k^3 + 3k^2 + 4k + 2$ for $k \neq 0$, we cannot prove the statement $P(k + 1)$:

$$\sum_{i=1}^{k+1} 2i^2 = k^3 + 3k^2 + 4k + 2.$$

We showed $P(n)$ holds for $n = 1$. In fact, it holds for $n = 2$ as well:

$$\sum_{i=1}^{2} 2i^2 = 2^3 + 2$$

or $\quad 2 \cdot 1^2 + 2 \cdot 2^2 = 2^3 + 2$

or $\qquad 2 + 8 = 8 + 2.$

The statement, however, breaks down at $n = 3$:

$$\sum_{i=1}^{3} 2i^2 \overset{?}{=} 3^3 + 3$$

or $\quad 2 \cdot 1^2 + 2 \cdot 2^2 + 2 \cdot 3^2 \overset{?}{=} 3^3 + 3$

or $\qquad 2 + 8 + 18 \overset{?}{=} 27 + 3$

but $\qquad\qquad 28 \neq 30$ ◼

EXAMPLE 3

Another *incorrect* conjecture is that

$$\sum_{i=1}^{n} 2i = (n - 1)(n + 2) \qquad \text{for } n \in \mathbb{I}^{+}.$$

$P(1)$ in Examples 1 and 2, as in many examples, seems so straightforward that we may be tempted to omit its verification and start immediately showing $P(k) \Rightarrow P(k + 1)$. Verify this implication, and then show $P(1)$ does not hold.

Solution

Assume $P(k)$:

$$\sum_{i=1}^{k} 2i = (k - 1)(k + 2)$$

or $\quad 2\cdot 1 + 2\cdot 2 + 2\cdot 3 + \cdots + 2\cdot k = (k - 1)(k + 2).$

Prove $P(k + 1)$: $\qquad \sum_{i=1}^{k+1} 2i = ((k + 1) - 1)((k + 1) + 2)$

or

$$2\cdot 1 + 2\cdot 2 + 2\cdot 3 + \cdots + 2\cdot k + 2(k + 1) = k(k + 3).$$

By the induction hypothesis we have

$$\sum_{i=1}^{k} 2i = (k - 1)(k + 2).$$

Adding the $(k + 1)$ term, $2(k + 1)$, to both sides, we have

$$\sum_{i=1}^{k} 2i + 2(k + 1) = (k - 1)(k + 2) + 2(k + 1)$$

$$\sum_{i=1}^{k+1} 2i = k^2 + k - 2 + 2k + 2$$

$$= k^2 + 3k = k(k + 3).$$

Thus, $P(k) \Rightarrow P(k + 1)$.

We shouldn't be so quick to claim victory in the proof, however, because $P(1)$ fails—we can't even get on the ladder! The statement $P(1)$ is

$$\sum_{i=1}^{1} 2i = (1 - 1)(1 + 2)$$

But $2\cdot 1 \neq 0\cdot 3$. In fact, $\sum_{i=1}^{n} 2i = (n - 1)(n + 2)$ **never** holds! ■

Many times we wish to prove a statement $P(n)$ is true for all integers $n \geq 0$ (or $n \geq r$). In this case, the first step of induction is to prove $P(0)$ (or $P(r)$) instead of proving $P(1)$. This is the situation in the program verification example that follows.

EXAMPLE 4

Verify that the segment below will actually produce $Z + N \cdot X$, where $Z = 0$ and N is any nonnegative integer. Because $N \geq 0$, the first step of induction will be to prove $P(0)$ instead of $P(1)$. The statement $P(N)$ which we will prove for any nonnegative integer follows: If the program reaches the top of the following loop with Y having a value of N, then the program eventually exits the segment with $ANS = Z + N \cdot X$.

Z and X are assigned initial values. Y is assigned an initial value of N, a nonnegative integer.

```
WHILE Y <> 0 DO
        BEGIN
            Z := Z + X;
            Y := Y - 1
        END;
ANS := Z;
```

Solution

1. Prove $P(0)$: If the program reaches the top of the loop with $Y = 0$, then the program exits the segment with $ANS = Z + 0 \cdot X$.

   Since $Y = 0$, the program exits the loop immediately with the values of X, Y, and Z unchanged. Therefore, $ANS = Z = Z + 0 \cdot X$.

2. Assume $P(K)$: If the program reaches the top of the loop with $Y = K$, it exits the segment with $ANS = Z + K \cdot X$.

   Prove $P(K + 1)$: If the program reaches the top of the loop with $Y = K + 1$, it exits the segment with $Z = Z + (K + 1) \cdot X$.

   Suppose the program reaches the top of the loop with $Y = K + 1$. It executes the body of the loop and returns to the top of the loop with the following values of the variables:

   value of $X = X' = X$ unchanged

   value of $Y = Y' = (K + 1) - 1 = K$

   value of $Z = Z' = Z + X$.

   Since we are now at the top of the loop with the value of Y being K, we can use the induction hypothesis, $P(K)$, to say that the program exits the segment with

   $ANS = Z' + K \cdot X'$

   $= (Z + X) + K \cdot X$

   $= Z + (K \cdot X + X)$

   $= Z + (K + 1)X.$

We have proved the assertion. Therefore, if Z has an initial value of 0, we have shown that this program segment will produce an answer of $N \cdot X$. $\square$ ■

Many of the properties we covered in set theory or logic for two sets or statements can be generalized to n sets or statements, respectively. Verification of these extended properties again depends on mathematical induction. In Chapter 2 we used the law of syllogism, $((p_1 \Rightarrow p_2) \wedge (p_2 \Rightarrow q)) \Rightarrow (p_1 \Rightarrow q)$, extensively. The proof of the general law is presented in the next example.

EXAMPLE 5

Prove using mathematical induction that in the algebra of propositions for statements $p_1, p_2, p_3, \ldots, p_n$, and q, $n \in \mathbb{N}$, $n \geq 2$,

$$((p_1 \Rightarrow p_2) \wedge (p_2 \Rightarrow p_3) \wedge \cdots \wedge (p_n \Rightarrow q)) \Rightarrow (p_1 \Rightarrow q).$$

Proof

We start the proof using p_1, p_2, q and one conjunction.

1. Prove $P(2)$: $((p_1 \Rightarrow p_2) \wedge (p_2 \Rightarrow q)) \Rightarrow (p_1 \Rightarrow q)$ law of syllogism
2. Assume $P(k)$:

$$((p_1 \Rightarrow p_2) \wedge (p_2 \Rightarrow p_3) \wedge \cdots \wedge (p_k \Rightarrow q)) \Rightarrow (p_1 \Rightarrow q).$$

Prove $P(k + 1)$:

$$((p_1 \Rightarrow p_2) \wedge (p_2 \Rightarrow p_3) \wedge \cdots \wedge (p_k \Rightarrow p_{k+1})$$
$$\wedge (p_{k+1} \Rightarrow q)) \Rightarrow (p_1 \Rightarrow q).$$

By the associative property in the algebra of propositions (see Exercise 29), we can write

$$(p_1 \Rightarrow p_2) \wedge (p_2 \Rightarrow p_3) \wedge \cdots \wedge (p_k \Rightarrow p_{k+1}) \wedge (p_{k+1} \Rightarrow q)$$
$$\equiv ((p_1 \Rightarrow p_2) \wedge (p_2 \Rightarrow p_3) \wedge \cdots \wedge (p_k \Rightarrow p_{k+1})) \wedge (p_{k+1} \Rightarrow q)$$
$$\equiv (p_1 \Rightarrow p_{k+1}) \wedge (p_{k+1} \Rightarrow q) \qquad \text{by the induction hypothesis applied to the first grouping of statements with } p_1, p_2, \ldots, p_k, \text{ and } p_{k+1}$$

$$\Rightarrow (p_1 \Rightarrow q) \qquad \text{by the law of syllogism with statements } p_1, p_{k+1}, \text{ and } q \quad ■$$

EXERCISES 6.4

Prove using mathematical induction that each of the formulas in Exercises 1–17 is true for all indicated integers n.

1. $\displaystyle\sum_{i=1}^{n} i = \frac{n(n+1)}{2}$, $\quad n \geq 1$ $\quad$ 2. $\displaystyle\sum_{i=1}^{n} i^3 = \frac{n^2(n+1)^2}{4}$, $\quad n \geq 1$

3. $2^{2n} - 1 \equiv 0 \pmod{3}$, $\quad n \geq 0$

4. $\displaystyle\sum_{i=1}^{n} i^2 = 2C(n+1, 3) + C(n+1, 2)$, $\quad n \geq 1$

5. $\displaystyle\sum_{i=1}^{n}(4i - 2) = 2n^2$, $\quad n \geq 1$

6. $2^n < n!$, $\quad n \geq 4$

7. $2^n > n$, $\quad n \geq 0$

8. $\displaystyle\sum_{i=1}^{n} i(i+1) = \frac{n(n+1)(n+2)}{3}$, $\quad n \geq 1$

9. $\displaystyle\sum_{j=0}^{n} \frac{1}{(j+1)(j+2)} = \frac{n+1}{n+2}$, $\quad n \geq 0$

10. $\displaystyle\sum_{i=0}^{n} r^i = \frac{(r^{n+1} - 1)}{r - 1}$, $\quad r \neq 1, n \geq 0$ (finite geometric series)

11. $\displaystyle\sum_{i=0}^{n} 2i = n(n+1)$, $\quad n \geq 0$ $\quad$ 12. $\displaystyle\sum_{i=0}^{n}(2i+1) = (n+1)^2$, $\quad n \geq 0$

13. $n^3 - 4n + 6$ is divisible by 3, $n \geq 0$. *Note:* Recall that x is divisible by 3 if there exists an integer y such that $x = 3y$.

14. $n(n^2 + 8)$ is divisible by 3.

15. $n + 1 < n^2$, $\quad n \geq 2$. $\qquad$ 16. $3n < n^2 - 1$, $\quad n \geq 4$.

17. $n - 2 < (n^2 - n)/12$, $\quad n > 10$.

18. **a.** For the *incorrect* statement $P(n)$, $\sum_{i=1}^{n} i = n^2$, show $P(1)$ holds but not the implication $P(k) \Rightarrow P(k+1)$.
    b. Find an n for which $P(n)$ is false.

19. Find a counterexample for the following *incorrect* statement: $x + y$ is a factor of $x^n + y^n$ for all positive integers n.

20. Find a counterexample for the following *incorrect* statement:

$$\sum_{i=1}^{n} i^2 = n^3 \text{ for all positive integers } n.$$

21. Let V be a vocabulary in a formal grammar. Suppose A and B are subsets of V. Define A^n as the set of all strings of length n with terminal elements from A. For example, if A contains all the letters of the alphabet, A^4 consists of all 4-letter strings; "book" and "xyzw" are two strings in A^4. By induction prove for all positive integers n that if A is a subset of B, then A^n is a subset of B^n.

22. **a.** Prove for any $N \in \mathbb{N}$ and Z having any numeric value that if the program reaches the top of the following segment with Y having a value of N, then the program exits the segment with $ANS = Z \cdot X^N$.

```
WHILE (Y <> 0) DO
    BEGIN
        Z := Z * X;
        Y := Y - 1
    END;
ANS := Z;
```

b. What value must Z have initially for the segment to yield an ANS of X^N? □

23. **a.** Prove for any $N \in \mathbb{N}$ that if the Pascal program reaches the top of the following loop with Y having a value of N, then the program exits the segment with $ANS = Z \cdot N!$.

```
WHILE (Y <> 0) DO
    BEGIN
        Z := Z * Y;
        Y := Y - 1
    END;
ANS := Z;
```

b. Give the initial value of Z to produce a final answer of N!. □

24. **a.** Prove for any $N \in \mathbb{N}$ that if the Pascal program reaches the top of the following loop with Y having a value of N, then the program exits the segment with $ANS = X + N(N + 1) / 2$.

```
WHILE (Y <> 0) DO
    BEGIN
        X := X + Y;
        Y := Y - 1
    END;
ANS := X;
```

b. Give the initial value of X to produce a final answer of $N(N + 1) / 2$. □

25. **a.** Prove for any $N \in \mathbb{N}$ that if the Pascal program reaches the top of the following loop with Y having a value of N, then the program exits the segment with $ANS = X + N$.

```
WHILE (Y <> 0) DO
    BEGIN
        X := X + 1;
        Y := Y - 1
    END;
ANS := X;
```

b. Give the initial value of X to produce a final answer of N. □

26. Prove by mathematical induction that if S is a set with n elements then $\mathcal{P}(S)$ has 2^n elements. *Note:* For a set with $k + 1$ elements, separate out one of the elements, giving a set with k elements and a set with one

element. Consider forming the subsets of the $(k + 1)$-element set from the subsets of the k-element set and the 1-element set.

27. Assume the following theorem: If $p_1, p_2, \ldots, p_k$ are the first k primes, then $(p_1 \cdot p_2 \cdots p_k) + 1$ is a prime. Using this theorem prove there exists infinitely many primes. *Note:* Let statement $P(n)$ be that there exist more than n primes.

28. Prove by mathematical induction that any integer greater than 1 can be written as a product of primes, $p_1 \cdot p_2 \cdots p_n$. *Note:* Let $P(k)$ be the statement that any integer m with $1 < m \le k$ can be written as a product of primes. Prove $P(k + 1)$ using proof by contradiction.

29. Prove the general associative property for conjunction:
$$(p_1 \wedge p_2 \wedge \cdots \wedge p_{n-1}) \wedge p_n = p_1 \wedge (p_2 \wedge \cdots \wedge p_{n-1} \wedge p_n).$$

30. Use mathematical induction to prove one of the generalized distributive properties: For any sets $A, B_1, B_2, \ldots, B_n$, we have
$$A \cap (B_1 \cup B_2 \cup \cdots \cup B_n) = (A \cap B_1) \cup (A \cap B_2) \cup \cdots \cup (A \cap B_n).$$

31. Show a general proof by cases holds: In the algebra of propositions for statements $p_1, p_2, p_3, \ldots, p_n$, and q, $n \in \mathbb{N}$, $n \ge 2$,
$$(p_1 \vee p_2 \vee \cdots \vee p_n \Rightarrow q) \Longleftrightarrow (p_1 \Rightarrow q) \wedge (p_2 \Rightarrow q) \wedge \cdots \wedge (p_n \Rightarrow q).$$

For Exercises 32–34 first write the general property, and then prove it is true using mathematical induction.

32. The commutative property for union in the algebra of sets.

33. DeMorgan's law, $(A \cap B)' = A' \cup B'$, in the algebra of sets.

34. The idempotent property for disjunction in the algebra of propositions.

Chapter 6   REVIEW

Terminology/Algorithms	Pages	Comments
Array	186	Data structure containing items of same type
Index	186	Indicates particular array element
Dimension	186	In one-dimensional array, maximum number of elements
Vector, v	187	$(v_1, v_2, \ldots, v_n)$
Standard position	187	Initial point at origin
Equal, x = y	188	$x_i = y_i$ for all i
Sum, x + y	189	$(x_1 + y_1, x_2 + y_2, \ldots, x_n + y_n)$

Terminology/Algorithms	Pages	Comments
Scalar	190	Real number
Scalar times vector, a**x**	190	$(ax_1, ax_2, \ldots, ax_n)$
Dot product, **x** $\cdot$ **y**	191	$x_1 y_1 + x_2 y_2 + \cdots + x_n y_n$
Norm or length, $\|\mathbf{v}\|$	192	$\sqrt{\mathbf{v} \cdot \mathbf{v}}$
Distance between points	192	$\|\mathbf{x} - \mathbf{y}\|$
Sigma notation, $\displaystyle\sum_{i=1}^{n} v_i$ or $\Sigma_{i=1}^{n} v_i$	195	$v_1 + v_2 + \cdots + v_n$
$\displaystyle\sum_{i=m}^{n} v_i$	199	$v_m + v_{m+1} + \cdots + v_n$
Finite geometric series	199	$\displaystyle\sum_{i=0}^{n-1} a^i, \quad a \neq 1$
Basic formulas with Σ	200	$\displaystyle\sum_{i=1}^{n} 1 \quad \sum_{i=1}^{n} (u_i + v_i) \quad \sum_{i=1}^{n} i$ $\displaystyle\sum_{i=m}^{n} v_i \quad \sum_{i=1}^{n} av_i \quad \sum_{i=0}^{n-1} a^i$
Polynomial of degree n	200	$a_0 + a_1 x + a_2 x^2 + \cdots + a_n x^n, \quad a_n \neq 0,$ $a_i \in \mathbb{R}, n \in \mathbb{N}$
Pi notation, $\displaystyle\prod_{i=m}^{n} v_i$ or $\Pi_{i=m}^{n} v_i$	201	$v_m \cdot v_{m+1} \cdots v_n$
Recursive function or Recurrence relation or Difference equation	205	x_n defined in terms of $x_0, x_1, \ldots, x_{n-1}$
Initial condition	205	First (few) value(s) of recursive function
Generating rule	205	How to compute values of recursive function from previous value(s)
Solve recursive relation	207	Find a nonrecursive definition
Backus-Normal form, BNF	209	Recursive way of formally defining a language
Formal grammar	210	Language defined with V, V_T, S, P
Mathematical induction	217	Method to prove a statement, $P(n)$, true for all integers, $n \geq r$
Induction hypothesis	218	$P(k)$ assumed for arbitrary k

REFERENCES

Chin, F. and M. Wong. "An Optimal Schedule for Printing and Binding." *International Journal of Computer Mathematics* 15: 117–127, 1984.

Dijkstra, E. W. *A Discipline of Programming.* Englewood Cliffs, NJ: Prentice-Hall, 1976.

Peter, R., and E. Horwood. *Recursive Functions in Computer Theory.* Chichester: Horwood, 1981.

Roberts, F. S. *Applied Combinatorics.* Englewood Cliffs, NJ: Prentice-Hall, 1984.

Rohl, J. S. *Recursion via Pascal.* Cambridge: Cambridge University Press, 1984.

Tropp, H. S., Ed. "FORTRAN Anecdotes," *Annals of the History of Computing* 6(1): 61, January 1984.

Tucker, A. *Applied Combinatorics*, 2nd ed. New York: Wiley, 1984.

Ward, M. *Induction, Recursion, and Programming*, New York: North-Holland, 1980.

Welch, G. W., "Howard Hathaway Aiken, the Life of a Computer Pioneer," *The Computer Museum Report*, 3–12, Spring 1985.

7 MATRICES

Tree. Matrices can be used to store the information in various graphs, such as trees, that are important to the study of computer science (see Chapter 8).

Matrices or Two-Dimensional Arrays

In Section 6.1 we considered the data structure of a one-dimensional array or vector. One example involved an array RATE which stored under that one name the rates of pay of five employees of a company. Quite often, however, more features need to be stored and manipulated. In such cases two-dimensional arrays may be helpful.

The name used in mathematics and in many computer science applications for a two-dimensional array is matrix. A **matrix** (plural, matrices) is a rectangular array of numbers. We can think of a matrix as a table of numbers, and in fact, **table** is the term employed in COBOL for matrix.

Applications involving matrices are prevalent in the area of computer graphics. Combinations of various intensities of the three primary colors, red, green, and blue, are used to create all the colors. For a particular design there exists a table storing various intensity levels, from 0 to 255, for each primary color. Table 7.1 has a few choices of colors that can be used in the picture.

TABLE 7.1
Table of Colors for a Particular Application

	Red	Green	Blue
1	255	0	0
2	0	255	0
3	0	0	255
4	255	255	0
5	96	77	13
6	0	120	151
7	0	0	0
8	255	255	255

Let us call this matrix of 8 different colors C. It has 2 dimensions, the first dimension indicating the number of colors that can be used in the picture (8) and the second dimension indicating the number of primaries (3) that are combined to create the various colors. We say that C is an eight-by-three, 8×3, matrix with 8 rows and 3 columns. The numbers found in the matrix indicate the intensity for that color and primary. To pick out a specific number we can use a notation that gives the row and column of the element. For example, 77, the intensity level for color 5 (brown) of the second primary (green) is in row 5, column 2. In most computer languages, including FORTRAN, this array element is indicated by $C(5, 2)$, while in others, such as Pascal, square brackets are used as in $C[5, 2]$. The no-

LANGUAGE
EXAMPLE

tation is derived from mathematical notation which can conveniently use subscripts, c_{52}. ☐ Another mathematical notation for the matrix C is $[c_{ij}]_{8\times3}$. C is the matrix, and c_{ij} indicates an arbitrary element on the ith row and jth column of the 8×3 matrix.

Sometimes we need to manipulate particular rows or columns. For instance, suppose we wanted to find the average intensity for the primary green in the table. First, we must add all the elements in column 2 and then divide by 8. Using the sigma notation discussed in Section 6.2 we would compute

$$\frac{\sum_{i=1}^{8} c_{i2}}{8}.$$

In a Pascal program segment with C declared as

```
C:  ARRAY[1..8, 1..3] OF 0..255;
```

we accomplish the desired result using a loop:

```
SUM := 0;
FOR  I := 1 TO 8 DO
     SUM := SUM + C[I,2];
COLAVG := SUM/8;
```

If we desired the average intensity for color 6, aqua, instead, we calculate

$$\frac{\sum_{j=1}^{3} c_{6j}}{3}$$

which in Pascal is

```
SUM := 0;
FOR  J := 1 TO 3 DO
     SUM := SUM + C[6,J];
ROWAVG := SUM/3;
```

The sum of all the elements can be computed by

$$\sum_{i=1}^{8}\sum_{j=1}^{3} c_{ij} \quad \text{or} \quad \sum_{j=1}^{3}\sum_{i=1}^{8} c_{ij}.$$

The former computes the sum of all the row totals: $255 + 255 + 255 + 510 + 186 + 271 + 0 + 765 = 2497$. This calculation can be generated with nested loops:

```
SUM := 0;
FOR  I := 1 TO 8 DO
      FOR  J := 1 TO 3 DO
            SUM := SUM + C[I,J];
```

We would switch the two FOR statements to imitate $\sum_{j=1}^{3} \sum_{i=1}^{8} c_{ij}$, which is the sum of all the column totals: $861 + 962 + 674 = 2497$. The result is the same regardless of which FOR statement comes first, but the distinction is important if you want to print row or column sums as you are going along. □

If we isolate the sixth row or the second column of C, we have one-dimensional arrays or row and column vectors:

$$[0, 120, 151] \quad \text{and} \quad \begin{bmatrix} 0 \\ 255 \\ 0 \\ 255 \\ 77 \\ 120 \\ 0 \\ 255 \end{bmatrix}.$$

In fact, these can be considered matrices themselves, the first being a 1×3 matrix and the second an 8×1. The operations on vectors discussed in Section 6.1 can be extended to matrices.

One such operation was multiplication by a scalar where each element of the vector was multiplied by a constant. If we want to calculate the percent of the maximum for each intensity in matrix C, we must divide each element by 255 and multiply by 100, or simply multiply by $100/255$. Here,

$$100/255 \cdot C = \begin{bmatrix} 100 & 0 & 0 \\ 0 & 100 & 0 \\ 0 & 0 & 100 \\ 100 & 100 & 0 \\ 38 & 30 & 5 \\ 0 & 47 & 59 \\ 0 & 0 & 0 \\ 100 & 100 & 100 \end{bmatrix}$$

with values rounded to integers, gives percent intensity for the primaries for each color.

DEFINITION

Let $C = [c_{ij}]$ be an $m \times n$ matrix. The **product of scalar a and matrix C** is the $m \times n$ matrix

$$aC = [ac_{ij}]$$

that is, each element of C is multiplied by a.

Suppose with the original table C you want to see what effect lowering each of the brightest intensities (255) by 20 and raising the least intensity (0) by 15 has on your picture. The resulting matrix R is the sum of the original matrix C and an adjustment matrix A. As with vectors, when adding matrices, the sum of corresponding elements is computed:

$$\begin{bmatrix} 255 & 0 & 0 \\ 0 & 255 & 0 \\ 0 & 0 & 255 \\ 255 & 255 & 0 \\ 96 & 77 & 13 \\ 0 & 120 & 151 \\ 0 & 0 & 0 \\ 255 & 255 & 255 \end{bmatrix} + \begin{bmatrix} -20 & 15 & 15 \\ 15 & -20 & 15 \\ 15 & 15 & -20 \\ -20 & -20 & 15 \\ 0 & 0 & 0 \\ 15 & 0 & 0 \\ 15 & 15 & 15 \\ -20 & -20 & -20 \end{bmatrix} = \begin{bmatrix} 235 & 15 & 15 \\ 15 & 235 & 15 \\ 15 & 15 & 235 \\ 235 & 235 & 15 \\ 96 & 77 & 13 \\ 15 & 120 & 151 \\ 15 & 15 & 15 \\ 235 & 235 & 235 \end{bmatrix} = R.$$

Notice, for example, that $c_{41} + a_{41} = 255 + (-20) = 235$. Since we are adding corresponding elements, the two matrices must be exactly the same size. It wouldn't make sense to add $C + B$ when

$$B = \begin{bmatrix} -20 & 18 & 15 & 30 \\ 15 & 20 & 15 & 85 \end{bmatrix}.$$

What would we add to $b_{14} = 30$ and $b_{24} = 85$?

> **DEFINITION**
>
> Let $A = [a_{ij}]$ and $C = [c_{ij}]$ be two $m \times n$ matrices. The **matrix sum** of A and C is an $m \times n$ matrix
> $$A + C = [a_{ij} + c_{ij}]$$
> that is, corresponding elements are added.

Also, observe that the order of addition is irrelevant: $R = C + A$, but we would get the same result by computing $A + C$. This property, called commutativity for matrix addition, is true because commutativity of addition holds in the set of real numbers. After all, we are adding corresponding elements. Not only does $c_{41} + a_{41} = 255 + (-20) = (-20) + 255 = a_{41} + c_{41}$, but $c_{ij} + a_{ij} = a_{ij} + c_{ij}$ for all i and j.

COMMUTATIVE PROPERTY FOR MATRIX ADDITION

> If A and C are $m \times n$ matrices, then $A + C = C + A$.

Another property of matrices that parallels one for real numbers is the associative property of addition. $(A + B) + C = A + (B + C)$

is true whether A, B, and C are real numbers or matrices of the same size. The property indicates that grouping does not matter; it is irrelevant whether we add A and B first or B and C.

EXAMPLE 1

Show the associative property holds for the following matrices:

$$A = \begin{bmatrix} 1 & 2 \\ 4 & 4 \end{bmatrix} \quad B = \begin{bmatrix} 3 & 1 \\ -1 & -4 \end{bmatrix} \quad C = \begin{bmatrix} 6 & 3 \\ 0 & 2 \end{bmatrix}$$

Solution

$$(A + B) + C = \left(\begin{bmatrix} 1 & 2 \\ 4 & 4 \end{bmatrix} + \begin{bmatrix} 3 & 1 \\ -1 & -4 \end{bmatrix} \right) + \begin{bmatrix} 6 & 3 \\ 0 & 2 \end{bmatrix}$$

$$= \begin{bmatrix} 4 & 3 \\ 3 & 0 \end{bmatrix} + \begin{bmatrix} 6 & 3 \\ 0 & 2 \end{bmatrix} = \begin{bmatrix} 10 & 6 \\ 3 & 2 \end{bmatrix}$$

$$A + (B + C) = \begin{bmatrix} 1 & 2 \\ 4 & 4 \end{bmatrix} + \left(\begin{bmatrix} 3 & 1 \\ -1 & -4 \end{bmatrix} + \begin{bmatrix} 6 & 3 \\ 0 & 2 \end{bmatrix} \right)$$

$$= \begin{bmatrix} 1 & 2 \\ 4 & 4 \end{bmatrix} + \begin{bmatrix} 9 & 4 \\ -1 & -2 \end{bmatrix} = \begin{bmatrix} 10 & 6 \\ 3 & 2 \end{bmatrix} \quad \blacksquare$$

ASSOCIATIVE PROPERTY FOR MATRIX ADDITION

If A, B, and C are $m \times n$ matrices, then $(A + B) + C = A + (B + C)$.

It is not uncommon to need to **zero out** a matrix, or assign each matrix element to 0, before using it in a program. For example, perhaps a questionnaire with 20 questions and 5 possible choices for each question has been distributed to a number of people. A 20×5 array, Q, can be used to tally the number of responses to each choice of each question. But since each element of Q will be a sum, each must be initialized to be 0. At that initial point the matrix Q is a **zero matrix** or a matrix containing only 0's. This zero matrix, $\mathbf{0}_{20 \times 5}$, is special in that if it is added to any other 20×5 matrix, A, the result is A: $A + \mathbf{0} = A = \mathbf{0} + A$. Thus, the $m \times n$ zero matrix $\mathbf{0}_{m \times n}$ behaves in the set of all $m \times n$ matrices like 0 behaves in the set of real numbers.

The similarity continues as we find a parallel to additive inverses of real numbers. For any real number a there is always another number b such that $a + b = 0$; b is the additive inverse of a. In fact, as we know, $b = -a$. If $a = -7$, then $b = -(-7) = 7$. The same property is true for matrices. For

$$A = \begin{bmatrix} 6 & -3 \\ -4 & 2 \end{bmatrix}$$

there exists a 2×2 matrix B, namely

$$-1 \cdot A = \begin{bmatrix} -6 & 3 \\ 4 & -2 \end{bmatrix} = -A$$

such that

$$A + B = \begin{bmatrix} 6 & -3 \\ -4 & 2 \end{bmatrix} + \begin{bmatrix} -6 & 3 \\ 4 & -2 \end{bmatrix} = \begin{bmatrix} 0 & 0 \\ 0 & 0 \end{bmatrix} = \mathbf{0}_{2 \times 2}.$$

PROPERTY OF ADDITIVE INVERSES FOR MATRICES

For any $m \times n$ matrix A there exists an $m \times n$ matrix B, called the **additive inverse** of A, such that

$$A + B = \mathbf{0}_{m \times n}$$

Moreover, $B = -A$, where $-A = (-1)A$.

LANGUAGE EXAMPLES

In many programming languages operations on matrices must be performed in loops, element by element. Some languages, however, such as SAS, APL, and MACSYMA allow the user to write the operations using matrices with mathematical notation. For example, in MACSYMA, ZERO[20, 5] is a 20×5 zero matrix. Also if A and B are appropriately defined matrices of the same size,

    A + B;

yields the sum, and

    100/255 * A;

gives the expected scalar times matrix A. In APL A[17;] gives the 17th row, A[;2] the 2nd column, and $\lceil$/A the maximum number in each row. If the language you are using for an application allows matrix operations, use them where appropriate instead of generating the results with loops.   ☐

EXERCISES 7.1

1. Consider the matrix $A = [a_{ij}] = \begin{bmatrix} 6 & 3 & -2 \\ 0 & -8 & 4 \end{bmatrix}$

   a. What is A's size?   **b.** Find a_{21}, a_{12}, a_{31}, and a_{13}.

   c. Compute $\sum\limits_{i=1}^{2} a_{i1}$.   **d.** Calculate $\sum\limits_{j=1}^{3} a_{ij}$ for $i = 1, 2$.

   e. Use part d to compute $\sum\limits_{i=1}^{2} \sum\limits_{j=1}^{3} a_{ij}$.

   Let $A = \begin{bmatrix} 6 & 3 & -2 \\ 0 & -8 & 4 \end{bmatrix}$ $B = \begin{bmatrix} -1 & 2 & 3 \\ -7 & 2 & 1 \end{bmatrix}$ and $C = \begin{bmatrix} 0 & -4 & 1 \\ 3 & 1 & -8 \end{bmatrix}$

Calculate the matrices in Exercises 2–14 using matrices A, B, C above.

2. $3A$ **3.** $3B$

4. $3A + 3B$ **5.** $A + B$

6. $3(A + B)$ (Compare Exercises 4 and 6)

7. $B + A$ (Compare Exercises 5 and 7)

8. $-A$ **9.** $\mathbf{0}_{2\times3}$

10. $B + C$ **11.** $(A + B) + C$ (Use Exercise 5)

12. $A + (B + C)$ (Use Exercise 10; compare Exercises 11 and 12)

13. $2(3A)$ (Use Exercise 1)   **14.** $6A$ (Compare Exercises 13 and 14)

15. Consider an M × N matrix C.
  a. Write an algorithm to print row sums.
  b. Revise your algorithm to print the sum of all the elements of the matrix, also.
  c. Revise again to print row averages.
  d. Revise to print the average of all elements. ☐

16. Repeat Exercise 15 for column sums. ☐

$$\text{Suppose } A = [3 \quad 5] \quad and \quad B = \begin{bmatrix} 1 & 1 \\ 3 & 4 \\ -1 & 4 \end{bmatrix}.$$

If possible, compute the matrices in Exercises 17–19.

17. $2A$ **18.** $A + A$ **19.** $A + B$

20. Write an algorithm to zero out the M × N matrix A.

21. Suppose D is declared to be an M × N array.
  a. Write a Pascal segment to read values into D where the data is typed a row at a time; that is, in the data, the numbers for the first row are followed by those of the second row, etc.
  b. How would the program segment change if the data is typed a column at a time?
  Note: If data is stored a column at a time, in what is called **column-major order,** the entire array in FORTRAN can be read as a result of the simple statement, READ, D. ☐

22. Suppose D is an M × N array that already has values read into it. Let E be declared an M × N matrix. Write an algorithm to assign E = D.

23. Write an algorithm to compute C = A + B where A, B, and C have been declared to be M × N matrices and where A and B already have values. ☐

24. How many elements are in matrix A if it is of the size
  a. 20×5 **b.** $m \times n$ **c.** 5×5 **d.** $n \times n$

25. The memory of the computer can be thought of as a one-dimensional array of cells, MEM, with the address of each cell being the index.

Two-dimensional arrays must be placed into one-dimensional memory. The most common method and that employed with Pascal arrays is to store the elements a row at a time in what is called **row-major order.** Consider array A declared as having 5 rows, numbered 1 through 5, and 3 columns, 1 through 3:

VAR A: ARRAY[5,3] OF INTEGER;

Suppose each element of array A is stored in a memory cell starting at address 342. Thus, A[1, 1] is stored in MEM[342].

a. What is the highest memory location A uses?

In which MEM element is each of the array elements in parts b–g stored?

b. A[2, 1]   **c.** A[4, 1]   **d.** A[I, 1]

e. A[2, 3]   **f.** A[4, 2]   **g.** A[I, J]

What array element of A is stored in each of the following MEM elements?

h. MEM[343]   **i.** MEM[349]   **j.** MEM[359]   ☐

26. In FORTRAN applications, two-dimensional arrays are stored in memory in column-major order, a column at a time. Memory can be thought of as a one-dimensional array MEM (see Exercise 25). Consider array A declared as a 5×3 array of integers

INTEGER A(5,3).

Assume each element of the array can be stored in one memory cell starting at memory location 342. Thus, A(1, 1) is stored in MEM(342).

a. What is the highest memory location A uses?

In which MEM element is each of the array elements in parts b–g stored?

b. A(1, 2)   **c.** A(1, 3)   **d.** A(1, J)

e. A(3, 1)   **f.** A(5, 2)   **g.** A(I, J)

What array element of A is stored in each of the following MEM elements?

h. MEM(343)   **i.** MEM(349)   **j.** MEM(359)   ☐

27. Consider the square matrix

$$A = \begin{bmatrix} 1 & 5 & 3 \\ 5 & 4 & 6 \\ 3 & 6 & 8 \end{bmatrix} = [a_{ij}]_{3 \times 3}.$$

a. Find a_{21} and a_{12}.

b. Verify that $a_{ij} = a_{ji}$ for all i and j. In this case we say that A is a **symmetric matrix.**

c. Suppose B is a symmetric matrix. Fill in the blanks.

$$B = \begin{bmatrix} -7 & — & — & — \\ 2 & 3 & — & — \\ -1 & 4 & -4 & — \\ 6 & 5 & 0 & -3 \end{bmatrix}$$

d. If we were storing B in the computer, we would only need to store the given elements in a vector, $(-7, 2, 3, -1, 4, -4, 6, 5, 0, -3)$. B has $4 \cdot 4 = 16$ elements, but the stored vector has only $1 + 2 + 3 + 4 = 10$ elements. In general, if B is an $n \times n$ symmetric matrix, how many elements must be stored in the vector?

e. -7, 3, -4, and -3 are the elements b_{ii} with $i = 1, 2, 3, 4$. These are called the **diagonal elements.** Suppose the diagonal elements of an $n \times n$ symmetric matrix are known to be 0 so that they do not need to be stored. How many elements must be stored in the vector?

28. Recall that $\mathbb{B} = \{0, 1\}$. Let M be the set of all 2×2 matrices with elements from $\mathbb{B}$.
    a. How many matrices are in M?
    b. How many functions are there from M to $\mathbb{B}$?
    How many of these are
    c. Onto functions?
    d. One-to-one functions?
    Note: Refer to Section 5.1.

29. Let M be the set of all $m \times n$ matrices with elements from $\mathbb{B}$.
    a. How many matrices are in M?
    b. How many functions are there from M to $\mathbb{B}$?
    How many of these are
    c. Onto functions?
    d. One-to-one functions?

30. Suppose in a small class of 4 students there are 3 tests. The grades are as follows:

    Jane   97, 85, 84   Jim   83, 90, 71
    Julia   76, 72, 64   Joe   93, 89, 95

    Arrange the grades in a matrix.

31. During a Christmas shopping spree a couple buys 3 men's shirts at \$21 each, 4 ties at \$12 each, 2 pocketbooks for \$16 each, 5 ornaments at \$3 each, and a tree for \$40. Write their total charges as the dot product of an amount vector and a cost vector.

32. Prove the generalized commutative property for matrix addition using mathematical induction:
    For any positive integer n, if $A_1, A_2, \ldots, A_n$ are matrices of the same size, then
    $$A_1 + A_2 + \cdots + A_n = A_n + \cdots + A_2 + A_1.$$

33. Consider the matrix
    $$T = \begin{bmatrix} 0 & 50 & 20 \\ 100 & 150 & 120 \\ 90 & 170 & 200 \end{bmatrix}.$$
    For any 3×3 matrix M with elements from $\mathbb{N}$, apply the function f to each element, where $f: \mathbb{N} \to \mathbb{B}$ is defined as
    $$f(m_{ij}) = \begin{cases} 0, & \text{if } m_{ij} < \text{corresponding threshold value, } t_{ij} \\ 1, & \text{if } m_{ij} \geq \text{corresponding threshold value, } t_{ij}. \end{cases}$$
    T is called a **threshold matrix** and each t_{ij} is a threshold value. An element of the matrix M is mapped to 1 if and only if it is at least as big as the corresponding threshold value. Fill in the blanks for the image of the elements of the following matrix M:
    $$M = \begin{bmatrix} 110 & 112 & 100 \\ 100 & 70 & 75 \\ 90 & 80 & 90 \end{bmatrix} \to \begin{bmatrix} — & 1 & — \\ — & 0 & — \\ — & — & — \end{bmatrix}.$$

Note: In APL the image matrix can be quickly calculated using the statement $M \geq T$. *Note:* See Exercise 35 for an application of threshold matrices.   ☐

34. Repeat Exercise 33 for matrix

$$M = \begin{bmatrix} 10 & 10 & 10 \\ 75 & 75 & 75 \\ 100 & 100 & 100 \end{bmatrix}.$$

35. Much has been done in recent years in the area of digitizing pictures. A spacecraft could take a black-and-white picture in space. A computer on board would analyze the degree of grayness of each dot or **pixel** of the picture and assign it a value for intensity, say from 0 (white) to 255 (black). The numbers for the entire picture would be transmitted to earth and the picture reconstructed. One method of reconstruction is called **dithering.** Each digitized pixel is compared with an individual threshold value to determine if a dot will or will not be placed at that point on the reconstructed picture. There are no gray dots in the reconstructed picture; a black dot is either present or not present at each position, depending on the presence of a 1 or 0 in the corresponding position of the final matrix. To accomplish this procedure a threshold matrix, called a **dither matrix,** is needed. Much experimentation has been done in dithering to find the best threshold matrix to help produce a clear, apparently continuous image using black dots on a white background. The construction of one dither matrix is presented in this problem. Let

$$D_2 = \begin{bmatrix} 0 & 2 \\ 3 & 1 \end{bmatrix} \quad \text{and} \quad V_2 = \begin{bmatrix} 1 & 1 \\ 1 & 1 \end{bmatrix}.$$

To develop the dither matrix, find the following matrices:
a. $4D_2$ **b.** $4D_2 + 2V_2$ **c.** $4D_2 + 3V_2$ **d.** $4D_2 + V_2$
e. Construct the 4×4 matrices D_4 and V_4 with the 2×2 matrices you've calculated placed in the indicated positions.

$$D_4 = \begin{bmatrix} 4D_2 & 4D_2 + 2V_2 \\ 4D_2 + 3V_2 & 4D_2 + V_2 \end{bmatrix} \qquad V_4 = \begin{bmatrix} V_2 & V_2 \\ V_2 & V_2 \end{bmatrix}$$

f. Now calculate the dither matrix $16D_4 + 8V_4$. This is the threshold matrix that will be used in reconstructing the picture.
g. Consider the 4×4 matrix M containing pixel intensities transmitted from space.

$$M = \begin{bmatrix} 100 & 145 & 100 & 178 \\ 111 & 60 & 250 & 102 \\ 40 & 200 & 20 & 73 \\ 254 & 198 & 223 & 204 \end{bmatrix}.$$

With function f defined as in Exercise 33, find the image of M after applying f to every point.
h. Draw the picture in a 4×4 array. *Note:* If the picture were larger, the same dither matrix could be used by applying that threshold matrix in a checkerboard fashion over the entire picture.

7.2

Multiplication of Matrices

The ability to multiply matrices allows us to solve many problems in computer science and mathematics. In computer graphics the rotation, shifting, and enlargement of pictures is accomplished with multiplication of matrices. Modern cryptography, the coding and decoding of messages, is possible only by the marriage of computers and mathematics with matrices playing a significant role. After coving topics in Boolean algebra in Chapter 9, we will be able to study some encoding techniques and methods of correcting errors in the transmission of messages. In Chapter 8 we will represent a computer network by a matrix and decide on the existence of paths through the network by matrix multiplication.

The operation of matrix multiplication is built upon the concept of a dot product of vectors. Suppose a college pays a nearby university for time-sharing use of its large computer facilities. The college must keep track of the amount of money charged to various accounts for working on the computer system. Suppose use is broken down into time-on-line (in hours), CPU time (in minutes), and pages of printout generated at the local facility (in thousands of pages). For example, the combined student account for the month might be summarized by the vector (4560, 68.6, 19.5), meaning for the month the students were on-line a total of 4560 hours; they used 68.6 minutes of CPU time and had 19,500 pages printed. Suppose each account is charged at the following rates for the resources: $0.17 per hour on-line, $18 per minute of CPU time, and $40 per thousand pages of printout. This information, too, can be summarized in a 3-tuple, (0.17, 18, 40). To compute the total amount charged on the student account, calculate the dot product of the two vectors.

$$(4560, 68.6, 19.5) \cdot (0.17, 18, 40)$$
$$= 4560 \cdot 0.17 + 68.6 \cdot 18 + 19.5 \cdot 40 = \$2790.00$$

As pointed out in Section 6.1, we could write the second vector as a column vector. Then, we are actually multiplying a 1×3 matrix by a 3×1 matrix to find a 1×1 matrix

$$[4560, 68.6 \ 19.5] \begin{bmatrix} 0.17 \\ 18 \\ 40 \end{bmatrix} = [2790].$$

The students' account is not the only one to be considered. The registrar's account might generate the vector (320, 98.8, 12); the admissions account, (320, 81.2, 35); and the financial affairs account, (160, 53.1, 8.2). Each must be dotted with (0.17, 18, 40) to evaluate the amount charged to that account. To save room and avoid the repetition of writing (0.17, 18, 40) over and over, let us put all the account vectors together in 4×3 matrix to be multiplied by (0.17, 18, 40), expressed as 3×1 matrix.

$$
\begin{array}{l}
\text{student} \\
\text{registrar} \\
\text{admissions} \\
\text{financial}
\end{array}
\begin{bmatrix}
4560 & 68.6 & 19.5 \\
320 & 98.8 & 12 \\
320 & 81.2 & 35 \\
160 & 53.1 & 8.2
\end{bmatrix}
\begin{bmatrix}
0.17 \\
18 \\
40
\end{bmatrix}
=
\begin{bmatrix}
2790 \\
2312.8 \\
2916 \\
1311
\end{bmatrix}.
$$

By dotting each account vector with (0.17, 18, 40) we find that the accounts should be charged as follows:

student $\$2790.00 = 4560 \cdot 0.17 + 68.6 \cdot 18 + 19.5 \cdot 40$
registrar $\$2312.80 = 320 \cdot 0.17 + 98.8 \cdot 18 + 12 \cdot 40$
admissions $\$2916.00 = 320 \cdot 0.17 + 81.2 \cdot 18 + 35 \cdot 40$
financial $\$1311.00 = 160 \cdot 0.17 + 53.1 \cdot 18 + 8.2 \cdot 40.$

Suppose the school used the university's facilities by dialing up their computer so that each account also must be charged a separate surcharge of $0.50 per hour for telephone use. The amount of on-line time is the only term to be considered. Again, the computation can be accomplished with a dot product. For the student account we have

$$
\begin{bmatrix} 4560 & 68.6 & 19.5 \end{bmatrix}
\begin{bmatrix} 0.50 \\ 0 \\ 0 \end{bmatrix} = \begin{bmatrix} 2280 \end{bmatrix}.
$$

Notice how the 0's in the second and third positions eliminated the effect of CPU time and printout. Certainly, we can take the account matrix and multiply by

$$
\begin{bmatrix} 0.50 \\ 0 \\ 0 \end{bmatrix} \quad \text{as well as by} \quad \begin{bmatrix} 0.17 \\ 18 \\ 40 \end{bmatrix}.
$$

But since both calculations need to be done at the end of the month, why not perform them together? Dot each row of the first matrix (account matrix) by each column of the second matrix (rate matrix) to get a resulting matrix with the charges for the month by account. There are four accounts, thus four rows, one for each account, in the resulting charges matrix. There are two types of rates, one for the university computer center and one for the telephone company. Thus, there exist two columns in the charges matrix, two types of charges for each account. Four rows in the first matrix along with two columns in the second yield a 4×2 resultant matrix:

$$
\begin{array}{c}
\quad\quad\quad \text{on-line} \quad \text{CPU} \quad \text{printout} \quad\quad\quad \text{U.} \quad \text{telephone} \\
\quad\quad\quad\quad | \quad\quad\quad | \quad\quad\quad | \quad\quad\quad\quad\quad\quad | \quad\quad\quad |
\end{array}
$$

$$
\begin{array}{r}
\text{student} \\
\text{registrar} \\
\text{admissions} \\
\text{financial}
\end{array}
\begin{bmatrix}
4560 & 68.6 & 19.5 \\
320 & 98.8 & 12 \\
320 & 81.2 & 35 \\
160 & 53.1 & 8.2
\end{bmatrix}
\begin{bmatrix}
0.17 & 0.5 \\
18 & 0 \\
40 & 0
\end{bmatrix}
\begin{array}{l}
\text{on-line} \\
\text{CPU} \\
\text{printout}
\end{array}
$$

$$
\begin{array}{c}
\quad\quad \text{U.} \quad\quad \text{telephone} \\
\quad\quad | \quad\quad\quad\quad |
\end{array}
$$

$$
=
\begin{bmatrix}
2790 & 2280 \\
2312.8 & 160 \\
2916 & 160 \\
1311 & 80
\end{bmatrix}
\begin{array}{l}
\text{student} \\
\text{registrar} \\
\text{admissions} \\
\text{financial}
\end{array} .
$$

In the charges matrix the third-row, first-column element indicates that the admissions account should be charged \$2916 by the university. The fourth-row, second-column element indicates that the financial affairs account is responsible for \$80 of telephone surcharge. For the dot products to be possible, the number of columns in the first matrix and the number of rows in the second have to be identical; here both are 3.

DEFINITION

Let $A = [a_{ij}]_{m \times q}$ be an $m \times q$ matrix and $B = [b_{ij}]_{q \times n}$ a $q \times n$ matrix. The **matrix product** of A and B is an $m \times n$ matrix AB or $A \cdot B = C = [c_{ij}]_{m \times n}$, where c_{ij} is the dot product of the ith row of A and the jth column of B.

EXAMPLE 1

For

$$
A = \begin{bmatrix} 8 & 5 & 3 & -4 \\ -5 & 1 & 0 & 1 \end{bmatrix}_{2 \times 4}
\quad \text{and} \quad
B = \begin{bmatrix} 2 & -6 \\ 7 & 1 \\ 4 & 3 \\ -9 & -2 \end{bmatrix}_{4 \times 2}
$$

evaluate $A \cdot B$ and $B \cdot A$.

Solution

$$
A \cdot B = \begin{bmatrix} 8 & 5 & 3 & -4 \\ -5 & 1 & 0 & 1 \end{bmatrix}
\begin{bmatrix} 2 & -6 \\ 7 & 1 \\ 4 & 3 \\ -9 & -2 \end{bmatrix}
= \begin{bmatrix} 99 & -26 \\ -12 & 29 \end{bmatrix}
$$

because

$$8 \cdot 2 \quad + 5 \cdot 7 + 3 \cdot 4 + -4 \cdot -9 = \quad 99$$
$$8 \cdot -6 + 5 \cdot 1 + 3 \cdot 3 + -4 \cdot -2 = -26$$
$$-5 \cdot 2 \quad + 1 \cdot 7 + 0 \cdot 4 + \quad 1 \cdot -9 = -12$$
$$-5 \cdot -6 + 1 \cdot 1 + 0 \cdot 3 + \quad 1 \cdot -2 = \quad 29$$

$$B \cdot A = \begin{bmatrix} 2 & -6 \\ 7 & 1 \\ 4 & 3 \\ -9 & -2 \end{bmatrix} \begin{bmatrix} 8 & 5 & 3 & -4 \\ -5 & 1 & 0 & 1 \end{bmatrix}$$

$$= \begin{bmatrix} 46 & 4 & 6 & -14 \\ 51 & 36 & 21 & -27 \\ 17 & 23 & 12 & -13 \\ -62 & -47 & -27 & 34 \end{bmatrix}. \quad \blacksquare$$

Notice that we were able to evaluate both AB and BA but that the products were not equal, not even the same size. Thus, the commutative law of multiplication does *not* hold for matrices. The associative law for multiplication, however, is true.

ASSOCIATIVE PROPERTY OF MATRIX MULTIPLICATION

Let A, B, and C be matrices so that $(AB)C$ is defined. Then $A(BC)$ is defined and

$$(AB)C = A(BC).$$

Another property that holds for real numbers and matrices is the distributive law. This property for numbers allows us to distribute a product through a sum as in $3(x + y) = 3x + 3y$.

DISTRIBUTIVE PROPERTY FOR MATRICES

Let A be an $m \times q$ matrix, and let B and C be $q \times n$ matrices. Then

$$A(B + C) = AB + AC.$$

EXAMPLE 2

Multiply $A\mathbf{I}_2$ and $\mathbf{I}_3A$, where

$$A = \begin{bmatrix} 6 & 5 \\ 1 & -3 \\ 2 & -8 \end{bmatrix} \qquad \mathbf{I}_2 = \begin{bmatrix} 1 & 0 \\ 0 & 1 \end{bmatrix} \qquad \text{and} \qquad \mathbf{I}_3 = \begin{bmatrix} 1 & 0 & 0 \\ 0 & 1 & 0 \\ 0 & 0 & 1 \end{bmatrix}.$$

Solution

$A\mathbf{I}_2 = A$ and $\mathbf{I}_3 A = A$. The $n \times n$ square matrix $\mathbf{I}_n$ with 1's along the diagonal, the line of elements from the top left to the bottom right, and zero's everywhere else is called the identity matrix. When multiplication is possible, $\mathbf{I}_n$ performs a role comparable to that of the number 1 in the set of real numbers. ▪

DEFINITION

An $n \times n$ matrix is called a **square matrix**.

DEFINITION

In an $n \times n$ square matrix B, the **diagonal** is the set of elements $\{b_{11}, b_{22}, \ldots, b_{nn}\}$.

DEFINITION

An $n \times n$ matrix $\mathbf{I}_n$ with 1's along the diagonal and 0's everywhere else is called an **identity matrix**.

PROPERTY OF MULTIPLICATIVE IDENTITY

Let A be an $m \times n$ matrix, and let $\mathbf{I}_m$ and $\mathbf{I}_n$ be $m \times m$ and $n \times n$ identity matrices, respectively. Then

$$\mathbf{I}_m A = A \quad \text{and} \quad A\mathbf{I}_n = A.$$

In a computer graphics figure, there are three basic manipulations you might want to accomplish: **scaling** (enlarging or shrinking), **translation** (moving up/down, right/left), and **rotation** with one point fixed. Taking an individual vector with endpoint (x, y), each of the operations can be performed on (x, y) by matrix multiplication.

For example, suppose we have the vector $(2, 3)$ as pictured in Figure 7.1a. Doubling the length of the vector can be accomplished by multiplying the vector by the scalar 2 as in $2(2, 3) = (4, 6)$, shown in Figure 7.1b. Notice the norm or length of $(2, 3)$ is $\sqrt{4 + 9} = \sqrt{13}$, while the length of $(4, 6)$ is $\sqrt{16 + 36} = \sqrt{52} = 2\sqrt{13}$. The same result can be achieved by multiplying $2\mathbf{I}_2$ times $(2, 3)$, expressed as a column matrix:

$$2\mathbf{I}_2 \begin{bmatrix} 2 \\ 3 \end{bmatrix} = 2 \begin{bmatrix} 1 & 0 \\ 0 & 1 \end{bmatrix} \begin{bmatrix} 2 \\ 3 \end{bmatrix} = \begin{bmatrix} 2 & 0 \\ 0 & 2 \end{bmatrix} \begin{bmatrix} 2 \\ 3 \end{bmatrix} = \begin{bmatrix} 4 \\ 6 \end{bmatrix}.$$

We will see the utility of working with $(2, 3, 1)$ instead of $(2, 3)$ when translating. The extra 1 is useful in translations, and so we append it for all transformations of vectors. Thus, for scaling we use

Figure 7.1

(a) Vector (2, 3); (b) vector 2(2, 3); (c) vector (4, 6) translated to initial point of (1, −4)

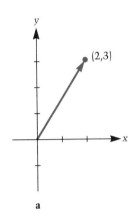

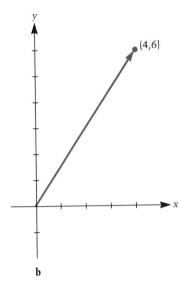

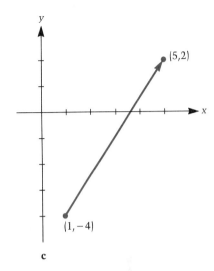

a
b
c

$$\begin{bmatrix} 2 & 0 & 0 \\ 0 & 2 & 0 \\ 0 & 0 & 1 \end{bmatrix} \begin{bmatrix} 2 \\ 3 \\ 1 \end{bmatrix} = \begin{bmatrix} 4 \\ 6 \\ 1 \end{bmatrix}.$$

Translation of a figure means moving the drawing parallel to the axes. Let's say we want to move every point of Figure 7.1b to the right 1 unit and down 4 units. We get new coordinates for every point (x', y') such that

$$x' = x + 1 \quad \text{and} \quad y' = y - 4.$$

Therefore, (4, 6) is translated to $(4 + 1, 6 - 4) = (5, 2)$ as seen in Figure 7.1c. Such an operation can be performed by a matrix product, if we multiply as follows

$$\begin{bmatrix} 1 & 0 & 1 \\ 0 & 1 & -4 \\ 0 & 0 & 1 \end{bmatrix} \begin{bmatrix} 4 \\ 6 \\ 1 \end{bmatrix} = \begin{bmatrix} 5 \\ 2 \\ 1 \end{bmatrix}.$$

Notice how the extra row in the left matrix and the 1 in (5, 2, 1) resulted in adding the value of 1 to $x(4)$ and -4 to the value of y (6). Ignore the bottom 1 of the result to get the answer (5, 2).

We can perform one operation followed by another by representing the point (x, y) as $(x, y, 1)$ and multiplying matrices. Taking the original vector (2, 3), doubling, and then translating, we have

$$\begin{bmatrix} 1 & 0 & 1 \\ 0 & 1 & -4 \\ 0 & 0 & 1 \end{bmatrix}\begin{bmatrix} 2 & 0 & 0 \\ 0 & 2 & 0 \\ 0 & 0 & 1 \end{bmatrix}\begin{bmatrix} 2 \\ 3 \\ 1 \end{bmatrix} = \begin{bmatrix} 1 & 0 & 1 \\ 0 & 1 & -4 \\ 0 & 0 & 1 \end{bmatrix}\begin{bmatrix} 4 \\ 6 \\ 1 \end{bmatrix} = \begin{bmatrix} 5 \\ 2 \\ 1 \end{bmatrix}$$

or the endpoint of the resulting vector is (5, 2).

In a drawing there may be many vectors, not just (2, 3), to scale and translate. Therefore, instead of performing the same two matrix products for each point, use the associative law and multiply the two 3 × 3 matrices once:

$$\begin{bmatrix} 1 & 0 & 1 \\ 0 & 1 & -4 \\ 0 & 0 & 1 \end{bmatrix}\begin{bmatrix} 2 & 0 & 0 \\ 0 & 2 & 0 \\ 0 & 0 & 1 \end{bmatrix} = \begin{bmatrix} 2 & 0 & 1 \\ 0 & 2 & -4 \\ 0 & 0 & 1 \end{bmatrix}.$$

Then we can scale and translate any number of vectors in half the time. To scale and translate (2, 3) and (−7, 4) we multiply:

$$\begin{bmatrix} 2 & 0 & 1 \\ 0 & 2 & -4 \\ 0 & 0 & 1 \end{bmatrix}\begin{bmatrix} 2 \\ 3 \\ 1 \end{bmatrix} = \begin{bmatrix} 5 \\ 2 \\ 1 \end{bmatrix} \qquad \text{giving (5, 2)}$$

$$\begin{bmatrix} 2 & 0 & 1 \\ 0 & 2 & -4 \\ 0 & 0 & 1 \end{bmatrix}\begin{bmatrix} -7 \\ 4 \\ 1 \end{bmatrix} = \begin{bmatrix} -13 \\ 4 \\ 1 \end{bmatrix} \qquad \text{giving (−13, 4).}$$

Actually we could put two (or m) points together as columns in a second 3 × 2 (or 3 × m) matrix and perform the scaling and translation all at once:

$$\begin{bmatrix} 2 & 0 & 1 \\ 0 & 2 & -4 \\ 0 & 0 & 1 \end{bmatrix}\begin{bmatrix} 2 & -7 \\ 3 & 4 \\ 1 & 1 \end{bmatrix} = \begin{bmatrix} 5 & -13 \\ 2 & 4 \\ 1 & 1 \end{bmatrix}.$$

Matrix multiplication can also rotate, as demonstrated in Exercises 65–69.

EXERCISES 7.2

Let $A = \begin{bmatrix} 1 & 2 \\ 3 & 4 \end{bmatrix}$ $B = \begin{bmatrix} 6 & 2 & 0 \\ 0 & -1 & 4 \end{bmatrix}$ $C = \begin{bmatrix} -2 & 1 & -5 \\ 7 & 1 & 0 \end{bmatrix}.$

Where possible, perform the indicated operation or answer the question in each of the Exercises 1–19.

1. AB　　　　　　2. AC　　　　　3. $AB + AC$ (Use Exercises 1 and 2)

4. $B + C$　　　　　　　　　　　5. $A(B + C)$ (Use Exercise 4)

6. Compare Exercises 3 and 5. What property does the comparison illustrate?

7. BA **8.** BC

9. $3(AC)$ (Use Exercise 2)   **10.** $3A$

11. $(3A)C$ (Use Exercise 10)   **12.** How do Exercises 9 and 11 compare?

13. $B\mathbf{I}_2$ **14.** $(3\mathbf{I}_2)A$

15. How do Exercises 10 and 14 compare?

16. $A^2 = A \cdot A$ **17.** B^2

18. $\mathbf{0}_{2\times 2} \cdot A$ **19.** $B \cdot \mathbf{0}_{3\times 3}$

$$Let\ D = \begin{bmatrix} 0 & 1 \\ 1 & 0 \end{bmatrix} \qquad E = \begin{bmatrix} 3 & 0 \\ 0 & 0 \end{bmatrix} \qquad F = \begin{bmatrix} 0 & 0 \\ 4 & 0 \end{bmatrix}$$

Where possible, perform the indicated operation or answer the question in each of the Exercises 20–39.

20. DA

21. DB

22. What does multiplication by D on the left accomplish?

23. AD

24. What does multiplication by D on the right accomplish?

25. Find a matrix G such that BG switches the first and second columns of B.

26. Does CG switch the first and second columns of C?

27. EA

28. EC

29. What does multiplication by E on the left accomplish?

30. FA

31. FB

32. What does multiplication by F on the left accomplish?

33. AE

34. What does multiplication by E on the right accomplish?

35. AF

36. What does multiplication by F on the right accomplish?

37. $(AD)B$ (Use Exercise 23)   **38.** $A(DB)$ (Use Exercise 21)

39. Compare the results of 37 and 38. What property does the comparison illustrate?

Using the matrix A *as defined above, where possible, perform the indicated operation or answer the question in each of the Exercises 40–48.*

40. Find the matrix H such that $HA = \begin{bmatrix} 5 \cdot 1 & 5 \cdot 2 \\ 7 \cdot 3 & 7 \cdot 4 \end{bmatrix}$.

41. Find the matrix J such that $AJ = \begin{bmatrix} 1 \cdot 4 & 2 \cdot 9 \\ 3 \cdot 4 & 4 \cdot 9 \end{bmatrix}$.

42. $A + I_2$ **43.** $A + 3I_2$ **44.** $[6 \quad 1]A$ **45.** $A\begin{bmatrix} 6 \\ 1 \end{bmatrix}$

46. $A\begin{bmatrix} x \\ y \end{bmatrix}$ **47.** $\begin{bmatrix} x \\ y \end{bmatrix}A$ **48.** $A[6 \quad 1]$

Using the matrices A–F *as defined above, answer the questions in Exercises 49–52.*

49. Which matrices A–F are square?

50. Give the diagonal elements of A.

51. Give the diagonal elements of B.

52. Give the diagonal elements of D.

In Exercises 53–55, various matrix products along with the sizes of the matrices are indicated. Fill in the missing dimensions.

53. $A_{6\times4}\, B_{4\times7} = C_{\underline{\quad}\times\underline{\quad}}$

54. $A_{1\times\underline{\quad}}\, B_{9\times\underline{\quad}} = C_{\underline{\quad}\times2}$

55. $A_{\underline{\quad}\times3}\, B_{\underline{\quad}\times\underline{\quad}}\, C_{2\times\underline{\quad}} = D_{4\times5}$

56. a. Suppose in a graphics drawing you want to translate every vector to the left 3 units and up 2 units. What matrix would you see?
   b. Give the matrix to scale your drawing by $1/2$.
   c. Write the one matrix that would translate and then scale.
   d. Find where the following points will be moved using only one matrix product: (8, 6), (−4, 3), (−7, −5), (2/3, 0), (−1, 4/5).

57. a. Suppose in a graphics drawing you want to translate every vector to the right 4 units and up 1 unit. What matrix would you use?
   b. Give the matrix to scale your drawing by 5.
   c. Write the one matrix that would scale and then translate.
   d. Find where the following points will be moved using only one matrix product: (8, 6), (−4, 3), (−7, −5), (2/3, 0), (−1, 4/5).

58. a. Write the following points as columns in a matrix M: (4, 5), (−3, 1), (2, 0), (0, 4).
   b. Perform the matrix product $P_1 M$ where $P_1 = \begin{bmatrix} 1 & 0 \\ 0 & 0 \end{bmatrix}$.
   c. Write the columns of $P_1 M$ from part b as ordered pairs.
   d. Plot the points of parts a and c. *Note:* Multiplication on the left by P_1 projects the points onto the x axis.
   e. Find a matrix P_2 such that $P_2 M$ projects the points onto the y axis.
   f. Perform the multiplication $P_2 M$.
   g. Plot the points of part a along with the points projected onto the y axis. *Note:* Recall the projection function of Section 5.2 is a mapping to a designated coordinate.

59. Suppose $A = \begin{bmatrix} 2 & 1 \\ 0 & -3 \end{bmatrix}$.

    a. Find $A^2 = A \cdot A$.   **b.** Find $A^3 = A^2 \cdot A$.

    c. What must be true about a matrix B for B^2 to exist?

60. Consider $A = \begin{bmatrix} 1 & -2 & 4 & 0 \\ 8 & 5 & 7 & 3 \end{bmatrix}$ and $B = \begin{bmatrix} 0 & 6 & -3 \\ 5 & 2 & 8 \\ 3 & 1 & 1 \\ -7 & 9 & -9 \end{bmatrix}$.

    a. Write the second row of A as a vector.

    b. Write the third column of B as a vector.

    c. Express the c_{23} element of $C = AB$ as a dot product. (Use parts a and b.)

    d. Write c_{23} as a sum of products.

61. Let $A = [a_{ij}]_{m \times q}$, $B = [b_{ij}]_{q \times n}$, and $C = [c_{ij}]_{m \times n}$ be matrices of the indicated sizes with $C = AB$. Thus, the c_{ij} element is the dot product of the ith row of A and jth column of B. Suppose $m \geq 2$ and $n \geq 3$. The second row of A can be written as a vector, $\mathbf{a} = (a_{21}, a_{22}, a_{23}, \ldots, a_{2q})$.

    a. Write the third column of B as a vector.

    b. Write c_{23} as a dot product.

    c. Write c_{23} as a sum of products.

    d. Write c_{23} using sigma notation.

    e. Write the ith row of A as a vector.

    f. Write the jth column of B as a vector.

    g. Write c_{ij} as a dot product.

    h. Write c_{ij} as a sum of products.

    i. Write c_{ij} using sigma notation. *Note:* This notation is used in many books to define matrix product.

62. Write $\begin{bmatrix} 7x - 3y \\ -2x + 5y \end{bmatrix}_{2 \times 1}$ as a matrix product of a 2×2 matrix of numbers and a 2×1 matrix of variables.

63. Let $A = \begin{bmatrix} 3 & 0 \\ 0 & 5 \end{bmatrix}$. Find

    a. A^2 **b.** A^3 **c.** A^n for any positive integer n.

64. Prove the generalized distributive property for matrices using mathematical induction:

Let A be an $m \times q$ matrix, and for any positive integer $t > 1$ let B_1, $B_2, \ldots, B_t$ be $q \times n$ matrices. Then

$$A(B_1 + B_2 + \cdots + B_t) = AB_1 + AB_2 + \cdots + AB_t.$$

Exercises 65–69 require some knowledge of trigonometry and are optional. The following matrix product rotates a vector (x, y) *clockwise through an angle σ about the origin, as indicated in Figure 7.2.*

$$\begin{bmatrix} x' \\ y' \\ 1 \end{bmatrix} = \begin{bmatrix} \cos \sigma & \sin \sigma & 0 \\ -\sin \sigma & \cos \sigma & 0 \\ 0 & 0 & 1 \end{bmatrix} \begin{bmatrix} x \\ y \\ 1 \end{bmatrix}$$

Figure 7.2

Rotation of (x, y) clockwise about the origin through angle σ, or rotation of axes counterclockwise through σ

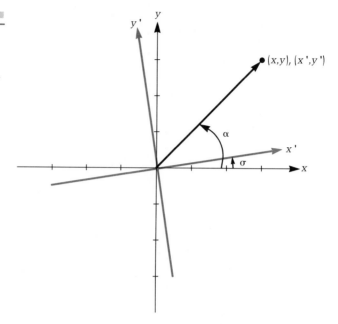

65. **a.** Find the matrix R to rotate each point (x, y) in the plane 30° clockwise about the origin. *Note:* See the bottom of page 251.
    b. Find a matrix T_1 to translate axes so that $(-1, 4)$ is on the new origin.
    c. Find a matrix T_2 to move every point 1 unit to the left and 4 units up.
    d. Compute $T_2\,RT_1$. *Note:* The result is a matrix for a clockwise rotation of 30° about the fixed point $(-1, 4)$.
    e. To where is $(2, 3)$ mapped by the rotation of part d?
    f. Using $T_2\,RT_1$ as one matrix, find to where the following points would be mapped in a clockwise rotation of 30° about $(-1, 4)$: $(5, 1)$, $(-3, 0)$, $(-1, 4)$, $(0, 0)$, $(1, -5)$.
    g. Plot the points of part f and see if your answers seem reasonable.

66. Find a 3 × 3 matrix to do each of the following:
    a. Scale by 3.
    b. Rotate every point 60° clockwise about the origin.
    c. Translate every point 5 units to the right and 2 units up.
    d. Translate every point so that $(7, 3)$ is the new origin.
    e. Rotate every point 75° counterclockwise about the origin. *Note:* Use a negative angle for counterclockwise rotation.

67. Use Exercise 66 to help you find the one matrix in each case to do the following:
    a. Scale by 3, then rotate every point 60° clockwise about the origin.
    b. Rotate every point 60° clockwise about the origin and then scale by 3.
    c. Rotate every point 60° clockwise about $(7, 3)$ and then scale by 3.
    d. Scale by 3 and then rotate every point 60° clockwise about $(7, 3)$.
    e. Rotate every point 75° counterclockwise about $(-5, -2)$ and then scale by 3.

  f. Scale by 3 and then rotate every point 75° counterclockwise about $(-5, -2)$.

68. a. Find the one matrix that would be used to rotate every point 45° counterclockwise about $(1, 2)$.

  b. To where would the points $(-3, -1)$ and $(3, 5)$ be mapped by the rotation?

69. In this example we will prove that counterclockwise rotation of the axes by an angle σ about the origin moves (x, y) to (x', y'), where

$$x' = x \cos \sigma + y \sin \sigma \quad (1)$$
$$y' = -x \sin \sigma + y \cos \sigma \quad (2)$$

  a. Draw right triangles in Figure 7.2 to help justify:

$$
\begin{array}{llll}
x = r \cos \alpha & (3) & y = r \sin \alpha & (5) \\
x' = r \cos (\alpha - \sigma) & (4) & y' = r \sin (\alpha - \sigma) & (6) \\
\;\; = r \cos (\alpha + (-\sigma)) & & \;\; = r \sin (\alpha + (-\sigma)) &
\end{array}
$$

  b. Use the following trigonometric identities to prove equations (1) and (2) above.

$$\cos (\beta + \gamma) = \cos \beta \cos \gamma - \sin \beta \sin \gamma \quad (7)$$
$$\sin (\beta + \gamma) = \sin \beta \cos \gamma + \sin \gamma \cos \beta \quad (8)$$
$$\cos (-\beta) = \cos (\beta) \quad (9) \qquad \sin (-\beta) = -\sin \beta \quad (10)$$

7.3

Systems of Linear Equations

The following are examples of linear equations:

$$6x = 1$$
$$5x + 7y = 3$$
$$-2x + \pi y + \sqrt{3} z = 8$$
$$1/2\, x_1 + 33.2x_2 + 15x_3 + 10x_4 = 33/4.$$

They derive their name from the fact that when they have only one or two variables as in the first two examples, their graphs are straight lines in 2-space. Interestingly, with three variables, the graph is a plane in 3-space. Though the general **linear equation** is

$$\sum_{i=1}^{n} a_i x_i = a_1 x_1 + a_2 x_2 + \cdots + a_n x_n = c$$

where $a_i, c \in \mathbb{R}$, $i = 1, 2, \ldots, n$, we will consider only equations with three or fewer variables.

  The geometric interpretation of these equations with at most three variables as lines or planes indicates that they will be useful in computer graphics. Systems of several linear equations, however, arise in a variety of applications in which matrices and the computer can be used as tools in their solution. In fact, the first electronic computer, the ABC, named after its designers Atanasoff and Berry, was built specifically to solve systems of 29 equations in 29 unknowns, a formidable task without the aid of a computer. ☐

HISTORICAL NOTE

Let's see how we can use matrices to solve an applied problem involving a system of equations. A computer programmer has been given an accounting program and a transactions program to complete in 60 working days. She estimates that 4 times as much time must be spent developing the accounting program as the transactions program. How much time should she budget for each project?

You might jump immediately to an answer of 12 days for the transactions program and $4 \cdot 12 = 48$ days for the accounting program by dividing 60 by 5 (4 parts for one program and 1 part for the other) and distributing the days proportionally. To examine the method of solving systems of equations with matrices, however, let us approach the problem more slowly.

For any word problem, first assign meanings to the unknowns:

x, number of days to be spent on accounting program

y, number of days to be spent on transactions program.

Now, go through the statement of the problem a sentence or phrase at a time. The first sentence says she has a total of 60 days, split between the two projects, to work. Thus, we have the linear equation

$$x + y = 60.$$

The second sentence yields the equation

$$x = 4y.$$

Be careful! 4 is multiplied times y, not x. The object of the phrase "as much time as," or the noun after the last "as," is what is multiplied by 4.

This system of linear equations can be written as

$$x + y = 60$$
$$x - 4y = 0.$$

Matrices can also be used to express the system:

$$\begin{bmatrix} 1 & 1 \\ 1 & -4 \end{bmatrix} \begin{bmatrix} x \\ y \end{bmatrix} = \begin{bmatrix} 60 \\ 0 \end{bmatrix}.$$

Often x and y are ignored (after all, they are just "dummy" variables), and the matrix equation is written as what is called an **augmented matrix**

$$\left[\begin{array}{cc|c} 1 & 1 & 60 \\ 1 & -4 & 0 \end{array} \right].$$

There are three equation (row) operations that you can perform on a system of equations (augmented matrix) to yield an **equivalent system,** one that has the same solution:

1. Multiply an equation (row) by a nonzero scalar.
2. Replace an equation (row) by the sum of it and another equation (row).
3. Exchange equations (rows).

Let's solve the system in equation and matrix form using these operations in a method called **Gaussian elimination,** named after Carl Friedrich Gauss, the prodigy mentioned in Section 6.2, who became the greatest mathematician of the 19th century. ☐

Eliminate x in the 2nd equation.
Multiply the 2nd equation by -1:

$$\begin{aligned} x + \ y &= 60 \\ -x + 4y &= \ \ 0 \end{aligned} \qquad \left[\begin{array}{cc|c} 1 & 1 & 60 \\ -1 & 4 & 0 \end{array}\right].$$

Replace the 2nd equation by the sum of the 1st and 2nd equations:

$$\begin{aligned} x + \ y &= 60 \\ 5y &= 60 \end{aligned} \qquad \left[\begin{array}{cc|c} 1 & 1 & 60 \\ 0 & 5 & 60 \end{array}\right].$$

Eliminate y in the 1st equation.
Multiply the 2nd equation by $1/5$:

$$\begin{aligned} x + y &= 60 \\ y &= 12 \end{aligned} \qquad \left[\begin{array}{cc|c} 1 & 1 & 60 \\ 0 & 1 & 12 \end{array}\right].$$

Replace the 1st equation by the sum of the 1st equation and -1 times the 2nd equation:

$$\begin{aligned} x \ \ &= 48 \\ y &= 12 \end{aligned} \qquad \left[\begin{array}{cc|c} 1 & 0 & 48 \\ 0 & 1 & 12 \end{array}\right].$$

Plugging $x = 48$ and $y = 12$ into both original equations we check that the answer $(x, y) = (48, 12)$ is correct:

$$48 + 12 = 60 \qquad \text{and} \qquad -48 + 4 \cdot 12 = 0.$$

Observe that working with matrices instead of the full equations uses an identical process with less effort.

The same two equations could be used in a computer graphics design. The graphs, pictured in Figure 7.3, are straight lines; and (48, 12) is the point of intersection. Two lines in the same plane can intersect in a point, a line (the two are identical), or be parallel.

Figure 7.3

Graph of $\begin{cases} x + y = 60 \\ \quad x = 4y \end{cases}$

Note: (48, 12) is the point of intersection for the two lines. Since two points determine a line, to graph we only need to find one other point on each line

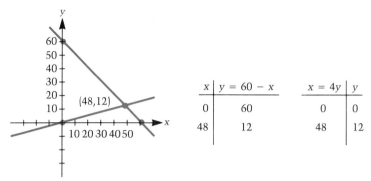

x	$y = 60 - x$
0	60
48	12

$x = 4y$	y
0	0
48	12

Many graphics applications involve three-dimensional figures projected onto a two-dimensional surface such as a computer screen or printer paper. Three linear equations in three unknowns can represent three planes that intersect in a point, a line, a plane, or not at all.

EXAMPLE 1

Solve the following system of equations using matrices:

$$\begin{cases} -2x + 7y - 52z = -93 \\ \quad x - \quad y + 10z = \quad 22 \, . \\ \quad x - 3y + 26z = \quad 48 \end{cases}$$

Solution

For easy reference we will label the rows as "r" followed by a reference number:

$$\begin{array}{c} r1 \\ r2 \\ r3 \end{array} \left[\begin{array}{ccc|c} -2 & 7 & -52 & -93 \\ 1 & -1 & 10 & 22 \\ 1 & -3 & 26 & 48 \end{array} \right].$$

Exchange rows 1 and 3 to get the number 1 in the top, left position:

$$\begin{array}{c} r3 \\ r2 \\ r1 \end{array} \left[\begin{array}{ccc|c} ① & -3 & 26 & 48 \\ 1 & -1 & 10 & 22 \\ -2 & 7 & -52 & -93 \end{array} \right].$$

Get 0's below the circled number which is called the **pivot number:**

$$\begin{array}{rrrrr} -1 \times r3 & -1 & 3 & -26 & -48 \\ + r2 & \underline{1} & \underline{-1} & \underline{10} & \underline{22} \\ & 0 & 2 & -16 & -26. \end{array}$$

Note: Though subtracting one row from the other works, you will be less likely to make careless errors if you always multiply one row by -1 and add instead.

Divide by 2 to simplify:

$$
\begin{array}{lrrrr}
r4 & 0 & 1 & -8 & -13 \\
\\
2 \times r3 & 2 & -6 & 52 & 96 \\
+\ r1 & \underline{-2} & \underline{7} & \underline{-52} & \underline{-93} \\
r5 & 0 & 1 & 0 & 3
\end{array}
$$

$$
\begin{array}{l}
r3 \\
r4 \\
r5
\end{array}
\left[
\begin{array}{rrr|r}
1 & -3 & 26 & 48 \\
0 & ① & -8 & -13 \\
0 & 1 & 0 & 3
\end{array}
\right].
$$

Now using the 2nd row, $r4$, get 0 below the pivot number:

$$
\begin{array}{lrrrr}
-1 \times r4 & 0 & -1 & 8 & 13 \\
+\ r5 & \underline{0} & \underline{1} & \underline{0} & \underline{3} \\
& 0 & 0 & 8 & 16.
\end{array}
$$

Divide by 8:

$$
\begin{array}{lrrrr}
r6 & 0 & 0 & 1 & 2
\end{array}
$$

$$
\begin{array}{l}
r3 \\
r4 \\
r6
\end{array}
\left[
\begin{array}{rrr|r}
1 & -3 & 26 & 48 \\
0 & 1 & -8 & -13 \\
0 & 0 & ① & 2
\end{array}
\right].
$$

Now start from the 3rd row, $r6$, and go up. Get 0's above the pivot number which is in $r6$:

$$
\begin{array}{lrrrr}
8 \times r6 & 0 & 0 & 8 & 16 \\
+\ r4 & \underline{0} & \underline{1} & \underline{-8} & \underline{-13} \\
r7 & 0 & 1 & 0 & 3
\end{array}
$$

$$
\begin{array}{lrrrr}
-26 \times r6 & 0 & 0 & -26 & -52 \\
+\ r3 & \underline{1} & \underline{-3} & \underline{26} & \underline{48} \\
r8 & 1 & -3 & 0 & -4
\end{array}
$$

$$
\begin{array}{l}
r8 \\
r7 \\
r6
\end{array}
\left[
\begin{array}{rrr|r}
1 & -3 & 0 & -4 \\
0 & ① & 0 & 3 \\
0 & 0 & 1 & 2
\end{array}
\right].
$$

Going up to the 2nd row, $r7$, we get 0 above the pivot:

$$
\begin{array}{lrrrr}
3 \times r7 & 0 & 3 & 0 & 9 \\
+\ r8 & \underline{1} & \underline{-3} & \underline{0} & \underline{-4} \\
r9 & 1 & 0 & 0 & 5
\end{array}
$$

$$r9\begin{bmatrix} 1 & 0 & 0 & | & 5 \\ 0 & 1 & 0 & | & 3 \\ 0 & 0 & 1 & | & 2 \end{bmatrix}.$$

Hence, the answer is $(x, y, z) = (5, 3, 2)$. The three planes of the system intersect in a single point, $(5, 3, 2)$. ■

EXAMPLE 2

Use Gaussian elimination on the following matrix, $A = [a_{ij}]$:

$$r1\begin{bmatrix} 0 & 3 & | & 4 \\ 7 & -2 & | & 1 \end{bmatrix}.$$

Solution

Because there is a nonzero number below $a_{11} = 0$, exchange the rows.

$$r2\begin{bmatrix} \textcircled{7} & -2 & | & 1 \\ 0 & 3 & | & 4 \end{bmatrix} \quad \begin{matrix} 1/7 \times r2 = r3 \\ 1/3 \times r4 = r4 \end{matrix} \quad \begin{bmatrix} 1 & -2/7 & | & 1/7 \\ 0 & \textcircled{1} & | & 4/3 \end{bmatrix}.$$

With 1 in $r4$ as the pivot, we need to eliminate $-2/7$ in $r3$. One way is to multiply $r4$ by $2/7$ and add. To avoid fractions as long as possible, however, you may find it easier to use $r2$ and $r1$, which are just multiples of $r3$ and $r4$, respectively. Multiply $r2$ by 3 and $r1$ by 2 and add:

$$\begin{array}{lrrr} 3 \times r2 & 21 & -6 & 3 \\ 2 \times r1 & \underline{0} & \underline{6} & \underline{8} \\ & 21 & 0 & 11. \end{array}$$

Replacing $r3$ with this row divided by 21, we have

$$\begin{bmatrix} 1 & 0 & | & 11/21 \\ 0 & 1 & | & 4/3 \end{bmatrix}. \quad ■$$

Algorithm For Gaussian Elimination

1. Continue from top to bottom:
   a. Start with the top unprocessed row. If on a lower row there is a nonzero number to the left of the first non-zero number on the unprocessed row, exchange rows.
   b. The leftmost nonzero element of the top unprocessed row is the **pivot.** If it is not 1, divide the row by this number.
   c. By row operations get 0's below the pivot.
2. Continue from bottom to top:
   a. Start at the bottom unprocessed row. The leftmost nonzero element is the pivot. By row operations get 0's above the pivot.

Some languages such as SAS or MACSYMA and its micro-computer counterpart, muMATH, have built-in functions to solve a system of linear equations and/or perform row reduction on a matrix. When not built-in, you can write a routine yourself, following the steps outlined above, to solve the system.

EXERCISES 7.3

Solve the systems of equations in Exercises 1–5 using Gaussian elimination on matrices.

1. $\begin{cases} x - 10y = 2 \\ 3x + 4y = 6 \end{cases}$

2. $\begin{cases} 2x + 7y = 1 \\ 9x - 3y = 2 \end{cases}$

3. $\begin{cases} 2x_1 - x_2 + x_3 = 13 \\ x_1 + 3x_2 - x_3 = 8 \\ 5x_1 - x_2 + 4x_3 = 37 \end{cases}$

4. $\begin{cases} 3x + 2y - 2z = -4 \\ -6x - 2y + 5z = 15 \\ 6x + 2y - 3z = -7 \end{cases}$

5. $\begin{cases} -7x + 7y - 2z = 13 \\ 4x - 14y + 4z = 6 \\ -3x + 21y - z = 8 \end{cases}$

6. Sketch the system in Exercise 1.

7. Sketch the system in Exercise 2.

In row reduction if you derive a row with 0's on the left and the right-most number is not 0, then the system has no solution. In this case, the set of lines or planes do not intersect in any common points. The lines are parallel; the planes do not intersect at all.

8. a. Solve $\begin{cases} 4x + y = 1 \\ 8x + 2y = -1 \end{cases}$.

   b. Sketch the graph of the system.

   c. What is true about the lines?

9. a. Solve $\begin{cases} x_1 - 3x_2 = 1/2 \\ -2x_1 + 6x_2 = -1 \end{cases}$.

   b. Sketch the graph of the system.

   c. What is true about the lines?

10. a. Solve $\begin{cases} x + y - 5z = 1 \\ 2x - 2y + z = 3 \\ 3x - y - 4z = 4 \end{cases}$.

   b. How do the planes intersect?

11. a. Solve $\begin{cases} 8x_1 + 2x_2 - x_3 = 2 \\ -x_1 + 5x_2 + 8x_3 = 7 \\ x_1 + x_2 + x_3 = 6 \end{cases}$.

   b. How do the planes intersect?

12. Suppose a 5-pound blend of coffee is to be made from three types of coffee: dark roast costing $5.72 per pound, Turkish at $6.28 a pound, and South American at $4.42 per pound. The resulting price is $5.47 per pound. If the combined total of the first two coffees is to be 2 pounds more than the amount of South American coffee, how much of each coffee will be used?

13. Suppose each student at a university is alloted $150 in the computing budget for a 14-week semester. A student is charged $0.17 for each hour on-line, $18 per minute of CPU time, and $40 for 1000 pages of printout. In trying to budget, a particular student decides to allow 10 seconds of CPU time for every hour on-line and 20 pages of printout for every minute of CPU time. Under these assumptions, how much on-line time, CPU time, and pages of printout will be used for the semester by the student? Per week?

14. 512K of memory is to be partitioned into three segments. The second segment is twice the length of the third and 2K more than the first. What is the length of each segment?

15. Suppose a computer center is given a grant of $20,000 to spend on microcomputers. The director decides to buy two brands of computer, brand A at $4000 each and brand B at $1200 each. The director wants twice as many of brand A as brand B.
   a. What is the best integer solution to the problem given these constraints?
   b. How much money is left over?

7.4

Matrix Inverses

As discussed in the last section, a system of equations like

$$\begin{cases} x_1 + x_2 = 60 \\ x_1 - 4x_2 = 0 \end{cases}$$

can be written as a matrix equation

$$\begin{bmatrix} 1 & 1 \\ 1 & -4 \end{bmatrix} \begin{bmatrix} x_1 \\ x_2 \end{bmatrix} = \begin{bmatrix} 60 \\ 0 \end{bmatrix}$$

or as $AX = B$, where

$$A = \begin{bmatrix} 1 & 1 \\ 1 & -4 \end{bmatrix} \quad X = \begin{bmatrix} x_1 \\ x_2 \end{bmatrix} \quad \text{and} \quad B = \begin{bmatrix} 60 \\ 0 \end{bmatrix}.$$

If A and B were numbers in $AX = B$ instead of matrices, and A was not 0, we could solve the problem immediately by dividing both sides by A or, equivalently, by multiplying by A^{-1}.

$$A^{-1}AX = A^{-1}B$$

$$X = A^{-1}B.$$

For example, if $2X = 6$, then

$$2^{-1} \cdot 2X = 2^{-1} \cdot 6$$

$$1/2 \cdot 2X = 1/2 \cdot 6$$

$$X = 3.$$

$1/2$ is the multiplicative inverse of 2 since $1/2 \cdot 2 = 1$, the multiplicative identity in $\mathbb{R}$.

For 2×2 matrices we have the multiplicative identity

$$\mathbf{I}_2 = \begin{bmatrix} 1 & 0 \\ 0 & 1 \end{bmatrix}.$$

Moreover, many, though not all, 2×2 matrices A have a multiplicative inverse, A^{-1}, such that $A^{-1} \cdot A = A \cdot A^{-1} = \mathbf{I}_2$. The multiplicative inverse of

$$A = \begin{bmatrix} 1 & 1 \\ 1 & -4 \end{bmatrix}$$

does exist and equals

$$A^{-1} = \begin{bmatrix} 4/5 & 1/5 \\ 1/5 & -1/5 \end{bmatrix}.$$

You might want to check that the product is the identity:

$$AA^{-1} = \begin{bmatrix} 1 & 1 \\ 1 & -4 \end{bmatrix}\begin{bmatrix} 4/5 & 1/5 \\ 1/5 & -1/5 \end{bmatrix} = \begin{bmatrix} 1 & 0 \\ 0 & 1 \end{bmatrix} = A^{-1}A.$$

DEFINITION

Let A be an $n \times n$ matrix. If there exists an $n \times n$ matrix B such that

$$AB = BA = \mathbf{I}_n,$$

then B is called the **multiplicative inverse** of A. Write $B = A^{-1}$.

Knowing A^{-1} we can solve $AX = B$ for X:

$$X = A^{-1}B = \begin{bmatrix} 4/5 & 1/5 \\ 1/5 & -1/5 \end{bmatrix}\begin{bmatrix} 60 \\ 0 \end{bmatrix} = \begin{bmatrix} 48 \\ 12 \end{bmatrix}.$$

LANGUAGE
EXAMPLE

HISTORICAL
NOTE

Thus, $x_1 = 48$ and $x_2 = 12$. Once we have A^{-1}, we can easily solve $AX = B$ for any B. Later in this section we will see how matrices and their inverses can be used for coding and decoding messages.

But how do we find A^{-1}? If we were using the language APL, the answer would be simple since it has a built-in operation for matrix inversion. The statement

$$\text{INV} \leftarrow \boxed{\div}A$$

takes the inverse of A and assigns it to INV. □ APL is essentially the product of one man's effort, that of Kenneth Iverson. It was introduced in the early 1960s mainly to solve problems that use vectors and arrays extensively. □

To find the inverse mathematically we use the same Gaussian elimination technique covered in the last section. Given

$$A = \begin{bmatrix} 1 & 1 \\ 1 & -4 \end{bmatrix}$$

we want to find A^{-1} such that $AA^{-1} = \mathbf{I}_2$. If

$$A^{-1} = \begin{bmatrix} a_{11} & a_{12} \\ a_{21} & a_{22} \end{bmatrix}$$

we are solving

$$\begin{bmatrix} 1 & 1 \\ 1 & -4 \end{bmatrix}\begin{bmatrix} a_{11} & a_{12} \\ a_{21} & a_{22} \end{bmatrix} = \begin{bmatrix} 1 & 0 \\ 0 & 1 \end{bmatrix}.$$

Notice that by definition of matrix product

$$\begin{bmatrix} 1 & 1 \\ 1 & -4 \end{bmatrix}\begin{bmatrix} a_{11} \\ a_{21} \end{bmatrix} = \begin{bmatrix} 1 \\ 0 \end{bmatrix}$$

and

$$\begin{bmatrix} 1 & 1 \\ 1 & -4 \end{bmatrix}\begin{bmatrix} a_{12} \\ a_{22} \end{bmatrix} = \begin{bmatrix} 0 \\ 1 \end{bmatrix}.$$

But we solved these kinds of equations in the last section with Gaussian elimination on the augmented matrices like

$$\begin{bmatrix} 1 & 1 & | & 1 \\ 1 & -4 & | & 0 \end{bmatrix} \quad \text{and} \quad \begin{bmatrix} 1 & 1 & | & 0 \\ 1 & -4 & | & 1 \end{bmatrix}.$$

What a waste of time it would be, however, to do the elimination twice. The left matrix is the same in both cases, so the steps would be identical for both problems. Therefore, put everything together and perform the elimination on the following matrix:

$$\begin{bmatrix} 1 & 1 & | & 1 & 0 \\ 1 & -4 & | & 0 & 1 \end{bmatrix}.$$

Initially, the identity matrix is on the right. After elimination, I_2 is on the left, and A^{-1} is on the right:

$$\begin{bmatrix} 1 & 0 & | & 4/5 & 1/5 \\ 0 & 1 & | & 1/5 & -1/5 \end{bmatrix}.$$

The same procedure works for any square matrix $A_{n \times n}$. Place A on the left and the identity matrix I_n on the right, and perform the elimination to form A^{-1} on the right. As mentioned above, however, A^{-1} does not exist for every square matrix. Moreover, it never exists for a nonsquare matrix, $A_{n \times m}$, where $n \neq m$. Even for real numbers a multiplicative inverse does not exist for every element—there is no x such that $0 \cdot x = 1$. We will discover that a matrix does not have an inverse if in the row reduction we obtain a row of 0's to the left of the middle bar.

EXAMPLE 1

Find A^{-1}, if it exists, for

$$A = \begin{bmatrix} 1 & 2 & 3 \\ -1 & 4 & 5 \\ 0 & 3 & 4 \end{bmatrix}.$$

Solution

$$\begin{matrix} r1 \\ r2 \\ r3 \end{matrix} \begin{bmatrix} 1 & 2 & 3 & | & 1 & 0 & 0 \\ -1 & 4 & 5 & | & 0 & 1 & 0 \\ 0 & 3 & 4 & | & 0 & 0 & 1 \end{bmatrix}$$

$$\begin{matrix} r1 \\ r1 + r2 = r4 \\ r3 \end{matrix} \begin{bmatrix} 1 & 2 & 3 & | & 1 & 0 & 0 \\ 0 & 6 & 8 & | & 1 & 1 & 0 \\ 0 & 3 & 4 & | & 0 & 0 & 1 \end{bmatrix}$$

$$\begin{matrix} r1 \\ 1/6\ r4 = r5 \\ r3 \end{matrix} \begin{bmatrix} 1 & 2 & 3 & | & 1 & 0 & 0 \\ 0 & 1 & 4/3 & | & 1/6 & 1/6 & 0 \\ 0 & 3 & 4 & | & 0 & 0 & 1 \end{bmatrix}$$

$$\begin{matrix} r1 \\ r5 \\ -3\ r5 + r3 \end{matrix} \begin{bmatrix} 1 & 2 & 3 & | & 1 & 0 & 0 \\ 0 & 1 & 4/3 & | & 1/6 & 1/6 & 0 \\ 0 & 0 & 0 & | & -1/2 & -1/2 & 1 \end{bmatrix}.$$

At this point we realize that A^{-1} does not exist. Since there are all 0's on the left in the third row, there is no way to get I_3 on the left part of the augmented matrix. ∎

Modern cryptography involves a great deal of mathematics and use of computers. One great achievement of cryptography occurred during World War II when the British intelligence broke the German code being produced by a machine called Enigma. The British team was headed by the brilliant mathematician Alan Turing who made his impact on computer science in other ways as we will see in Chapter 8. □

One method of encoding involves matrices. First, map every character that can be used via a one-to-one function into the set of nonnegative integers. This map could be performed in any number of ways. As we know from work with permutations, if we plan to use 29 characters (the alphabet, blank, comma, and period), there are 29! ways to map to the integers 0 through 28. Taking one of the more obvious mappings, we have

blank	A	B	C	D	E	F	G	H	I	J	K	L	M	N
0	1	2	3	4	5	6	7	8	9	10	11	12	13	14

O	P	Q	R	S	T	U	V	W	X	Y	Z	.	,
15	16	17	18	19	20	21	22	23	24	25	26	27	28

Suppose the message is

SEND HELP AT ONCE

It would be mapped to

19 5 14 4 0 8 5 12 16 0 1 20 0 15 14 3 5

We could use a less obvious mapping and send the message as the corresponding numbers. With enough messages, however, the code could be cracked because letters occur with different frequencies in English. For example, the letter "E" occurs most frequently, so that number that is used the most in a message, here 5, has a good chance of being "E."

To proceed with the encoding, after mapping we group the numbers into vectors of 2, 3, 4, or any number of elements. Suppose we group them into 3-tuples, padding with the code for blank, 0:

(19, 5, 14), (4, 0, 8), (5, 12, 16),   (0, 1, 20), (0, 15, 14), (3, 5, 0)

Now take any matrix A with elements from the integers, carefully picked such that A^{-1} also has integer elements. For example, let

$$A = \begin{bmatrix} 1 & 2 & 3 \\ 0 & 1 & -2 \\ 1 & 3 & 2 \end{bmatrix}.$$

We can verify that

$$A^{-1} = \begin{bmatrix} 8 & 5 & -7 \\ -2 & -1 & 2 \\ -1 & -1 & 1 \end{bmatrix}.$$

A will be used for coding and A^{-1} for decoding. Arrange the 3-tuples as columns in a matrix M and compute AM:

$$AM = \begin{bmatrix} 1 & 2 & 3 \\ 0 & 1 & -2 \\ 1 & 3 & 2 \end{bmatrix} \begin{bmatrix} 19 & 4 & 5 & 0 & 0 & 3 \\ 5 & 0 & 12 & 1 & 15 & 5 \\ 14 & 8 & 16 & 20 & 14 & 0 \end{bmatrix}$$

$$= \begin{bmatrix} 71 & 28 & 77 & 62 & 72 & 13 \\ -23 & -16 & -20 & -39 & -13 & 5 \\ 62 & 20 & 73 & 43 & 73 & 18 \end{bmatrix}.$$

We would like to send the message as letters, so apply the $\bmod_{29}$ function to each number to get the matrix M' and then map those numbers back to the character set.

$$M' = \begin{bmatrix} 13 & 28 & 19 & 4 & 14 & 13 \\ 6 & 13 & 9 & 19 & 16 & 5 \\ 4 & 20 & 15 & 14 & 15 & 18 \end{bmatrix}.$$

Thus, the encoded message is

MFD,MTSIODSONPOMER

Consider the problem of deciphering this message if you didn't even know the size of the vectors, much less the matrix A.

If you were the correct contact, however, you would reverse the process to decode. After mapping the letters to numbers and placing them in a matrix M', you would start to undo the coded message with the product $A^{-1}M'$:

$$A^{-1}M' = \begin{bmatrix} 8 & 5 & -7 \\ -2 & -1 & 2 \\ -1 & -1 & 1 \end{bmatrix} \begin{bmatrix} 13 & 28 & 19 & 4 & 14 & 13 \\ 6 & 13 & 9 & 19 & 16 & 5 \\ 4 & 20 & 15 & 14 & 15 & 18 \end{bmatrix}$$

$$= \begin{bmatrix} 106 & 149 & 92 & 29 & 87 & 3 \\ -24 & -29 & -17 & 1 & -14 & 5 \\ -15 & -21 & -13 & -9 & -15 & 0 \end{bmatrix}.$$

Then applying the $\bmod_{29}$ function once more, you get a matrix of integers from 0 to 28 which is identical to M:

$$M'' = \begin{bmatrix} 19 & 4 & 5 & 0 & 0 & 3 \\ 5 & 0 & 12 & 1 & 15 & 5 \\ 14 & 8 & 16 & 20 & 14 & 0 \end{bmatrix}.$$

Now, mapping back to the characters the message is decoded.

A note on the use of your calculator. Store the modulo number, 29, in the memory. Since $\mod_{29}(x)$ is the positive remainder from $x \div 29$, repeated subtraction or addition of 29 until you get a number between 0 and 28 will accomplish the same result. For example, $106 - 3 \cdot 29 = 19$ and $-24 + 29 = 5$.

To review, the steps for matrix encoding were as follows:

1. Find a matrix A with inverse A^{-1} such that both A and A^{-1} have only integer elements.
2. Map the n characters to the integers 0 through $(n - 1)$.
3. Separate the numbers into vectors, and place the vectors as columns in a matrix M.
4. Calculate AM.
5. Evaluate $\mod_n(x)$ for each element x in AM.
6. Take the numbers out of the matrix a column at a time.
7. Map the integers to the characters.

To decode the message the steps are reversed:

1. Map the characters to the integers.
2. Put the numbers in the matrx M' by columns.
3. Calculate $A^{-1}M'$.
4. Apply the function $\mod_n$ to each element.
5. Take out the numbers a column at a time.
6. Map the numbers to the characters.

There are many techniques of coding and decoding, but this method nicely illustrates the utility of matrices and their inverses.

EXERCISES 7.4

Find A^{-1} *where possible for each of the matrices* A *in Exercises 1–8.*

1. $\begin{bmatrix} 1 & 2 \\ 1 & 3 \end{bmatrix}$ 2. $\begin{bmatrix} 4 & 5 \\ 7 & 9 \end{bmatrix}$ 3. $\begin{bmatrix} 6 & -10 \\ -3 & 5 \end{bmatrix}$

4. $\begin{bmatrix} 6 & 4 & -5 \\ 11 & 3 & 1 \end{bmatrix}$ 5. $\begin{bmatrix} -1 & 1 & 0 \\ 1 & 1 & 1 \\ 0 & 1 & 1 \end{bmatrix}$

6. $\begin{bmatrix} 1 & 4 & -1 \\ 0 & 3 & 5 \\ 2 & 3 & 7 \end{bmatrix}$ 7. $\begin{bmatrix} 3 & 0 & 8 \\ 0 & 5 & 5 \\ 3 & -8 & 0 \end{bmatrix}$ 8. $[7 \quad 4 \quad 8 \quad -6]$

9. a. Use A^{-1} from Exercise 2 to solve the following system of equations, $AX = B$:

$$4x_1 + 5x_2 = -3$$
$$7x_1 + 9x_2 = \quad 4.$$

b. With one matrix product solve $AX = B$ for B equal to each of the following:

$$\begin{bmatrix} 6 \\ 1 \end{bmatrix}, \begin{bmatrix} 1 \\ 0 \end{bmatrix}, \begin{bmatrix} -7 \\ 4 \end{bmatrix}, \begin{bmatrix} 5 \\ 8 \end{bmatrix}, \begin{bmatrix} -9 \\ 2 \end{bmatrix}.$$

10. With A equal to the matrix in Exercise 6, use A^{-1} to solve for X in $AX = B$:

$$B = \begin{bmatrix} 10 & -7 & 0 & 1 \\ 2 & 0 & 1 & -5 \\ 4 & 3 & 0 & -5 \end{bmatrix}.$$

11. Using the matrix and character mapping for the cryptography example in this section, decode the following message:

L ELRARSHMKJVNG

12. a. Use the matrix from Exercise 1 and the character mapping to the integers given in this section to encode the message

RANDOLPH IS A SPY

b. Check your work by decoding your answer of part a.

13. a. Use the matrix from Exercise 5 and the character mapping below to encode the message

13 RUE LORAINE

b	A	B	C	D	E	F	G	H	I	J	K	L	M	N	O
0	5	10	15	20	25	30	35	1	6	11	16	21	26	31	36

P	Q	R	S	T	U	V	W	X	Y	Z	0	1	2	3	4
2	7	12	17	22	27	32	3	8	13	18	23	28	33	4	9

5	6	7	8	9
14	19	24	29	34

b. Check your work by decoding your answer of part a.

Chapter 7   REVIEW

Terminology/Algorithms	Pages	Comments
Matrix, $C = [c_{ij}]_{m \times n}$	232	2-dimensional array of numbers, m rows and n columns
Element, c_{ij}	233	ith-row, jth-column element
Multiplication by scalar, aC	234	Multiply each element of matrix C by a.
Addition, $A + C$	235	Add corresponding elements
+ commutative	235	$A + C = C + A$
+ associative	236	$(A + B) + C = A + (B + C)$

Terminology/Algorithms	Pages	Comments
Zero matrix, $0_{m \times n}$	236	Matrix of all zeros
Additive inverse, $-A$	237	$A + (-A) = 0$
Matrix product, $A_{m \times q} B_{q \times n} = C_{m \times n}$	244	c_{ij} = dot product of ith row of A and jth column of B
· Associative	245	$(A \cdot B) \cdot C = A(B \cdot C)$
Distributive property	245	$A(B + C) = AB + AC$
Square matrix	246	$n \times n$ matrix
Diagonal	246	$\{b_{11}, b_{22}, \ldots, b_{nn}\}$ in square matrix
Identity matrix, I_n	246	$n \times n$, 1's on diagonal, 0's elsewhere $I_m A_{m \times n} = A$, $A I_n = A$
Scaling	246	Multiplication by scalar or by $a I$
Translation	246	Move parallel to axes
Linear equation	253	$\sum_{i=1}^{n} a_i x_i = c$
Augmented matrix	254	$[A_{m \times n} \mid B_{m \times 1}]_{m \times (n+1)}$
Equivalent system	255	Same solution
Row operations	255	Multiplication by scalar; replace with sum of rows; exchange rows
Gaussian elimination	255	Algorithm
Pivot number	256	Eliminate numbers above (below) it
Multiplicative inverse, A^{-1}	261	$A^{-1} A = A A^{-1} = I$
Find A^{-1}	263	Gaussian elimination on $[A \mid I]$

REFERENCES

Agnew, J., and R. C. Knapp, *Linear Algebra with Applications*, Monterey, CA: Brooks/Cole, 1978.

Grotch, S. L. "Three-Dimensional Graphics for Scientific Data Display and Analysis," Lawrence Livermore National Laboratory, University of California, UCRL–90099, Livermore, 1983.

Levine, S. R. *Introduction to Computer Graphics*, Tutorial at SIGGRAPH 85, 1985.

Roberts, F. S. *Applied Combinatorics*, Englewood Cliffs, NJ: Prentice-Hall, 1984.

Tucker, A. B., Jr. *Programming Languages*, 2nd ed. New York: McGraw-Hill, 1986.

8 GRAPH THEORY

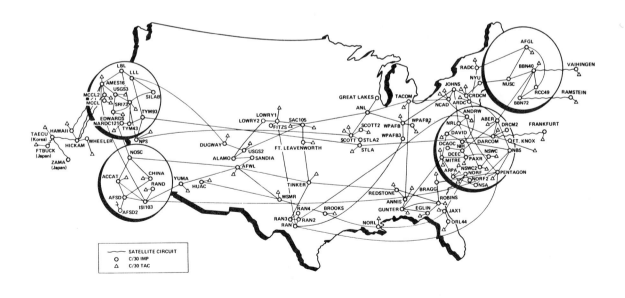

MILNET. This graph illustrates part of the data defense computer network. Communication between any two nodes in the network is possible.

8.1

Graphs

The area of mathematics called graph theory has numerous applications in computer science. In fact, in this book we have already used a tree graph to illustrate the Fundamental Counting Principle in Figure 3.1 of Section 3.1, a directed graph of the partially ordered set $(\mathcal{P}(U), \subseteq)$ in Figure 4.3 of Section 4.2, and bipartite graphs for the functions in Figures 5.1 and 5.2 of Section 5.1. Graphs are useful from two major standpoints—to help us understand and represent structures in computer science and to aid our application of computer science to other fields such as business and the sciences.

By a graph we do not mean the graphs such as that of $f(x) = x^2$ in Figure 4.1 of Section 4.1, but a set of points with arcs connecting some of the points, as illustrated in Figure 8.1.

Figure 8.1

Examples of graphs

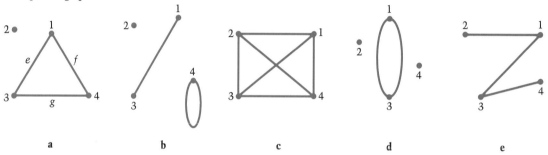

| a | b | c | d | e |

DEFINITION

A **graph** $G = (V, E)$ consists of a set V of **vertices** (singular, **vertex**) or **nodes** or points and a set E of **edges** or arcs connecting pairs of points. Denote an edge e between nodes u and v as (u, v) or as (v, u). (Here, (u, v) is not an ordered pair.)

We will need a number of terms to speak the language of graph theory. In Figure 8.1a points 1 and 3 are adjacent because there is an edge e connecting them. We say that edge e, which can be written as $(1, 3)$ or as $(3, 1)$, is incident to point 1 and to point 3. Nodes 1 and 2 with no joining edge are not adjacent. Moreover, vertex 2 is isolated, having degree 0 or no incident lines, while point 1 has degree 2 or is the endpoint of two edges. In Figure 8.1b, the degree of point 4 is also 2 since this vertex appears twice as an endpoint.

DEFINITION

Two vertices u and v of a graph are **adjacent** if there exists an edge (u, v) connecting them. An edge e is **incident** to vertex v if v is an endpoint of e. The **degree** of a vertex v, $\deg(v)$, is the number of times v is an endpoint of an edge. If $\deg(v) = 0$, then v is called an **isolated point.**

Since each edge is incident to two vertices, an edge (u, v) will contribute 1 to the degree of u and 1 to the degree of v. Thus, as stated in the next theorem, if we add together the degrees of all the vertices, we will end up with a number that is twice the number of edges.

THEOREM

Let $G = (V, E)$ be a graph with set of vertices, $V = \{v_1, v_2, \ldots, v_n\}$. Then

$$\sum_{i=1}^{n} \deg(v_i) = 2 \cdot n(E).$$

This theorem can be used to show that in a graph, an even number of vertices have odd degree. Notice in Figure 8.1a, 0 vertices have odd degree; in Figure 8.1e, two nodes, nodes 2 and 4, have degree 1; and in Figure 8.1c, each of the four nodes has degree 3. We will apply this observation to Euler paths and, in turn, to coding theory in Section 8.4.

THEOREM

In a graph there are an even number of nodes having odd degree.

Proof
Let $G = (V, E)$ be a graph. By the previous theorem, the sum of all the degrees of the nodes is $2 \cdot n(E)$, an even number. Subtracting all of the even degrees, we still have an even number, $2d$ (even − even = even). Thus, the resulting sum of all the odd degrees is $2d$. There must be an even number of these odd degrees for the result to be $2d$, an even number (odd + odd = even; odd + even = odd). Thus, there are an even number of nodes with odd degree.

Graphs can be used as models in numerous applications, from computer networks to projecting time tables; finding optimum paths through such graphs can be essential to solving many problems. Within graph a of Figure 8.1 we have a cycle or path that goes from point 1 along e to point 3, on g to point 4, back to 1 along edge f: 1, e, 3, g, 4, f, 1. We can, of course, think of this path, which is three edges in length, as starting with node 3 or with 4. Graph 8.1c has a number of cycles including the cycle of length 4 from node 1 to 2 to 3 to 4 to 1. The path with vertices 1, 2, 3, 4, 2, 1, is not a cycle because an edge and a vertex other than 1 are repeated. We do, however, have a circuit, or a path that ends where it began. Every cycle is a circuit, but not vice versa. The edge circling back to vertex 4 in graph 8.1b is called a loop.

DEFINITION

> In a graph G, a **path** from vertex v_0 to v_n along edges e_0 to e_{n-1} is the sequence
>
> $$v_0, e_0, v_1, e_1, \ldots, v_{n-1}, e_{n-1}, v_n$$
>
> where $e_i = (v_i, v_{i+1})$ for $i = 0, 1, \ldots, n - 1$. A path of n edges is said to be of **length** n. The path is a **circuit** if $v_0 = v_n$. The path is a **cycle** if $v_0 = v_n$ and no edge or vertex, except v_0, is repeated. Edge e is a **loop** if both endpoints are the same vertex.

Notice from the definition that a path with $n + 1$ distinct vertices, $v_0, v_1, \ldots, v_{n-1}, v_n$, has n distinct edges, $e_0, e_1, \ldots, e_{n-1}$. In a cycle, however, though all of the edges are distinct, the first vertex is repeated at the end. Thus, a cycle of length n

$$v_0, e_0, v_1, e_1, \ldots, v_{n-1}, e_{n-1}, v_n = v_0$$

has n edges, $e_0, e_1, \ldots, e_{n-1}$, and n vertices, $v_0 = v_n, v_1, \ldots, v_{n-1}$. As you will see in the algorithm of Exercise 13, in a path from a node u to a node v, by eliminating circuits you can always find a path from u to v of length less than the number of vertices in the graph.

Graphs display various properties. For instance, graphs c and e of Figure 8.1 are connected because there is a way to get from any point to any other point following the edges. Only graph 8.1c is also complete, with every point connected to every other point directly by an edge. Graph 8.1b, containing a loop, and Graph 8.1d with two edges joining nodes 1 and 3 are not simple. Suppose each of the graphs in Figure 8.1 illustrates four interconnected computers. There is a way of relaying information from a computer to any other one in the connected graphs 8.1c and 8.1e. In the complete graph 8.1c all computers can communicate directly with one another. Graph 8.1d provides a backup line for communication between two of the computers.

DEFINITION

A graph is **connected** if there exists a path from any vertex to any other vertex.

DEFINITION

A graph that does not have a loop or two edges joining the same points is **simple**.

DEFINITION

A graph is **complete** if it is simple and if each point is adjacent to every other point.

Notice in graph c of Figure 8.1 that each of the four points has three incident edges connecting that node to the remaining nodes. In other words, each node has degree 3. Therefore, the sum of the degrees of all the points is

$$3 + 3 + 3 + 3 = 4 \cdot 3 = 12.$$

Since this value is twice the number of edges, as we can observe, $n(E)$ is 6. The next theorem generalizes our observation.

THEOREM

Let $G = (V, E)$ be a complete graph with n vertices. The number of edges in G is

$$n(E) = \frac{n(n-1)}{2}.$$

Proof

Since G is complete, each vertex is adjacent to the remaining vertices. Thus, the degree of each of the n vertices is $n - 1$, and we have the sum of the degrees of all of the vertices being $n(n - 1)$. By the first theorem in this section, this sum is equal to twice the number of edges in G:

$$n(n - 1) = 2n(E).$$

Therefore, we have

$$n(E) = \frac{n(n-1)}{2}.$$

Several types of graphs are of particular importance in computer science. Trees, such as graph a of Figure 8.2 which can look much like an upside-down version of a live tree, are of such significance that we will consider them in a separate section. These graphs, as well as being connected, contain no cycles.

Figure 8.2

Examples of graphs

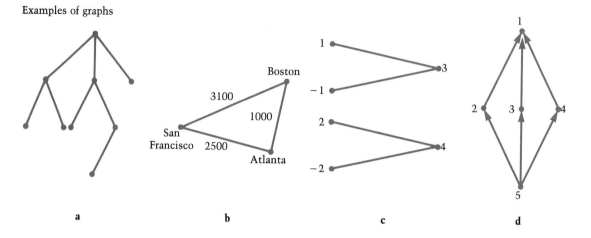

a   b   c   d

DEFINITION

A graph is **acyclic** if it contains no cycles.

DEFINITION

A **tree** is a connected, acyclic graph.

Many of the illustrations of graphs in previous chapters employed directed graphs where the edges were arrows as in Figure 8.2d. In such graphs we can expand the concept of degree. There are three edges coming out of vertex 5, so its out-degree is 3; but observe that its in-degree is 0.

DEFINITION

A **directed graph** or **digraph** $G = (V, E)$ consists of a set of vertices V and a set of directed arcs or edges E. Edge e can be written as an ordered pair (u, v) where u is the **initial point** and v is the **terminal point** of the edge. The **in-degree** of u is the number of edges directed into u while the **out-degree** is the number of edges pointing outward from u.

Network graphs attach a number to each edge. Perhaps these numbers represent distances between cities on a schematic map as in Figure 8.2b. Or perhaps they represent bit values (0 or 1) on the lines of a combinational circuit like those to be considered in Chapter 9. Applications involving paths through networks will also be discussed in Section 8.4.

DEFINITION

A **network** is a graph or digraph where edges are labeled with numbers which are not necessarily distinct.

Bipartite graphs like graph 8.2c have their vertices split into two sets with edges only between vertices in different sets. This graph could be used to picture the function {(1, 3), (−1, 3), (2, 4), (−2, 4)} from a domain of {1, −1, 2, −2} to a range of {3, 4}. As illustrated in Exercise 14, a bipartite graph can also serve as a model of the allocation of computer resources.

DEFINITION

A **bipartite graph** is a graph or digraph with vertices partitioned into two sets, V_1 and V_2, where arcs are only between vertices in different sets.

EXERCISES 8.1

1. Answer the following questions about Figure 8.3:
   a. How many arcs are incident to node 1?
   b. How many points are adjacent to node 1?
   c. Give the degree of node 1.
   d. Repeat parts a, b, and c for node 3.
   e. Find a path of length 1 from point 1 to 4.
   f. Find a path of length 2 from point 1 to 4.
   g. Find two paths of length 3 from vertex 2 to vertex 6.

Figure 8.3

Graph for Exercise 1

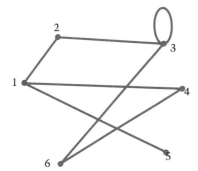

Figure 8.4

Graphs for Exercise 2

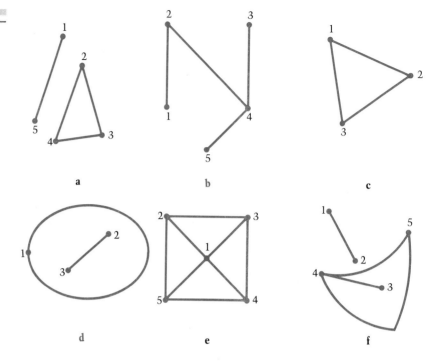

2. For each graph in Figure 8.4, indicate which terms apply: (i) connected, (ii) complete, (iii) loop-free, (iv) acyclic, (v) simple, (vi) tree.

3. How many edges are in a cycle with 35 distinct points?

4. Give all the cycles in Figure 8.1c.

5. Draw an undirected tree with
   a. Three points.
   b. Four points where the degree of each point is less than or equal to 2.
   c. Four points where one point has degree 3.
   d. Five points where the degree of each point is at most 2.
   e. Five points where one point has degree 3.
   f. Five points where one point has degree 4.

6. a. How would you partition the points of Figure 8.4b to verify that it is a bipartite graph?
   b. Is Figure 8.4a bipartite?   c. Is Figure 8.4d?

7. a. Draw a complete graph with two points.
   b. Using part a, add another point and make the new graph complete.
   c. Using part b, add a fourth point and make the new graph complete.
   d. Using part c, add a fifth point and make the new graph complete.

8. a. In a complete graph with n points, how many edges are incident to any point? (Refer to your answers in Exercise 7 for several examples.)
   b. Using sigma notation, give the number of edges in a complete graph with n points.

    c. Using a formula from Section 6.2, write your answer without sigma notation. *Note:* Notice how your answer agrees with the number given in the theorem concerning complete graphs in this section.

    d. Prove your answer to part b using mathematical induction. *Note:* You are giving an alternative proof to the one found in this section.

9. Let $G = (V, E)$ be a digraph. Express the sum of the out-degrees of the vertices in terms of $n(E)$.

10. a. How many edges are in a complete digraph that has edges going from every point to every other point?

    b. Prove your answer to part a.

11. In a connected graph an **articulation point** is a node whose removal, along with the elimination of all incident edges, will cause the graph to become disconnected. Find the articulations points, if any, in

    a. Figure 8.1e.

    b. Figure 8.1c.

    c. Figure 8.4b. *Note:* Such a point in a computer network might represent a computer that is essential for communications among all nodes of the system.

12. Let G be a simple graph with five nodes, labeled v_1, v_2, v_3, v_4, v_5.

    a. Find the maximum number of edges m that G can have. Using combinations, find how many such graphs G there can be with

    b. 0 edges   **c.** 1 edge   **d.** 2 edges

    e. 3 edges   **f.** m edges

    g. Using sigma and combination notation, give the number of simple graphs with five labeled nodes.

13. Fill in the justification for the following theorem: In a graph or digraph with n vertices, if there exists a path from point v_0 to point v_k with $v_0 \neq v_k$, then there exists a path of length less than n from v_0 to v_k.

Proof   Let $v_0, v_1, \ldots, v_k$ be the points of a path from v_0 to v_k of length k. If $k \geq n$, then at least one point v_i must appear more than once in the list (**a.** Why?) Cut out the section of path after v_i to the next appearance of v_i to form a shorter path. (**b.** Why is this possible?) Keep repeating the process until a path of length less than n is found.

    c. For the digraph in Figure 8.5 and the path of length 8 from point 1 to point 3 with points 1, 4, 2, 1, 4, 1, 1, 4, 3, follow the proof of the theorem to find a path of length less than or equal to 3.

Figure 8.5

Digraph for Exercise 13

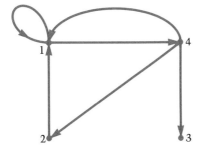

Figure 8.6

Graph for Exercise 14

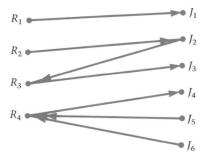

14. Consider a computer center with various resources such as several computers, printers, terminals, card readers, and card punches, and several jobs running on the system. A graph of the system at a particular moment might be as in Figure 8.6. An edge (R_i, J_k) indicates resource R_i has been allocated to job J_k, while edge (J_k, R_i) means job J_k is awaiting that resource.
   a. What kind of graph is this?
   b. What is the maximum reasonable out-degree for any node?
   c. Drawing one additional arrow will deadlock a subset of this system; i.e., all jobs in the subset will be waiting for resources that have been allocated to other jobs in the subset. What is that edge?

8.2

Representations

For a graph to be manipulated by your program, this structure must be stored in some convenient manner. Often, we use one- or two-dimensional arrays to represent graphs or digraphs. Such arrays, depending on the representation, can indicate such information as adjacent nodes, directions of edges in a digraph, data stored at nodes or along edges, and existence of paths between nodes.

A **linked list** is a data structure that can be thought of as a digraph; information is stored at each node and a pointer to the next node in the list is indicated by a directed arc. In the linked list illustrated in Figure 8.7a, the information stored at the head node is

Figure 8.7

A linked list and a doubly linked list

a

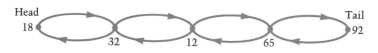

b

18. An arrow points to the next node, which contains the number 32, and another arrow links to the next node, and so forth. Ignoring direction of the edges, a linked list is a connected, simple graph with no cycles and with each node having an out-degree of 0 or 1. Taking into account direction, however, there is no path following the direction of the arrows from the tail to the head or to any other node. A **doubly linked list**, such as that in Figure 8.7b, has pointers from each node to the next and previous nodes. As an undirected graph it is connected with no loops, but it is not simple. Even with direction considered in graph 8.7b it is possible to find a path from any point to any other point in the list.

Figure 8.8

Three 5 × 1 arrays holding the information and pointers of the doubly linked list in Figure 8.7b

HEAD	INDEX	PREVIOUS	VALUE	NEXT
2	1	5	92	0
	2	0	18	4
	3	4	12	5
	4	2	32	3
	5	3	65	1

LANGUAGE EXAMPLE

Graphs can be stored and manipulated in the computer as arrays. As illustrated in Figure 8.8, a 5 × 3 two-dimensional array or three 5-element one-dimentional arrays can hold the doubly linked list of Figure 8.7b. Elements in the middle column contain the value of the node while elements in the first and third columns point to the previous and next node, respectively. Originally, the values are entered into the array in order, the first value in the first element of the middle column, second value in the second element, and so forth; but with additions and deletions this arrangement is lost and the references PREVIOUS and NEXT become essential. A variable HEAD has as its value the row of the first element in the linked list. Referring to Figure 8.8, we see that HEAD with a value of 2 points to the second row as the place to start on our list. The 0 in PREVIOUS[2], the first column element of row 2, also indicates that the middle column holds the first value in the list, 18. The last number in the row, 4, points to the fourth row for the next node value of 32. Since NEXT[4] = 3, the third row contains the information for the next node. The value contained there is 12, and NEXT[3] = 5 indicates the row with index 5 should subsequently be considered. We find VALUE[5] = 65 and NEXT[5] = 1. Since NEXT[1] = 0, we discover that row 1 holds information for our tail node. Wherever we are in the doubly linked list, we can always find the next node or retrace our steps to the previous node. For example, the fourth row indicates a pointer to the previous node, on row 2, and to the next node, on row 3. Deleting this PREVIOUS array from the figure, we have a model of the linked list in Figure 8.7a. ☐

Figure 8.9

Graph with adjacency and
connection matrices

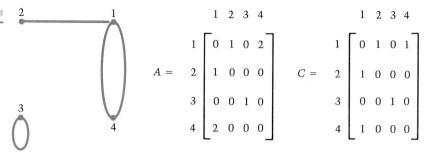

$$A = \begin{array}{c c} & \begin{array}{c c c c} 1 & 2 & 3 & 4 \end{array} \\ \begin{array}{c} 1 \\ 2 \\ 3 \\ 4 \end{array} & \left[\begin{array}{c c c c} 0 & 1 & 0 & 2 \\ 1 & 0 & 0 & 0 \\ 0 & 0 & 1 & 0 \\ 2 & 0 & 0 & 0 \end{array} \right] \end{array}$$

$$C = \begin{array}{c c} & \begin{array}{c c c c} 1 & 2 & 3 & 4 \end{array} \\ \begin{array}{c} 1 \\ 2 \\ 3 \\ 4 \end{array} & \left[\begin{array}{c c c c} 0 & 1 & 0 & 1 \\ 1 & 0 & 0 & 0 \\ 0 & 0 & 1 & 0 \\ 1 & 0 & 0 & 0 \end{array} \right] \end{array}$$

Adjacency and connection matrices can store graphs or digraphs where we are not interested in the values at the nodes but only in the graph itself. Take for example the graph and the digraph in Figures 8.9 and 8.10, respectively.

In the **adjacency matrix** the ij element indicates the number of edges between node i and node j. For instance, since there are two edges connecting points 1 and 4 in Figure 8.9, elements a_{14} and a_{41} are both 2. Note that the adjacency matrix for an undirected graph is symmetric about the diagonal while one for a directed graph usually is not (see Exercise 27, Section 7.1). Notice in Figure 8.10 that there are three arrows from node 4 to points 1, 2, and 3, resulting in 1's in elements b_{41}, b_{42}, and b_{43}. There is, however, only one arrow to node 4, causing b_{14} to be 1.

Figure 8.10

Digraph and its adjacency
matrix, which is also its
connection matrix

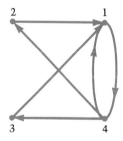

$$B = [b_{ij}] = \begin{array}{c c} & \begin{array}{c c c c} 1 & 2 & 3 & 4 \end{array} \\ \begin{array}{c} 1 \\ 2 \\ 3 \\ 4 \end{array} & \left[\begin{array}{c c c c} 0 & 0 & 0 & 1 \\ 1 & 0 & 0 & 0 \\ 1 & 0 & 0 & 0 \\ 1 & 1 & 1 & 0 \end{array} \right] \end{array}$$

In matrix B of Figure 8.10, b_{ij} gives the number of arcs or paths of length 1 from point i to point j. Interestingly, the ij element of B^2 is the number of paths of exactly length 2 from i to j:

$$B^2 = \begin{bmatrix} 0 & 0 & 0 & 1 \\ 1 & 0 & 0 & 0 \\ 1 & 0 & 0 & 0 \\ 1 & 1 & 1 & 0 \end{bmatrix} \begin{bmatrix} 0 & 0 & 0 & 1 \\ 1 & 0 & 0 & 0 \\ 1 & 0 & 0 & 0 \\ 1 & 1 & 1 & 0 \end{bmatrix} = \begin{bmatrix} 1 & 1 & 1 & 0 \\ 0 & 0 & 0 & 1 \\ 0 & 0 & 0 & 1 \\ 2 & 0 & 0 & 1 \end{bmatrix} .$$

Observe in the graph that there are paths of length 2 from point 1 to 1, from 1 to 2, from 1 to 3, from 2 to 4, from 3 to 4, and from 4 to 4. There are two ways to get from node 4 to node 1 in two steps. But why does squaring give the number of paths of length 2? To illustrate, how can you get from point 4 to point 1? By looking at row 4 we see there is one way each to get from node 4 to nodes 1, 2, and 3. Now considering column 1, there are three ways to get to point 1—from nodes 2, 3, and 4. Thus, there are two paths from 4 to 1, one through point 2 and one through point 3. The dot product of the fourth row and first column gives the 4, 1 element of B^2 and

$$[1 \ 1 \ 1 \ 0] \begin{bmatrix} 0 \\ 1 \\ 1 \\ 1 \end{bmatrix} = \overset{4 \to 1}{1 \cdot 0} + \overset{4 \to 2}{1 \cdot 1} + \overset{4 \to 3}{1 \cdot 1} + \overset{4 \to 4}{0 \cdot 1} = 2.$$

$$\underset{1 \to 1 \qquad 2 \to 1 \qquad 3 \to 1 \qquad 4 \to 1}{}$$

In general recall that the ij element of B^2 is the dot product of the ith row of B, which contains the number of ways from point i to every other point, and the jth column with the number of paths to point j.

To obtain the number of paths of length 2 or less we want the number of paths of length 1 or 2. For each ij position we have a disjoint-or situation and thus should add:

$$B + B^2 = \begin{bmatrix} 0 & 0 & 0 & 1 \\ 1 & 0 & 0 & 0 \\ 1 & 0 & 0 & 0 \\ 1 & 1 & 1 & 0 \end{bmatrix} + \begin{bmatrix} 1 & 1 & 1 & 0 \\ 0 & 0 & 0 & 1 \\ 0 & 0 & 0 & 1 \\ 2 & 0 & 0 & 1 \end{bmatrix} = \begin{bmatrix} 1 & 1 & 1 & 1 \\ 1 & 0 & 0 & 1 \\ 1 & 0 & 0 & 1 \\ 3 & 1 & 1 & 1 \end{bmatrix}.$$

To generalize, the ij element of B^n gives the number of paths of exactly length n from vertex i to vertex j while the ij element of

$$\sum_{k=1}^{n} B^k$$

gives the number of paths of length less than or equal to n.

The **connection matrix** only indicates existence of an edge from one point to another, not the number of such edges. Matrix C in Figure 8.9 as well as B in Figure 8.10 are connection matrices. If we are not interested in quantities but in existence, we perform addition and multiplication in the two-element Boolean algebra [$\mathbb{B}$, +, ·, ', 0, 1]. In this algebra, which is discussed in detail in Chapter 9, addition and multiplication is evaluated as in the real number system with one exception; we define $1 + 1$ as 1. Remember, we are not counting edges between each pair of points, but only seeing if at least one exists. C^2 indicates whether there is a path of length 2 from point i to point j, for $i,j = 1, 2, 3, 4$:

$$C^2 = \begin{bmatrix} 0 & 1 & 0 & 1 \\ 1 & 0 & 0 & 0 \\ 0 & 0 & 1 & 0 \\ 1 & 0 & 0 & 0 \end{bmatrix} \begin{bmatrix} 0 & 1 & 0 & 1 \\ 1 & 0 & 0 & 0 \\ 0 & 0 & 1 & 0 \\ 1 & 0 & 0 & 0 \end{bmatrix} = \begin{bmatrix} 1 & 0 & 0 & 0 \\ 0 & 1 & 0 & 1 \\ 0 & 0 & 1 & 0 \\ 0 & 1 & 0 & 1 \end{bmatrix}.$$

Notice that dotting the first row and first column gives us $0 \cdot 0 + 1 \cdot 1 + 0 \cdot 0 + 1 \cdot 1 = 1 + 1 = 1$. Yes, there does exist a path of length 2 from point 1 to itself. One way is through point 2, another through point 4. Adding C and C^2 in $[\mathbb{B}, +, \cdot, ', 0, 1]$, we discover whether paths of length 1 or 2 exist between points:

$$C + C^2 = \begin{bmatrix} 0 & 1 & 0 & 1 \\ 1 & 0 & 0 & 0 \\ 0 & 0 & 1 & 0 \\ 1 & 0 & 0 & 0 \end{bmatrix} + \begin{bmatrix} 1 & 0 & 0 & 0 \\ 0 & 1 & 0 & 1 \\ 0 & 0 & 1 & 0 \\ 0 & 1 & 0 & 1 \end{bmatrix} = \begin{bmatrix} 1 & 1 & 0 & 1 \\ 1 & 1 & 0 & 1 \\ 0 & 0 & 1 & 0 \\ 1 & 1 & 0 & 1 \end{bmatrix}.$$

Notice the isolation of point 3. The zeros in the third row and column indicate there are no paths of length 1 or 2 between node 3 and the other nodes.

Matrices can also be used to store networks. We must represent the values associated with the edges as well as incidency. In the network matrix, the ij element indicates the value associated with the edge between node i and node j. We write ∞ where there is no corresponding edge and place 0 in each diagonal element. The numbers might indicate distances or travel times between locations. Or perhaps the values stand for the time needed to execute various steps in a project. The network for Figure 8.11 follows.

Figure 8.11

A network

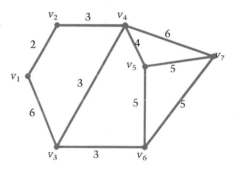

$$D = \begin{array}{c@{\;}c} & \begin{array}{ccccccc} 1 & 2 & 3 & 4 & 5 & 6 & 7 \end{array} \\ \begin{array}{c} 1 \\ 2 \\ 3 \\ 4 \\ 5 \\ 6 \\ 7 \end{array} & \left[\begin{array}{ccccccc} 0 & 2 & 6 & \infty & \infty & \infty & \infty \\ 2 & 0 & \infty & 3 & \infty & \infty & \infty \\ 6 & \infty & 0 & 3 & \infty & 3 & \infty \\ \infty & 3 & 3 & 0 & 4 & \infty & 6 \\ \infty & \infty & \infty & 4 & 0 & 5 & 5 \\ \infty & \infty & 3 & \infty & 5 & 0 & 5 \\ \infty & \infty & \infty & 6 & 5 & 5 & 0 \end{array} \right] \end{array}$$

EXERCISES 8.2

LANGUAGE
EXERCISES

1. Fill in the NEXT array along with the HEAD to store the linked list in Figure 8.12.

2. Fill in the PREVIOUS, VALUE, and NEXT arrays along with the HEAD to store the doubly linked list in Figure 8.13. □

For Exercises 3–5 give the adjacency matrix for each of the graphs from Figure 8.4 of Section 8.1.

3. Graph 8.4a.   4. Graph 8.4d.   5. Graph 8.4f.

Figure 8.12

Linked list for Exercise 1

Head   Tail

10 ──────▶ -4 ──────▶ 9 ──────▶ 6

HEAD	INDEX	VALUE	NEXT
——	1	-6	
	2	9	
	3	10	
	4	-4	

Figure 8.13

Doubly linked list for
Exercise 2

Head   Tail

10 ⟷ -4 ⟷ 9 ⟷ -6

HEAD	INDEX	PREVIOUS	VALUE	NEXT
——	1		-6	
	2		9	
	3		10	
	4		-4	

6. **a.** Give the adjacency matrix of the graph in Figure 8.14a.
   b. By matrix multiplication find the number of paths of length 2 between each pair of points. Verify your answers by observing the graph.
   c. Find the matrix of paths of length 1 or 2.
   d. Find the matrix of paths of length 3.
   e. Find the matrix of paths of length at most 3.

7. Repeat Exercise 6 for the graph in Figure 8.14b.

8. Give the connection matrix for the graph in Figure 8.14a.

9. Give the connection matrix for the graph in Figure 8.14b.

Figure 8.14

Graphs for Exercises 6–9

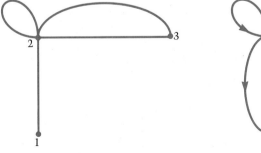
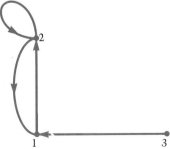

a   b

10. **a.** Give the connection matrix C of the graph in Figure 8.15.
   b. Find m such that

$$S = \sum_{i=1}^{m} C^i$$

has no zeros where arithmetic is in $[\mathbb{B}, +, \cdot, ', 0, 1]$. The resulting matrix illustrates that there exists a path of length 1, 2, . . . , or m from point v_i to point v_j for $i,j = 1, 2, 3, 4$.
   c. Find $C^1, C^2, \ldots, C^m$ and the sum S for this graph.

Figure 8.15

Graph for Exercise 10

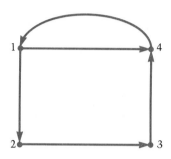

11. Recall the following theorem from Exercise 13 of Section 8.1: In a graph or digraph with n vertices, if there exists a path from point v_0 to point v_k with $v_0 \neq v_k$, then there exists a path of length less than n from v_0 to v_k. Fill in the blank in the following statement: Because of this theorem, for a digraph of n points and a connection matrix, C, we need to look only at C^m for m at most _____ to verify that there is a path from vertex v_i to v_j, $i \neq j$.

12. Consider a graph with n points and connection matrix C. The **reachability matrix** of the graph is a matrix

$$M = \sum_{i=0}^{n-1} C^i$$

where $C^0 = I_n$, the identity matrix, and where arithmetic is performed in $[\mathbb{B}, +, \cdot, ', 0, 1]$. Point v_j can be reached from point v_i if there exists a 1 in the ij position of M.
What is the reachability matrix
   a. Of Figure 8.15?
   b. Of the graph in Figure 8.9?
   c. Of Figure 8.14b?
   d. Suppose each of these graphs represents the routes of an airline. Is any city in these graphs not reachable from any of the other cities?

13. **a.** Give the network matrix D for the graph in Figure 8.16.
   b. For $i,j = 1, 2, 3, 4$, replace each d_{ij} with the smallest possible value for $d_{ik} + d_{kj}$, $k = 1, 2, 3, 4$. Assume ∞ is greater than any number. For example, d_{14} would be replaced by $d_{13} + d_{34} = 7 + 8 = 15$, the length of the shortest path from v_1 to v_4 is 15. The ij element of the resulting matrix is the minimum "distance" between vertices v_i and v_j, for $i,j = 1, 2, 3, 4$. *Note:* This approach for finding the length of the

Figure 8.16

Graph for Exercise 13

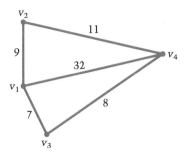

shortest path from v_i to v_j in a network, however, does not in general give us the path. The organized method for finding minimum "distances," called Floyd's algorithm, is easy to implement on the computer:

```
FOR K := 1 TO N DO
    FOR I := 1 TO N DO
        FOR J := 1 TO N DO
            IF (D[I,K] + D[K,J]) < D[I,J] THEN
                D[I,J] := D[I,K] + D[K,J];    □
```

14. Repeat Exercise 13, part b, for the network matrix associated with Figure 8.11.

8.3

Trees

The tree is probably the most important type of graph used in computer science. We have already used it to list the reports needed for each class by dorm (Chapter 3), to present an algorithm in hierarchical form (Chapter 2), and to enumerate the possibilities for statements in a truth table (Chapter 2). Trees are used extensively in a variety of areas including compiler design, searching and sorting methods, and artificial intelligence.

Since a tree is connected with no cycles, as an undirected graph there is always a unique path between any two vertices. Often one vertex is specially designated as the root, and all line segments are thought of as directed away from the root. Thus, the root is the only vertex with no arrows coming into it.

DEFINITION

A **rooted tree** is a connected digraph G with no cycles and with a unique vertex r, called the **root**, which has zero indegree.

Figure 8.17

Example of a tree drawn
with and without directed
arcs

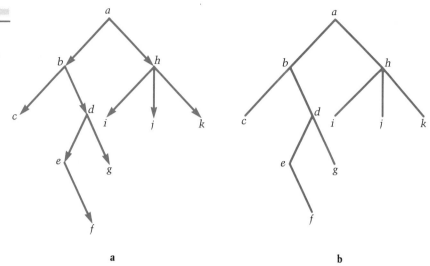

a   b

Notice that the tree in Figure 8.17 has 11 vertices and 10 edges. In fact, the number of edges is always one less than the number of vertices. The root has an in-degree of 0, while each of the other vertices has an in-degree of 1; that is, each of the n vertices, except the root, has exactly one edge coming into it.

THEOREM

A tree with n vertices has $n - 1$ edges.

Throughout this section we will be considering rooted trees and for brevity will call them trees. Often the root is drawn at the top of the diagram, and all arrows are pointing downward. When there is no ambiguity, arrows will be omitted from the sketches as in Figure 8.17b.

In that figure, directed line segments are understood from node b to d and from b to c. We say that b is the parent of c and d while c and d are the children of b. All the nodes adjacent to but below b, namely c, d, e, f, and g, are the descendants of b. The nodes b and a, which are above d along a path from the root to d, are the ancestors of d. Notice the similarity in terms to those used in describing a family tree. Considering b with some of its descendants, such as c, d, and e, and edges to them, we have a subtree with root b. Those nodes at the bottom that have no arrows going away from them (c, f, g, i, j, and k) are descriptively called leaves. The level of a node x is the number of edges that are on the path from the root to x. Thus, the root is at level 0. Two nodes, b and h, are at level 1; c, d, i, j, and k are at level 2; and node f is at level 4. With 4 being the greatest level, we say this tree has height 4.

DEFINITION

In a (rooted) tree if there is an edge from node u to node v, then say that u is the **parent** of v, and v is the **child** of u. All those nodes that can be reached by a path from node u are the **descendants** of u. The **ancestors** of v are those nodes found on the path from the root to v. Node v is a **leaf** node if its out-degree is zero.

DEFINITION

The **level** of node u is the length of the path from the root to u. The **height** of a tree is the maximum level.

DEFINITION

S is a **subgraph** of graph G if S is itself a graph and every node and edge of S is in G.

DEFINITION

S is a **subtree** of tree T if S is a tree which is a subgraph of T.

EXAMPLE 1

Figure 8.18

Tree for Example 1

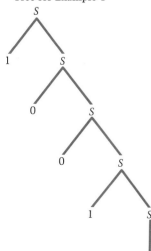

In Example 3 of Section 6.3 we considered a grammar with vocabulary $V = \{0, 1, S\}$, terminals $V_T = \{0, 1\}$, and productions

$$S \to 0S \qquad S \to 1S \qquad S \to 0 \qquad S \to 1.$$

We verified that 10011 is a word in the resulting language as follows:

$$S \to 1S \to 10S \to 100S \to 1001S \to 10011.$$

Verification of membership in the language can also be pictured with a tree as in Figure 8.18. Interior nodes are nonterminals, and leaves are the terminals that form the word. The children of a nonleaf node, read left-to-right, give the right side of the applied production. For instance, since we have $S \to 1S$ at the first step of the verification, the root contains S and its children are 1 and S. At the next step we have $S \to 0S$, so the nodes at level 2 contain 0 and S. Because the final production used is $S \to 1$, level 5 has only one leaf node. This graph is called a **parse tree** and the process of verifying membership in the language, a major task of the compiler, is called **parsing**.

The graph in Figure 8.18 is also an example of a binary tree, where every node has none, one, or two children. A parse tree does not have to be binary, but applications often result in these 2-ary trees. Various theorems covering properties of binary trees are presented in the exercises (see Exercises 21, 24, 25, and 26). ▮

> **DEFINITION** ▨▨▨▨▨▨▨▨▨▨▨▨▨▨▨▨▨▨▨▨▨▨▨▨▨▨▨▨▨▨▨▨
>
> A tree is a **binary tree** if every node has at most two children.

Using a binary tree to alphabetize a list of words is an efficient method of sorting. The first word in the list becomes the root. Then each new word W is sent to the left if it falls before the root alphabetically, or to the right if it falls after the root word. If W encounters another word, the same type of comparison is made to decide about left or right. Take for example the sentence

PLEASE HELP ME ALPHABETIZE THIS SENTENCE.

PLEASE is placed at the root as in Figure 8.19a. Since HELP < PLEASE alphabetically, a left branch is generated with HELP at the resulting node. ME is first compared to PLEASE and is sent to the

Figure 8.19 ▨▨▨▨▨▨▨▨▨▨

A sequence of trees generated as each new word is added to the alphabetizing tree

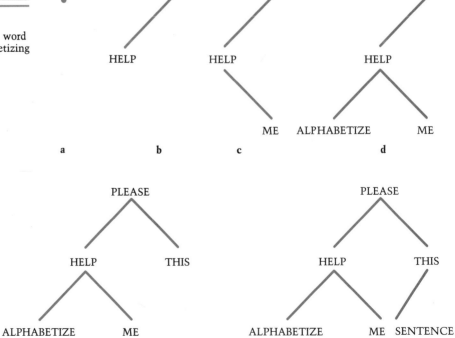

left. But HELP is encountered, so the process is repeated. Since ME > HELP, a right branch is sprouted. The process of adding branches to the tree as each word is added is illustrated in Figure 8.19.

Recursive Algorithm for Alpha (W, R), a Routine to Place Word W into a Tree with Root R for Alphabetizing

```
ALPHA(W, R):
     If the tree is empty (no word at the root), then
          R ← W and return.
     If W < R, then
          go along the left branch to node N
     else if W > R, then
          go along the right branch to node N
     else
          return.
     If node N is found, then
          ALPHA (W, N)
     else
          place a node containing W on end of branch.
```

We have the tree, but still the sentence is not alphabetized. To accomplish this task we travel through the tree in what is called inorder traversal. We go to the left as far as we can and write down the word we find, a process that is called **visiting the node.** Then we visit the parent. If this parent has a right subtree, repeat the process for the subtree. Continue this procedure for the uncovered part of the tree. The dotted line in Figure 8.20 along with the numbers indicates the inorder tree traversal and the order in which the nodes are visited to give the alphabetized list: ALPHABETIZE, HELP, ME, PLEASE, SENTENCE, THIS.

Figure 8.20

Inorder tree traversal indicated on tree

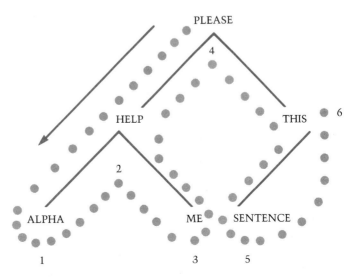

Recursive Algorithm, Inorder (R), for Inorder Traversal of Tree with Root R

> INORDER(R):
>   If the tree is empty, then
>   return.
>   If there is a node to the left of R, called LEFT (R), then
>   INORDER(LEFT(R)).
>   Visit node R.
>   If there is a node to the right of R, called RIGHT(R),
>   then INORDER(RIGHT(R)).

There are other sequences in which we could traverse the tree. For example, to traverse the tree in postorder, visiting of a node is accomplished after the visiting of all nodes on the left and right subtrees. Such a traversal is useful in converting an expression from infix to postfix notation as we did in Section 1.4.

EXAMPLE 2

Place the expression $a * b/(c + d * e)$ in a binary tree by taking each operation in reverse order of priority as a node of a subtree, splitting the expression into a left and a right part. Traverse the resulting tree in postorder to generate the expression in postfix notation.

Solution

First, write the expression with full parentheses:

$$((a * b)/(c + (d * e)))$$

Observe that / operates last on the subexpressions $(a * b)$ and $(c + (d * e))$. In $(c + (d * e))$, + operates last on c and $(d * e)$. The steps for generating the tree are given in Figure 8.21. Notice that operations end up on internal nodes while leaves contain the variables. The postorder traversal, indicated in Figure 8.22, yields the expression in postfix notation: $a\ b * c\ d\ e * +/.$ ∎

Figure 8.21

Steps to generate the parse tree for $a * b/(c + d * e)$

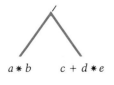

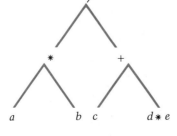

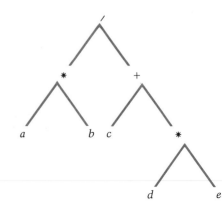

Figure 8.22

Postorder traversal of tree

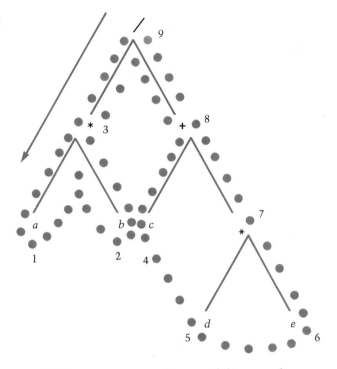

Solid computer graphics modeling can be accomplished using the same tree structure. As illustrated in Figure 8.23, primitive solids, such as blocks, cylinders, and cones, are combined using set operations to form more complex shapes. Notice how, once more, the internal nodes contain the operations while the leaves hold the operands. A subtree of a previously designed part can easily be attached using a set operation to become a component in another shape. Thus, the tree structure closely models the actual manufacturing process.

Figure 8.23

Solid geometry using a tree. Reprinted with permission from "The Build Group of Solid Modelers," Robin Hillyard, *IEEE Computer Graphics and Applications*, Vol. 2, No. 2, March 1982 (Copyright 1982 IEEE).

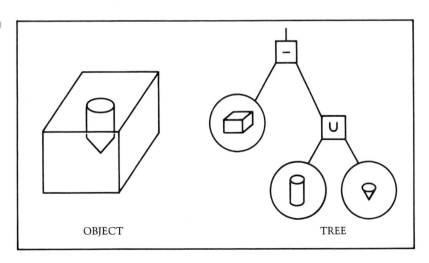

EXERCISES 8.3

1. In Figure 8.20, find
   a. The children of HELP.   **b.** The parent of ME.
   c. The ancestors of HELP.   **d.** The ancestors of ME.
   e. The descendants of THIS.   **f.** The level of THIS.
   g. The leaf nodes.

2. Fill in the reasons in the proof of the following theorem:
An undirected tree T with more than one vertex has more than one vertex of degree 1.

Proof   Pick a path P through T that has a maximum number of edges. P does not contain a cycle. (**a.** Why?) Thus, we may assume the path P goes from a vertex u to a different vertex v. If $\deg(u) > 1$ or $\deg(v) > 1$, then we could add another edge to P to form a new, longer path. But we then have a contradiction (**b.** Why?), so $\deg(u) = 1 = \deg(v)$.

3. How many edges are in a tree with
   a. 1 node?   **b.** 2 nodes?   **c.** 3?   **d.** 4?   **e.** 5?   **f.** n?
   g. Using the theorem in Exercise 2 and mathematical induction, prove the following theorem:

   A tree with n vertices has $n - 1$ edges.

   Note: From Exercise 2 you need only the fact that in a tree with more than one vertex, there is one vertex of degree 1.

4. Using the grammar of Example 1, parse the word 0010.

5. Using the productions
$$S \to S0 \qquad S \to 1S \qquad S \to 1$$
of a grammar, draw the parse tree for 11000.

6. Using the productions
$$S \to S0S \qquad S \to 1S1 \qquad S \to 0$$
of a grammar, draw the parse tree for   **a.** 10100.   **b.** 1110111.

7. Consider the grammar with vocabulary $V = \{0, 1, S, A\}$, terminals $V_T = \{0, 1\}$, start symbol S and productions
$$S \to 0SA \qquad A \to A1 \qquad S \to 0 \qquad A \to 1$$
   a. Draw a parse tree for 001.
   b. Draw two parse trees for 000111.

8. Consider the following simplified structure of an English sentence. (Recall that | means "or.")

   sentence → subject predicate
   predicate → verb direct-object
   subject → noun | adjective-phrase noun
   direct-object → noun | adjective-phrase noun
   adjective-phrase → adjective | particle adjective
   particle → A | AN | THE

There would also be productions to send each of the nonterminals such as "noun" and "verb" to all the appropriate English words. A similar structure would be used in artificial intelligence studies as part of a natural language interpreter. Parse the sentence

IDLE HANDS ARE THE DEVIL'S WORKSHOP

9. Repeat Exercise 8 using the sentence

WILLFUL WASTE MAKES WOEFUL WANT

10. Use a tree with inorder traversal to alphabetize the words in the sentence of Exercise 8.

11. Use a tree with inorder traversal to alphabetize the words in the sentence of Exercise 9.

12. Repeat Exercise 10 using the following sentence:

DUTY IS THE SUBLIMEST WORD IN THE ENGLISH LANGUAGE

Note: Notice how duplicate words can be detected and eliminated by this process.

13. Write an algorithm for postorder traversal.

14. Write an algorithm for **preorder traversal** where the node is visited first, then the left subtree and the right subtree.

LANGUAGE EXERCISE

15. We can store a tree in arrays by having a NODE array to store the value at the node and LEFT and RIGHT arrays to hold pointers to (or the index of) the left and right children, respectively. A value of 0 for a pointer indicates the corresponding child does not exist. Finish placing the binary tree of Figure 8.20 into such arrays:

	LEFT	NODE	RIGHT
1		PLEASE	
2		HELP	
3		ME	
4		ALPHABETIZE	
5		THIS	
6		SENTENCE	

LANGUAGE EXERCISES

16. Write the values of the arrays for the binary tree for the expression $a * b/(c + d * e)$ in Example 2, Figure 8.22 (see Exercise 15).

17. The following is a sample question from the *1984 AP Course Description in Computer Science**:

A certain binary tree T is represented as a two-dimensional 5×3 array A, with rows of A corresponding to nodes of T. The columns of A contain the following information:

column 1—the row index of the left child
column 2—the value stored at the node
column 3—the row index of the right child

---

*AP questions selected from *AP Course Description in Computer Science*. College Entrance Examination Board, 1984. Reprinted by permission of Educational Testing Service, the copyright owner of the sample questions.

A row index of 0 indicates a nonexistent child. If A consists of the entries

$$\begin{array}{ccc} 5 & 6 & 4 \\ 1 & 0 & 0 \\ 0 & 1 & 0 \\ 3 & 4 & 0 \\ 0 & 8 & 0 \end{array}$$

then T is given by which of the following diagrams? ☐

(A)
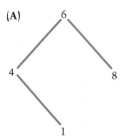

(B)
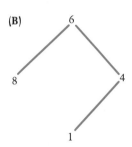

(C)
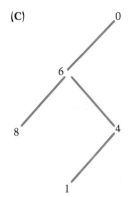

(D)
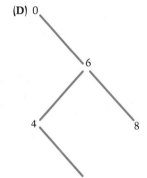

(E)
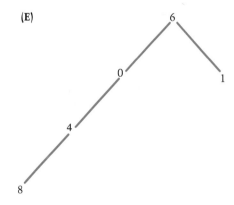

18. What is the best type of graph to illustrate play in a tennis tournament where the loser of each match is eliminated?

19. What is the best type of graph to illustrate play in high school football where each team plays every other team in the region once?

Use the following definitions in Exercises 20–23.

DEFINITION

A **full binary tree** is a binary tree where every node has 0 or 2 children but never just 1.

DEFINITION

An **internal node** of a tree is one that is not a leaf node.

20. Draw a full binary tree having
   a. No internal nodes.   **b.** One internal node.
   c. Draw two full binary trees having 2 internal nodes.
   d. Draw five full binary trees having 3 internal nodes.

21. Fill in the details of the proof of the following theorem: A full binary tree T with i internal nodes has $2i + 1$ nodes and $i + 1$ leaves.

Proof   Each of the i internal nodes has 2 children. (**a.** Why?) Thus, there are a total of ____**b**____ children in the tree. Only one node, ____**c**____, is not a child. Therefore, there are $2i + 1$ nodes in T. Consequently, there are $i + 1$ leaves (**d.** Why?)

22. Play in a tennis tournament can be represented by a full binary tree. The players' names are listed at the leaf nodes, while the internal nodes indicate matches. The winner of a match continues to the next match indicated by the tree, while the loser is eliminated. How many matches will be played in a tournament with 53 contestants? *Note:* See Exercise 21.

23. In a **full *m*-ary tree** every node has 0 or m children. Suppose such a tree has i internal nodes.
   a. How many nodes does the tree have in all?
   b. How many leaves does the tree have? *Note:* See Exercise 21.

24. a. For a binary tree, how many nodes are at level 0? What is the greatest number of nodes that can appear at the following levels?
   b. 1   **c.** 2   **d.** 3   **e.** n.

25. Prove by mathematical induction that in a binary tree there is a maximum of 2^n nodes at level n. *Note:* See your answers in Exercise 24 for specific examples.

26. What is the maximum number of nodes that can appear in a binary tree of height
   a. 0   **b.** 1   **c.** 2   **d.** 3   **e.** n
   Note: Use Σ notation and Exercise 25.
   f. Write your answer to part e without Σ notation.

8.4

Paths and Spanning Trees

Suppose we are given a network, a connected graph with non-negative numbers associated with each edge. And suppose we want to find a Hamilton path in the network, a path that goes through each node exactly once. Perhaps the network represents a map of cities that need to be visited by a traveling salesman, and he wishes to find the best route. He could list all possible routes, all possible Hamiltonian paths, and pick the best one. . . or could he? If he were visiting 26 cities, even with the use of a supercomputer that could calculate the total miles at a rate of a million routes per second, it would take over 4.9 billion centuries to check all possibilities! No

HISTORICAL
NOTE

efficient algorithm has ever been discovered to solve this problem. William R. Hamilton, the brilliant English mathematician, posed the question of how to determine the existence of such paths in the early 19th century, and the answer is still unknown.   □

DEFINITION

> A **Hamiltonian path** in a graph is a path that passes through each vertex, except perhaps the first vertex, exactly once. If this path is also a circuit, it is called a **Hamiltonian circuit.**

Hamiltonian paths are used in the development of a certain encoding scheme called a **Gray code.** Recall that 2^n characters can be encoded using n bits. A Gray code encodes integers so that in going from one integer to the next only one bit is changed. The code is useful when continuous or analog information from a physical system is converted to a digital reading for a digital computer. Say the computer is monitoring the position of a shaft. If the position were between marks 7 and 8, as we will see in Chapter 10, there would be four bits to change in the binary number system representation, from 0111 to 1000. With a Gray code, however, only one bit would change, perhaps from 0100 to 1100. Therefore, there is less chance of ambiguity with a Gray code.

EXAMPLE 1

Design a Gray code to encode the integers 0 through 3.

Solution

Since there are $4 = 2^2$ characters to be encoded, 2 bits will be needed. Draw a graph having four points labeled with the possible 2-bit strings, 00, 01, 10, 11, as in Figure 8.24a.

Now, draw an edge between any two nodes that differ by 1 bit. (See Figure 8.24b.) There is no edge connecting 00 and 11 since these strings disagree in more than one position. For the same reason, there is no edge between 10 and 01. We next draw a

Figure 8.24

Development of a graph for Example 1 to generate a Gray code to encode four characters

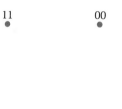

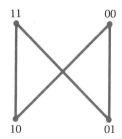

a

b

Figure 8.25

Hamiltonian paths indicated through the graph of Figure 8.24b.

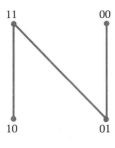

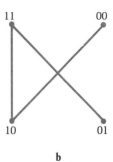

a   b

Hamiltonian path through the graph. One possibility, starting with 00, appears in Figure 8.25a.

From this graph, using a one-to-one mapping, we generated a Gray code for the integers 0, 1, 2, 3:

$00 \rightarrow 0$

$01 \rightarrow 1$

$11 \rightarrow 2$

$10 \rightarrow 3.$

Figure 8.25b shows another possible Hamiltonian path, and the corresponding Gray code follows:

$00 \rightarrow 0$

$10 \rightarrow 1$

$11 \rightarrow 2$

$01 \rightarrow 3.$

Of course, we are not forced to have 00 encode 0; any of the four 2-bit strings is adequate for the job. Once the code for 0 is established, there are only two choices to encode 1. After all, each vertex in graph 8.24b has degree 2. But once two vertices have been chosen, there is only one way to complete the Hamiltonian path. Therefore, there are $4 \cdot 2 = 8$ different Gray codes for 0, 1, 2, 3.

HISTORICAL NOTE

Though there is not a general theorem about the existence of a path through every vertex exactly once, a Hamiltonian path, there is an easy way to determine the existence of a path passing over every edge exactly once. This latter path, called an Euler path, was named after Leonhard Euler (pronounced oiler), the mathematician who in 1736 solved the existence question. Extremely productive in a variety of areas of mathematics and physics, he also established the notations for function, $f(x)$, and summation, Σ, we have been using. ☐

DEFINITION

> An **Euler path** in a graph G is a path that passes through each edge of G exactly once. If this path is also a circuit, it is called an **Euler circuit.**

Euler showed that such a path exists when either no nodes or exactly two nodes have odd degree. First, let us consider a constructive method for deriving an Euler circuit in the graph of Figure 8.26 where every node has even degree. Start with any point, say v_1. Travel through the graph, recording vertices as you go and being careful not to repeat any edge. Since each vertex has even degree, for every vertex we approach there is always an edge by which we can leave. Suppose this procedure leads us through vertices v_1, v_2, v_5, v_4, v_1 as pictured in Figure 8.26b.

Figure 8.26

Steps to find an Euler path through a graph with no vertices having odd degree

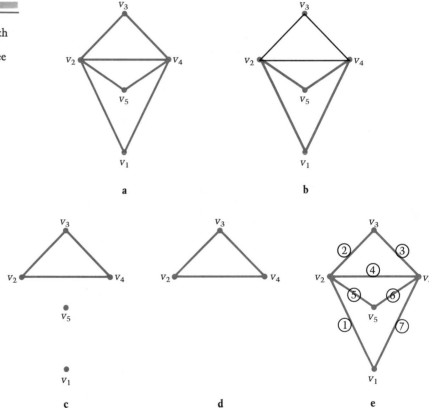

We haven't repeated any edge, but we haven't covered all of the edges either. Remove the edges that have been traversed leaving Figure 8.26c. Notice that v_1 and v_5, having degree 0, are isolated in this graph. Remove these isolated points as well, giving Figure 8.26d.

From the remaining subgraph choose a vertex that was also on the circuit we have already covered. In this example we have two candidates, v_2 and v_4. Starting with, say, v_2 we repeat the above procedure of a finding a circuit. Perhaps we would follow the path through v_2, v_3, v_4, v_2. Patch this path into the original path at v_2 to obtain a longer circuit covering edges only once:

v_1, **v_2, v_3, v_4, v_2,** v_5, v_4, v_1.

This process of attaching "side trips" will eventually result in an Euler circuit, as is the case already in our example. The order in which we covered the edges is depicted in Figure 8.26e.

Algorithm to Construct an Euler Circuit in a Graph G where Every Node Has Even Degree

1. Starting at any point, draw a circuit P that does not repeat any edge. Record in a list L the vertices covered by the path in order of appearance.
2. Repeat until you have an Euler circuit:
   a. Obtain a subgraph of G with all edges of P and subsequent isolated points deleted. Now, call this subgraph G.
   b. Pick a point v from P and G.
   c. Draw a circuit C starting at v in G as in step 1.
   d. Form a new circuit with the circuit C inserted into circuit P as a "side trip" from point v and back. Call this new circuit P. Revise the list L to include the points of the "side trip."

Suppose instead of none, our connected graph contains exactly two nodes of odd degree as in Figure 8.27a. For the time being insert an extra edge, (v_2, v_4), to form a graph where every node has even degree. Now, starting with v_2 or v_4 and edge (v_2, v_4), find an Euler circuit as above. One such circuit is

v_2, v_4, v_3, v_2, v_5, v_4, v_1, v_2.

Eliminating the first vertex and added edge we no longer have a circuit but still have an Euler path

v_4, v_3, v_2, v_5, v_4, v_1, v_2

Figure 8.27

Steps to find an Euler path
through a graph with two
vertices having odd degree

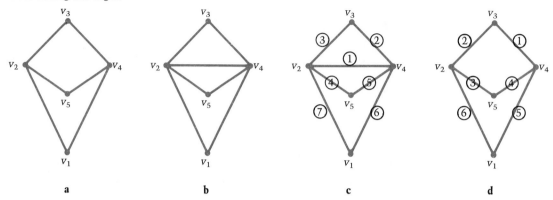

where we start at one vertex of odd degree and end at the other vertex
of odd degree. Thus, if all vertices have even degree, we can generate
an Euler circuit, starting and ending at the same point. If two vertices have odd degree, we cannot find an Euler circuit but still can
obtain an Euler path.

Algorithm to Construct an Euler Path in a Graph G where Exactly Two Nodes, u and v, Have Odd Degree

1. Draw an extra edge, (u, v), to form a graph G' in which
   every node has even degree.
2. Construct an Euler circuit as in the above algorithm,
   starting with edge (u, v).
3. Eliminate (u, v) from the resulting circuit.

As shown in Section 8.1 there are an even number of points with
odd degree, so exactly one point with odd degree is impossible in any
graph. Moreover, we cannot have more than two points with odd
degree and still have an Euler path. As seen by our construction, in
a graph having points of odd degree an Euler path must begin at one
point with odd degree and end at the other.

THEOREM

A connected graph has an Euler path if and only if exactly 0
or 2 nodes have odd degree.

THEOREM

> A connected graph has an Euler circuit if and only if every node has even degree.

A corresponding theorem for digraphs follows.

THEOREM

> A connected digraph has an Euler circuit if and only if each node has the same in-degree as out-degree.

One application of Euler circuits, also, involves finding a finite sequence of bits, called a **de Bruijn sequence,** and from this string producing a binary encoding scheme. Such a sequence is 00011101. When these bits are arranged in a circle as in Figure 8.28, each three consecutive bits are different and thus, can encode a different character. Thus, using this de Bruijn sequence, we could generate the following encoding scheme for 8 characters:

$$000 \rightarrow A \qquad 110 \rightarrow E$$
$$001 \rightarrow B \qquad 101 \rightarrow F$$
$$011 \rightarrow C \qquad 010 \rightarrow G$$
$$111 \rightarrow D \qquad 100 \rightarrow H$$

Figure 8.28

The de Bruijn sequence 00011101 presented in a circle

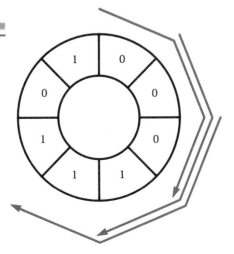

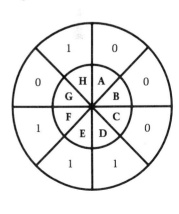

a   b

HISTORICAL NOTE

As mentioned in Chapter 3 this type of scheme was first developed by Baudot, the man for whom the term "baud" is named. The encoding system can be represented in a very succinct manner, requiring a sequence of only 2^n bits for the encoding of 2^n characters; each string of n bits in the circle encodes one character. This scheme was particularly helpful to telegraph operators who could dial a code by finding the corresponding letter on a circle as in Figure 8.28b. □

EXAMPLE 2

Find a de Bruijn sequence to encode 8 characters.

Solution

Since we have $8 = 2^3$ characters, the code for each character will contain $m = 3$ bits. Start a graph with $2^{m-1} = 2^2 = 4$ vertices, labeling each vertex with a different $(m - 1)$-bit, or 2-bit, string (see Figure 8.29a).

Figure 8.29

Development for Example 2 of an 8-bit de Bruijn sequence to encode 8 characters

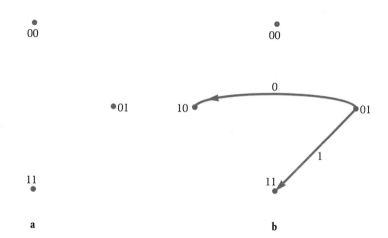

Now, draw two arrows from each node, labeled 0 and 1. For the directed edge labeled "0," indicate what would be the new 2-bit string if 0 was concatenated on the right while the leftmost bit was eliminated. Repeat the process in a similar fashion for the edge labeled "1." For instance, starting with the string 01, 0 moved into the right gives 010; then eliminating the leftmost bit results in the string 10. For 1 shifted into the right of 01, 011, we obtain a string 11. These arrows will be placed in the graph as indicated in Figure 8.29b. The completed graph is presented in Figure 8.30. Notice each of the $2^{m-1} = 2^2 = 4$ points has an out-degree of 2. Thus, there are $2 \cdot 2^{m-1} = 2^m = 2^3 = 8$ edges in the

Figure 8.30

Completion of graph in
Figure 8.29

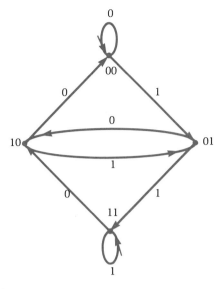

graph. To generate a de Bruijn sequence we produce an Euler circuit, recording the labels of the edges in the order in which they
occur. For example, one Euler circuit covers the nodes

01, 10, 00, 00, 01, 11, 11, 10, 01,

yielding the de Bruijn sequence from the edge labels

00011101. ∎

**Algorithm to Generate a de Bruijn Sequence for the Encoding
of 2^m Characters**

1. Start a graph with 2^{m-1} points, each labeled with a different $(m - 1)$-bit string.
2. For each vertex v labeled with the bit string
   $v_1 v_2 v_3 \ldots v_{m-1}$ do the following:
   a. Draw a directed edge, labeled 0, from v to a vertex la-
      beled with the string $v_2 v_3 \ldots v_{m-1} 0$.
   b. Draw a directed edge, labeled 1, from v to a vertex la-
      beled with the string $v_2 v_3 \ldots v_{m-1} 1$.
3. Draw an Euler circuit through the resulting graph, recording the labels for all the edges in order. This finite sequence of bits is a de Bruijn sequence.

Sometimes we are not concerned with a path through all the
vertices or along all the edges, but would like a tree connecting all

points in the graph, called a spanning tree. In the case of a network we may want such a tree to have a minimum total. For example, a major computer network connects many computer centers nationwide or even worldwide. We would like to have the best routing through the system to achieve the least possible delay in message relay.

DEFINITION

A **spanning tree** of a graph G is a subgraph which is a tree containing all the vertices of G.

DEFINITION

A **minimal spanning tree** of a network is a spanning tree that has the smallest possible number total.

There is an algorithm by Kruskal that can be applied in a reasonable amount of time to find a minimal spanning tree T for a network N of n vertices.

Kruskal's Algorithm for Finding a Minimal Spanning Tree

Repeat until there are $n - 1$ edges in T:
    Select for T an edge e of N that is not already in T, that has the smallest number associated with it, and whose addition to T will not form a cycle. (In case of a tie, pick any of the possible edges.)

EXAMPLE 3

Find a minimal spanning tree for the network in Figure 8.31 where the nodes indicate locations and the numbers indicate time in hours to get from one location to another.

Figure 8.31

Network for Example 3

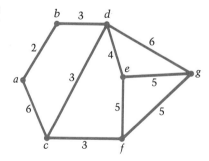

Figure 8.32

Steps of applying Kruskal's algorithm to the graph in Figure 8.31

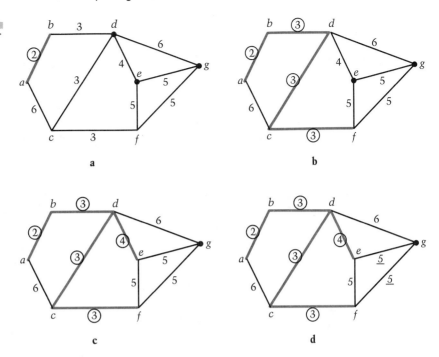

a

b

c

d

Solution

Since there are seven points in the network, we need to get a spanning tree with six edges. The first edge is obviously (a, b), since it has the smallest number associated with it. Circle the numbers on the graph as you use them, and refer to Figure 8.32 to follow the algorithm. The label 3 is used on three edges; pick any one, say (d, c). The minimum number is still 3, and in fact we can pick all edges labeled 3 since none will result in a cycle. Next take the edge labeled 4. Then pick one of the edges numbered 5. We must be careful, however, not to form a cycle. Thus, (e, f) is eliminated. Pick either (g, f) or (e, g). The two possible minimal spanning trees as pictured in Figure 8.33 contain edges (a, b), (b, d), (c, d), (c, f), (d, e), and (e, g) or (f, g) and have a total value $2 + 3 + 3 + 3 + 4 + 5 = 20$. ■

Figure 8.33

Two minimal spanning trees for Figure 8.31

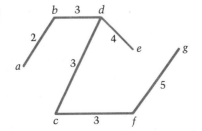

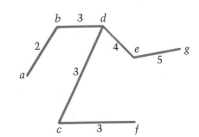

EXERCISES 8.4

If possible, find a Hamiltonian path through each of the graphs listed in Exercises 1–8. (Ignore edge labels.)

1. Figure 8.34a
2. Figure 8.34b
3. Figure 8.34c
4. Figure 8.31
5. Figure 8.26a
6. Figure 8.2a, Section 8.1
7. Figure 8.3, Section 8.1
8. Figure 8.5, Section 8.1

Figure 8.34

Three networks

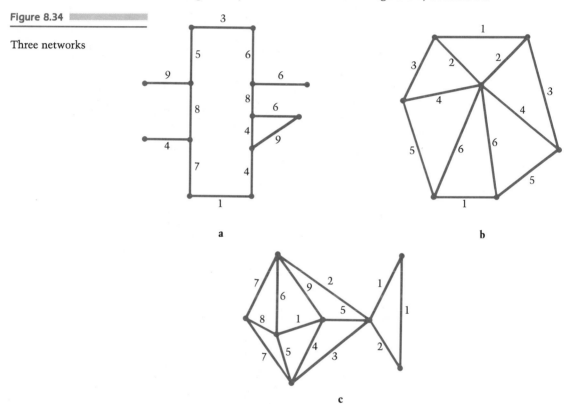

a

b

c

If possible, find an Euler path through each of the graphs listed in Exercises 9–16. If not possible, state why. (Ignore edge labels.)

9. Figure 8.31
10. Figure 8.34b
11. Figure 8.34c
12. Figure 8.24b
13. Figure 8.34a
14. Figure 8.2a, Section 8.1
15. Figure 8.3, Section 8.1
16. Figure 8.5, Section 8.1

Find a minimal spanning tree and its total numeric value for each network in Exercises 17–19.

17. Figure 8.34a   **18.** Figure 8.34b   **19.** Figure 8.34c

20. Using Figure 8.24b find all possible Gray codes to encode the integers 0, 1, 2, 3.

21. Why is it true that every complete graph contains a Hamiltonian circuit?

22. a. Calculate how long it would take a supercomputer to check routes for a salesperson to cover 26 cities, including the home city, if the machine can compute a million routes per second.
   b. For 20 cities.

23. Design two Gray codes for the integers 0 through 7.

24. Design two Gray codes for the integers 0 through 15.

25. There are codes where the bits of one letter's code do not appear at the first of another letter's representation. Thus, the codes for the letters can be of different lengths. The binary tree in Figure 8.35 represents part of such a code. For example, 00 encodes "e," and no other letter starts with 00. To find the code for "j," follow the path from the root to j writing the bits down from left to right as 1110. Find the code
   a. For "a."   **b.** For "o."
   c. Decode the string below by following the path from the root indicated by the bit string until arriving at a leaf. Write down the letter, and start the process over with the next bit in the string.

   0 0 0 1 1 0 0 1 0 1 0 0 1 1 0 0 1 0 1 0 1 1 1 0 1 1 0 0 0 0 1 1 1 0 1 1 1 1 0

Figure 8.35

Tree for Exercise 25

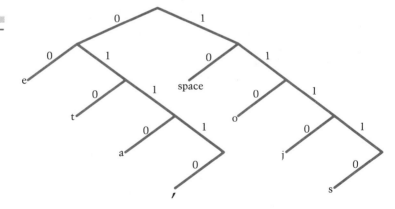

26. Draw a tree similar to the one in Figure 8.35 to depict the code found in Example 2.

27. a. Find a de Bruijn sequence for $m = 2$.
   b. Use this sequence to encode the first four letters in the alphabet.
   c. Draw a tree to display this code.

28. a. Find a de Bruijn sequence for $m = 4$.
   b. Use this sequence to encode the first 16 letters in the alphabet.
   c. Draw a tree to display this code.

Kruskal's algorithm can be used to find a spanning tree in any connected graph by considering each edge to have a label of 1. Find a spanning tree for each of the indicated graphs in Exercises 29–31.

29. Figure 8.4e, Section 8.1

30. Figure 8.16, Section 8.2

31. Figure 8.10, Section 8.2, considered as a graph, not digraph.

32. Does a tree with more than two leaves have an Euler path? Why or why not?

33. Euler's work was inspired by the following puzzle: A river with two islands in the middle flowed through the East Prussian city of Königsberg. Seven bridges connected four land masses as indicated in Figure 8.36. Was it possible to walk through the city, passing over each bridge exactly once?

To solve the problem, Euler drew a graph with each of the four land masses depicted as a point. Then the puzzle reduced to asking the existence of an Euler path through the graph. □
a. Draw such a graph.
b. Using this graph and Euler's theorem in this section, solve the puzzle.

HISTORICAL NOTE

Figure 8.36

Sketch for Königsberg bridge problem in Exercise 33

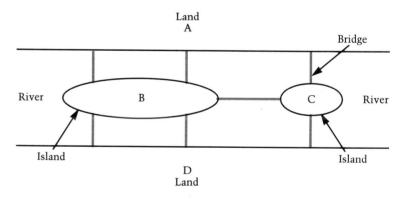

8.5

Finite-State Machines

Does this sound familiar? What is a device that can accept input, store information, process it, and produce output? "A computer," you say? Of course, you are correct; but there are other correct answers. A finite-state machine also works in this manner. Actually, this "machine" is not a piece of hardware at all but a mathematical model that can be used to simulate many I/O (input/output) computational devices including digital computers or compilers or even vending machines and elevators.

This model has an input/output unit, and, consequently, has a way of communicating with the world using a set of symbols. Let I be the set of input symbols and O, the set of output symbols. Often, though not always, $I = O$. Perhaps they are each $\mathbb{B}$ or the character set for Pascal. Or in the case of an elevator, I might be up, down, and floor selection, while O might be stops on particular floors. There is also a set of states S for our model. A state is like a snapshot of what is happening in the machine at a particular instant. In a computer with n bits, a state is a particular configuration of 0's and 1's, so there are 2^n possible states. An elevator might be in a state of going down to the first floor to pick up a passenger or in a state of stopping on the fifth floor on the way up.

Two functions are necessary in our model, one to get us to the next state and one to produce the output. The next state function or transition function, f_N, returns the next state based on the present state and input. For instance, if the elevator is in the state of moving up to the fifth floor and has the input of someone pressing the down button on the third floor, it goes to a state that says, "Remember, when coming back down from stopping on the fifth floor, to pick up someone on the third floor." With the same input but in a state of moving down from the fifth floor, the next state goes something like this, "Stop when you get to the third floor to pick up someone going down." The output function f_O depends on the state only. For instance, one state might automatically have the output of a stop on the third floor.

DEFINITION

> A **finite-state machine** is a 5-tuple $M = (S, I, O, f_N, f_O)$, where S is a finite set of states; I is a finite set of input symbols, called the **input alphabet**; O is a finite set of output symbols, called the **output alphabet**; f_N is a **next-state** or **transition function** with $f_N: S \times I \to S$; and f_O is an **output function**, with $f_O: S \to O$. The **initial state** of the machine is s_0.

It sounds like a lot is involved in a finite-state machine, but our directed graphs can come to the rescue, clarifying the situation. The **state graph** or **transition diagram** is a digraph having edges labeled with inputs and nodes with states. Since the output function f_O has domain S, output depends only on the state; and we can label a node with output along with the state. The next-state function f_N has domain $S \times I$; we're in a state and the input points us to the next state, a situation pictured by an arrow from one state to the next.

Figure 8.37

Finite-state machine for
Example 1

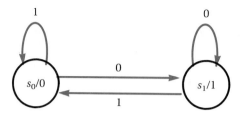

EXAMPLE 1

Given the finite-state machine in Figure 8.37 and input string
011001, what is the output? What does this machine do in
general?

Solution

From the graph we can determine the output and next state functions. The leftmost node is labeled with state s_0 and output 0,
while the other node indicates s_1 and output 1.

$$f_O(s_0) = 0 \qquad f_O(s_1) = 1$$

An input of 0 at state s_0 sends you to a next state of s_1, while an
input of 1 returns you to s_0.

$$f_N(s_0, 0) = s_1 \qquad f_N(s_0, 1) = s_0$$

Similarly, for state s_1 and input 0, the next state is also s_1; but s_1
with input 1 points to a next state of s_0.

$$f_N(s_1, 0) = s_1 \qquad f_N(s_1, 1) = s_0$$

Here the initial state s_0 produces an output of 0, regardless of the
input; but for many applications this initial output is ignored. Because of the input string, which is read left-to-right, the path followed through the graph has vertices $s_0, s_1, s_0, s_0, s_1, s_1, s_0$:

input   0   1   1   0   0   1

state $s_0 \rightarrow s_1 \rightarrow s_0 \rightarrow s_0 \rightarrow s_1 \rightarrow s_1 \rightarrow s_0.$

Or, writing the output below the state, we have

input   0   1   1   0   0   1

state s_0 s_1 s_0 s_0 s_1 s_1 s_0

output   0   1   0   0   1   1   0.

Ignoring the initial zero, we see that an input of 011001, gives an
output with every bit changed, 100110, called its complement.

EXAMPLE 2

Design a finite-state machine to shift a string of bits to the left one position, concatenating a 0 on the right.

Solution

As we discuss in Chapter 10, this process is an easy way to multiply a binary number by 2. Thus, the binary representation for 7, 111, is shifted to become 14, 1110. Our input and output alphabets will be {0, 1, b}, where "b" stands for "blank." We will just copy the number, bit by bit, until a blank is encountered, at which time a 0 will be produced. Following the state graph in Figure 8.38, we see that an input string of 111b gives an output string of 01110b. The leading 0 is of course irrelevant. We must indicate the next state for every possible input. Consequently, the edge from s_2 to s_3 and from s_3 to itself, labeled I, meaning the entire input alphabet {0, 1, b}, points to a state that will terminate the bit string output. ■

Figure 8.38

Finite-state machine for Example 2

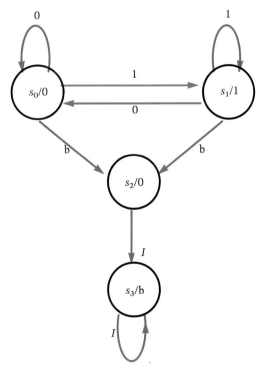

Suppose we want to design a finite-state machine that takes a string of n 1's

$$\underbrace{11 \ldots 1}_{n \text{ factors}}$$

abbreviated 1^n, and produces a string of n 1's followed by n 0's

$$1^n 0^n = \underbrace{11 \ldots 1}_{n} \underbrace{00 \ldots 0}_{n}$$

regardless of what positive integer value n has. Therefore, an input of 1 yields 10, while an input of $111 = 1^3$ is shifted three positions to the left, 111000. Moreover, $11111 \rightarrow 1111100000 = 1^5 0^5$. We could certainly have a machine mimic the 1's; but after that is complete, how does the machine know how many 0's to print? It can't go back and input the 1's again; and our machine has a limited memory which is realized by the states. We have a finite number of states to produce $1^n 0^n$ on input of 1^n for any of an infinite number of positive integers n. The finite-state machine simply is unable to simulate this process.

A better model was developed by Alan Turing in 1936. His **Turing machine** consists of a tape containing a finite string of characters from the alphabet, an unbounded read/write head, and a finite set of instructions. It operates much as a finite-state machine but has more memory and the additional ability to reread and even write over input. The Turing machine can model any procedure that can be performed on a general-purpose digital computer.

As you can see from the date, 1936, the work of this British mathematician was done before the invention of the first computer in the 1940s. Not only did his machine provide a simple mathematical model of a computer, but it originated an area important in computer science as well as mathematics, called **computability theory.** Turing showed that some functions cannot be computed; there are problems for which no algorithmic solution exists.

Alan Turing's concepts were published when he was only 25 years old. A little more than five years later he was heading the British team that broke the code of the German high command in World War II. Before the decade of the 1940s was over, he was working on the idea of a computer displaying intelligence, what we know now as artificial intelligence (AI). He wrote the first program for playing chess, a problem in AI that still poses challenges. In fact, each year at the Association for Computing Machinery (ACM) Conference, the North American Computer Chess Championship is held for the best chess programs to compete. ☐

Though the Turing machine is most interesting, we will not cover it here but will return to the finite-state machine. This time we will consider the finite-state machine not as a model of a computational device but as a recognizer such as in a compiler. As a **recognizer,** the machine will output a 1 when the input string belongs to a particular set. Since other output is irrelevant, output is omitted from the graph. A double circle around the **final** or **accepting state(s)** indicates the output of 1.

To make the discussion easier we need some notation for the strings. We have already used 1^n as an example of the notation

$$a^n = \underbrace{aa \ldots a}_{n \text{ factors}}$$

and $1^n 0^n$ indicated concatenation of the two strings 1^n and 0^n, so that ab means the string a followed by the string b. We write "string a or string b" as $a \vee b$. The empty string is named λ. a^* is the abbreviation for the set of strings that has any number of a's $\{\lambda, a, aa, aaa, aaaa, \ldots\}$.

EXAMPLE 3

Design a finite-state machine to recognize strings of the form $011^* = \{01, 011, 0111, 01111, \ldots\}$ with $I = \{0, 1\}$.

Solution

A string that starts with 0, 1, and then has any number of 1's stops in the final state s_2 with an output of 1 (see Figure 8.39). But if 1 is input at the first step or 0 at the second or any symbol other than 1 afterward, the machine goes to state s_3 and stays there. Consequently, s_3 is a **dead state.** ■

Figure 8.39

State graph for Example 3

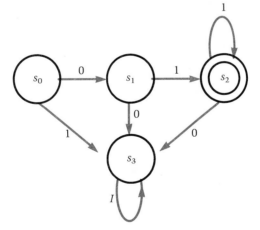

EXAMPLE 4

Design a finite-state machine to recognize $1^* \vee 1^* 011^*$.

Solution

Our machine of Example 3 must be altered to accept λ, 1, 11, 111, ... Thus, s_0 also becomes a final state as illustrated in Figure 8.40. ■

Figure 8.40

Finite-state machine for
Example 4

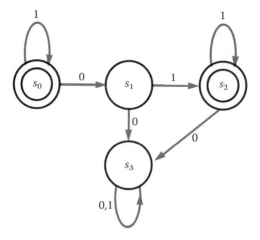

EXERCISES 8.5

1. Consider the finite-state machine in Figure 8.41. Give the output for input
   a. 111.   **b.** 10010.   **c.** 101101
   d. Describe in words what this machine does.

Figure 8.41

Finite-state machine for
Exercise 1

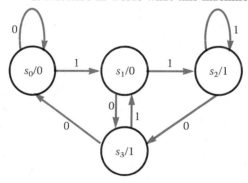

2. What strings are recognized by the finite-state machine in Figure 8.42?

3. Design a finite-state machine that outputs a 1 if an even number of 1's has been encountered and outputs a 0 otherwise. *Note:* Only considering the last output for a string, you have designed a machine to produce odd parity.

4. Describe in the notation of the section the set of strings recognized by the machine in Figure 8.43a.

5. Describe in the notation of the section the set of strings recognized by the machine in Figure 8.43b.

Figure 8.42

Finite-state machine for
Exercise 2

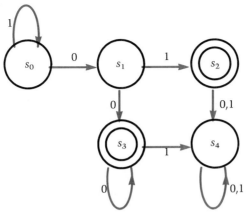

6. Design a finite-state machine to recognize $(01)^*$.

7. **a.** Design a finite-state machine to recognize only an even number of 1's.
   b. Write such a set in the notation of this section.

8. Repeat Exercise 7 for an odd number of 1's.

Indicate if each string in Exercises 9–13 belongs to the indicated set.

9. Is 011110100 in 01^*010^*?   **10.** Is 010011101 in $(0 \vee 1)^*$?

11. Is 010010 in $(0 \vee 1)00(0 \vee 1)$?   **12.** Is 1111110 in 11^*0?

13. Is 10001 in $(0^*1)^*$?

Figure 8.43

Machines for Exercises 4
and 5

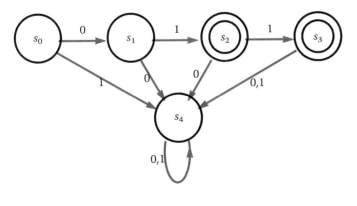

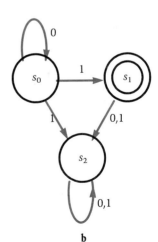

a   b

14. Design a finite-state machine to output a 0 for every leading input of 1. The first 0 input results on a 1 output. Afterward, output should be identical to input. Assume the leftmost bit is always 0. Ignoring the output of state s_0 and reading the input strings from right to left, we have the following examples:

$$01011000 \to 01011001 \quad \text{and} \quad 0100111 \to 0101000.$$

Note: We see in Chapter 10 that this process increments a binary number by 1.

15. Design a finite-state machine to recognize a two-character variable in BASIC where the first character is a letter and the second is a letter or digit or nothing. Let L be the set of letters, D the set of digits, and I the set of all characters in BASIC. □

16. Design a finite-state machine to recognize an integer, dd^*, where d is a digit.

17. Design a finite-state machine to recognize an expression which is the sum of two integers where exactly one blank must be on each side of the plus sign.

18. Repeat Exercise 17 where any number of blanks can be around the plus sign.

19. Design a finite-state machine to recognize a real number
    a. Of the form $dd^*.d^*$ of at least one digit, the decimal point and any number of digits.
    b. Of the form $dd^*.d^* \lor dd^* \lor .d^*$.

20. Design a finite-state machine to recognize a string ending with the letters of END in that order.

21. Design a finite-state machine to recognize a string containing a character string :=.

22. Design a finite-state machine to recognize a string starting with B.

23. Design a finite-state machine to recognize strings that start with 1 and have a 0 in the third position.

24. Design a finite-state machine to output 1 every time "$;$" is encountered.

25. Design a finite-state machine to output 1 every time the character string DO is encountered.

26. Design a finite-state machine to recognize $a^* b^*$.

27. Design a finite-state machine that multiplies a binary number by $2^2 = 4$ by shifting the number to the left two positions (see Example 2).

Chapter 8   REVIEW

Terminology/Algorithms	Pages	Comments
Graph	272	(V, E), vertices and edges
Vertex or node, v	272	Point
Edge, $e = (u, v)$	272	Arc connecting points u and v
Adjacent nodes	273	Edge connects
e incident to v	273	Vertex endpoint of edge
Degree of v, deg(v)	273	Number of times vertex is endpoint
Isolated point, v	273	Deg(v) = 0
$\sum_{i=1}^{n} \deg(v_i)$	273	$2 \cdot n(E)$
Number of nodes with odd degree	273	Even number
Path from v_0 to v_n	274	Sequence of vertices and edges
Length of path	274	Number of edges in path
Circuit	274	Path ends where began
Cycle	274	Path ends where began, no repeats
Loop	274	Edge ends where began
Connected graph	275	Path between any 2 vertices
Simple graph	275	At most 1 edge connecting vertices
Complete graph	275	All points adjacent, simple
$n(E)$ in complete graph with n vertices	275	$n(n - 1)/2$
Acyclic	276	No cycles
Tree	276	Connected graph, acyclic
Directed graph, digraph	276	(V, E), vertices and directed edges
Directed edge, $e = (u, v)$	276	Arrow from u to v

Terminology/Algorithms	Pages	Comments
Initial point u	276	Arrow starts at u
Terminal point v	276	Arrow points to v
In-degree of u	276	Number of edges into u
Out-degree of u	276	Number of edges from u
Network	277	Graph with labeled edges
Bipartite graph	277	Vertices partitioned into V_1 and V_2, arcs only between points of different sets
Array representation	280	For digraphs
Adjacency matrix	282	ij element = number of edges between ith and jth nodes
Connection matrix	283	ij element = 1 if edge between ith and jth nodes
Network matrix	284	ij element = value on edge between ith and jth nodes
Rooted tree	287	Tree with root
Root	287	Node with in-degree = 0
Number of edges in tree	288	$n - 1$ if n vertices
u parent of v	289	Edge from node u to node v in tree
v child of u	289	Edge from node u to node v in tree
Descendants of u	289	Nodes on paths from u in tree
Ancestors of v	289	Nodes on paths from root to v in tree
Leaf	289	Node with out-degree = 0 in tree
Level of node u	289	Length of path from root to u in tree
Height of tree	289	Maximum level
S subgraph of G	289	Graph contained in G
S subtree of T	289	Tree contained in T

Terminology/Algorithms	Pages	Comments
Parsing	289	Algorithm
Binary tree	290	At most 2 children for each node
Placing words in tree for alphabetizing	291	Algorithm
Inorder traversal	292	Algorithm
Hamiltonian path	298	Through every vertex once
Hamiltonian circuit	298	Hamiltonian path that is a circuit
Gray code	298	1 bit changes between representations of integers
Generate Gray code	298	Example. Hamiltonian path
Euler path	300	Through every edge once
Euler circuit	300	Euler path that is a circuit
Euler path in connected graph	302	0 or 2 nodes have odd degree
Euler circuit in connected graph	303	0 nodes have odd degree
Construct Euler path	301	Algorithm
Construct Euler circuit	300	Algorithm
De Bruijn sequence	303	2^n-bit string for encoding 2^n characters. Circular
Generate deBruijn sequence	304	Algorithm. Euler path
Spanning tree	306	Tree containing all nodes from a network
Minimal spanning tree	306	Spanning tree with smallest total
Kruskal's algorithm	306	Algorithm for minimal spanning tree
Finite-state machine	311	(S, I, O, f_N, f_O), S = finite set of states (see below)
Input alphabet, I	311	Finite set of input symbols
Output alphabet, O	311	Finite set of output symbols
Next-state function or transition function, f_N	311	$f_N: S \times I \to S$

Terminology/Algorithms	Pages	Comments
Output function, f_O	311	$f_O: S \rightarrow O$
Initial state	311	s_0
State graph or transition diagram	311	Digraph, edges labeled with inputs, nodes labeled with states
Design finite-state machine	313	Algorithm
Recognizer	314	Output 1 when string recognized
Final or accepting state	315	Has output of 1 in recognizer
a^n	315	String of n a's
ab	315	Concatenation of strings a and b
$a \vee b$	315	String a or string b
λ	315	Empty string
a^*	315	$\{\lambda, a, aa, aaa, aaaa, \ldots\}$
Dead state	315	Can't leave this state

REFERENCES

Charniak, E., and D. McDermott. *Introduction to Artificial Intelligence.* Reading, MA: Addison-Wesley, 1985.

Educational Testing Service. *1984 AP Course Description in Computer Science.* Princeton, NJ: College Board Publications, 1984.

Fleury, A. "The Discrete Structures Course: Making Its Purpose Visible." Presented at the ACM Annual Conference, 1985.

Hillyard, R. "The Build Group of Solid Modelers." *IEEE Computer Graphics and Applications* 2(2): 44, March 1982.

Kline, M. *Mathematical Thought from Ancient to Modern Times.* New York: Oxford University Press, 1972.

Ralston, A. "De Bruijn Sequences—A Model Example." *Mathematics Magazine* 55(3): 131–143, May 1982.

Roberts, F. S. *Applied Combinatorics*. Englewood Cliffs, NJ: Prentice-Hall 1984.

Tucker, A. *Applied Combinatorics*, 2nd ed. New York: Wiley, 1984.

Wand, M. *Induction, Recursion, and Programming*. New York: North-Holland, 1980.

Yarmish, R. "Determining the Reachability Matrix of a Digraph." *UMAP* 3(3): 357–375, 1982.

9 BOOLEAN ALGEBRA

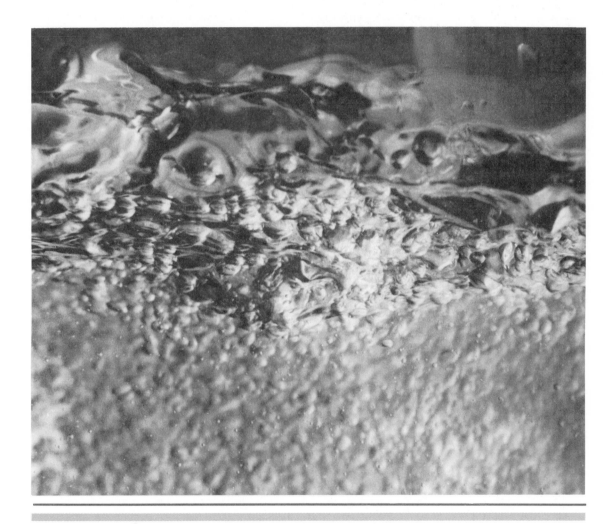

Fluorinert. A closeup of a splash of the fluid used to cool the circuitry in the CRAY-2 supercomputer is pictured (see Figure 6.2, Section 6.1).

Boolean Algebra Definition

HISTORICAL NOTE

The design of a computer, how data and programs are manipulated in the computer, and the logic of programs all have as their basis an algebra that has three operations and as few as two elements. It is surprising that so much depends on a system that may contain so few elements. When George Boole wrote of the algebra that now bears his name in *The Mathematical Analysis of Logic* in 1847, he was presenting an algebraic foundation for logic. But few appreciated Boole's genius. Born in England, the son of a poor shoemaker, George Boole was self-taught. Because of his work in logic he was given the position of professor of mathematics at Queens College, an honorable, but not particularly respected, position. Little did he or anyone else suspect that 100 years later his work would provide the theoretical basis for the design of electronic circuitry and, hence, of the computer. ☐

A Boolean algebra is a structure with a set of elements and three operations. The set may contain only two elements, such as 0 and 1. A switch in a computer has just two states, off and on, just as a condition in a program has but one of two values, FALSE and TRUE. The mathematical model for these structures is a Boolean algebra. There may, however, be more than two elements in the algebra. In Chapter 2, we considered the power set, $\mathcal{P}(U)$, of a universal set U. We saw that if U has n elements, then the collection of all subsets of U, $\mathcal{P}(U)$, has 2^n elements. Three basic set operations were covered, union, intersection, and complement. $\mathcal{P}(U)$ with these operations is, in fact, an example of a Boolean algebra. The basic properties we discussed in that context can be generalized to define a Boolean algebra. First, let us review the properties from set theory:

---

1. Commutative properties
   a. $A \cup B = B \cup A$ b. $A \cap B = B \cap A$
2. Associative properties
   a. $A \cup (B \cup C) = (A \cup B) \cup C$
   b. $A \cap (B \cap C) = (A \cap B) \cap C$
3. Distributive properties
   a. $A \cup (B \cap C) = (A \cup B) \cap (A \cup C)$
   b. $A \cap (B \cup C) = (A \cap B) \cup (A \cap C)$
4. Identities
   a. $A \cup \phi = A$ b. $A \cap U = A$
5. Complement
   a. $A \cup A' = U$ b. $A \cap A' = \phi$

$\mathcal{P}(U)$ with union, intersection, and complement and special elements ϕ and U, written $[\mathcal{P}(U), \cup, \cap, ', \phi, U]$, is but one example of a Boolean algebra; there are many others. In fact, as we discussed in Chapter 2, the algebra of propositions $[S, \vee, \wedge, \sim, F, T]$ with its 2^n statements and three operations of disjunction, conjunction, and negation has exactly the same structure. The formal definition of Boolean algebra follows. In reading through it, consider how $[\mathcal{P}(U), \cup, \cap, ', \phi, U]$ (or $[S, \vee, \wedge, \sim, F, T]$) is such an algebra with $+$ interpreted as $\cup$ $(\vee)$, $\cdot$ as $\cap$ $(\wedge)$, 0 as ϕ (F), and 1 as U(T). One note of caution, $+$ and $\cdot$ do not mean regular addition and multiplication; and 0 and 1 do not necessarily stand for the numbers 0 and 1. Some notation has to be used, and these symbols are convenient; but they are just symbols, having different meanings in different contexts. After all, don't $+$ and $*$ in the context of sets stand for union and intersection in Pascal?

DEFINITION

A **Boolean algebra**, $[A, +, \cdot, ', 0, 1]$, is a set A with two binary operations, $+$ and $\cdot$, a unary operation, $'$, and two distinct elements 0 and 1, satisfying the following properties for all $x, y, z \in A$:
1. **Commutative properties**
   a. $x + y = y + x$ b. $x \cdot y = y \cdot x$
2. **Associative properties**
   a. $x + (y + z) = (x + y) + z$ b. $x \cdot (y \cdot z) = (x \cdot y) \cdot z$
3. **Distributive properties**
   a. $x + (y \cdot z) = (x + y) \cdot (x + z)$ b. $x \cdot (y + z) = (x \cdot y) + (x \cdot z)$
4. **Identities**
   a. $x + 0 = x$ b. $x \cdot 1 = x$
5. **Complement**
   a. $x + x' = 1$ b. $x \cdot x' = 0$

The Boolean algebra containing only 0 and 1 is important to computer science.

EXAMPLE 1

Consider the set $\mathbb{B} = \{0, 1\}$ with the three operations $+, \cdot,$ and $'$ defined by the following tables:

$+$	0	1
0	0	1
1	1	1

$\cdot$	0	1
0	0	0
1	0	1

x	x'
0	1
1	0

To read the tables for $x + y$ and $x \cdot y$, take the value for x from the left and the value for y from the top. The value of $x + y$ or $x \cdot y$ can be found on the corresponding row and column. For example, $1 + 0$ is 1, the element in the second row and first column. x' is found by reading the corresponding row element so that $0'$ is 1, and $1'$ is 0. One interpretation of these operations is $x + y = \text{MAX}(x, y)$ and $x \cdot y = \text{MIN}(x, y)$.

This example is important, in part because the circuitry of a computer is built precisely on this Boolean algebra. The elements 0 and 1 can be interpreted as two states of a switch, off and on.

The ten properties can be verified to show that $[\mathbb{B}, +, \cdot, ', 0, 1]$ is in fact a Boolean algebra. Let's prove a couple of the properties by going through every possible situation.

Verify 1a, the commutative property for $+$:

$$x + y = y + x \qquad \text{for all } x \text{ and } y.$$

Solution   Note that there are two variables, x and y, and two possibilities for each variable. Thus, by the counting principle there are $2 \cdot 2 = 4$ cases we must consider. Use the table or the definition of $x + y = \text{MAX}(x, y)$ to check each situation:

$0 + 0 = 0 + 0$	is obvious since $x = y = 0$.
$0 + 1 = 1 = 1 + 0$	by the table.
$1 + 0 = 0 + 1$	is the same equality as the previous line.
$1 + 1 = 1 + 1$	since $x = y = 1$.

Verify 3a, the distributive property:

$$x + (y \cdot z) = (x + y) \cdot (x + z).$$

Solution   With three variables, x, y, z, and two choices for each variable, the counting principle implies that there are $2^3 = 8$ equalities to verify. We check that the left and right sides of the equality agree for all possibilities:

Left Side	Right Side
$0 + (0 \cdot 0) = 0 + 0 = 0$	$(0 + 0) \cdot (0 + 0) = 0 \cdot 0 = 0$
$0 + (0 \cdot 1) = 0 + 0 = 0$	$(0 + 0) \cdot (0 + 1) = 0 \cdot 1 = 0$
$0 + (1 \cdot 0) = 0 + 0 = 0$	$(0 + 1) \cdot (0 + 0) = 1 \cdot 0 = 0$
$0 + (1 \cdot 1) = 0 + 1 = 1$	$(0 + 1) \cdot (0 + 1) = 1 \cdot 1 = 1$
$1 + (0 \cdot 0) = 1 + 0 = 1$	$(1 + 0) \cdot (1 + 0) = 1 \cdot 1 = 1$
$1 + (0 \cdot 1) = 1 + 0 = 1$	$(1 + 0) \cdot (1 + 1) = 1 \cdot 1 = 1$
$1 + (1 \cdot 0) = 1 + 0 = 1$	$(1 + 1) \cdot (1 + 0) = 1 \cdot 1 = 1$
$1 + (1 \cdot 1) = 1 + 1 = 1$	$(1 + 1) \cdot (1 + 1) = 1 \cdot 1 = 1.$

Verification of some of the other properties will be left as exercises. ∎

Let's examine how the Boolean algebra structure can be applied to statements in logic or to conditions in a program. In both, statements and conditions that are either true (T) or false (F) can be combined to form compound statements or conditions that are also T or F. The logical connectors OR, AND, and NOT can be thought of as two binary operations and a unary operation, respectively. You might write

    IF (A > B) AND (C = 5) THEN....

If conditions or statements (A > B) and (C = 5) both have the value T, then the compound statement (A > B) AND (C = 5) is T.

EXAMPLE 2

Consider the set S of conditions which can be determined to be true (T) or false (F) with the three operations, OR, AND, NOT, defined as follows:

X	Y	X OR Y
F	F	F
F	T	T
T	F	T
T	T	T

X	Y	X AND Y
F	F	F
F	T	F
T	F	F
T	T	T

X	X'
F	T
T	F

[S, OR, AND, NOT, F, T] is a Boolean algebra. In fact, close comparison with Example 1 shows that this Boolean algebra with $S = \{T, F\}$ behaves the same as [$\mathbb{B}$, +, ·, ', 0, 1] with just a change in names of the elements and of the operations. Definitions of the operations here are displayed as truth tables with values written as in a one-dimensional array. In Example 1, the operation tables were pictured in a two-dimensional array format. We could have written the tables as in this example with a truth table format:

x	y	$x + y$
0	0	0
0	1	1
1	0	1
1	1	1

LANGUAGE EXAMPLE

The close connection between the algebra of sets and the Boolean algebra [$\mathbb{B}$, +, ·, ', 0, 1] is illustrated in the computer implementation of Pascal sets. Consider the Pascal declaration

    VAR
      A, B, C: SET OF 1..7;

Here, the universal set is established as $\{1, 2, 3, 4, 5, 6, 7\}$. Suppose we have the assignments

```
A := [1, 3, 6];
B := [3, 5, 6, 7];
```

Each set is represented as a string of 7 bits, $a_1a_2a_3a_4a_5a_6a_7$ and $b_1b_2b_3b_4b_5b_6b_7$, respectively. A position holds a 1 if the corresponding element is in the set; otherwise, the value held at the position is 0. Thus, the strings for A and B are

A:

1	0	1	0	0	1	0
1	2	3	4	5	6	7

or 1010010

and B:

0	0	1	0	1	1	1
1	2	3	4	5	6	7

or 0010111.

To evaluate $A \cup B$ for the Pascal assignment

```
C := A + B;
```

corresponding positions are added in $[\mathbb{B}, +, \cdot, ', 0, 1]$. For example, $a_1 + b_1 = 1 + 0 = 1$; $a_2 + b_2 = 0 + 0 = 0$; and $a_3 + b_3 = 1 + 1 = 1$. The bit 1 in a_i or b_i or both yields 1 for $a_i + b_i$. Similarly, if an element appears in set A or set B or both, that element is also in the union of A and B. The resulting bit string for the union is

C = A + B:

1	0	1	0	1	1	1
1	2	3	4	5	6	7

or 1010111

indicating C is assigned the set value $[1, 3, 5, 6, 7]$.

The intersection is computed by taking the product of corresponding positions, so that $a_1 \cdot b_1 = 1 \cdot 0 = 0$ and $a_3 \cdot b_3 = 1 \cdot 1 = 1$. Both a_i and b_i must be 1 for $a_i \cdot b_i$ to be 1. Similarly, an element x must be in both sets A and B to be in the intersection. Thus, for the assignment

```
C := A * B;
```

the associated bit string for C = [3, 6] is 0010010.

The test if [5, 6, 7] is a subset of B might be written as a condition in an IF statement.

```
IF [5, 6, 7] <= B THEN. . . .
```

The actual implementation involves testing the truth of the equality

$$\{5, 6, 7\} \cap B = \{5, 6, 7\}$$

or in Pascal of the condition

$$([5, \quad 6, \quad 7] * B = [5, \quad 6, \quad 7]).$$

A check of membership in a set, "is X in B?" or in Pascal, "(X IN B)," is accomplished by evaluating "is $\{X\}$ a subset of B?" or "(X <= B)," respectively. □

EXERCISES 9.1

1. Prove that $[\mathbb{B}, +, \cdot, ', 0, 1]$ of Example 1 satisfies the properties 1b, 2a, 4b, and 5a of a Boolean algebra.

2. Prove that $[S, \text{OR}, \text{AND}, \text{NOT}, F, T]$ of Example 2 satisfies properties 1a, 3a, 4a, and 5b of a Boolean algebra.

3. Let $D = \{1, 2, 3, 5, 6, 10, 15, 30\}$. Define the operations $+, \cdot, '$ for $x, y \in D$ as follows:

$$x + y = \text{lcm}(x, y) \qquad \text{(least common multiple of } x \text{ and } y)$$
$$x \cdot y = \text{gcd}(x, y) \qquad \text{(greatest common divisor of } x \text{ and } y)$$
$$x' = 30/x.$$

Thus, $6 + 15 = \text{lcm}(6, 15) = 30$, $6 \cdot 15 = \text{gcd}(6, 15) = 3$, and $6' = 30/6 = 5$.

a. Give the tables for $+$, $\cdot$, and $'$ written as two-dimensional arrays.

b. What are the identities for $+$ and $\cdot$?

c. Verify property 5a of a Boolean algebra for this exercise. Actually, D with these operations and appropriate identities is a Boolean algebra.

In Exercise 3, we could have started with the three primes, 2, 3, and 5, and found all possible greatest common divisors and least common multiples to generate D. Another way to consider the formation is to find all possible products of the elements, $2^a \cdot 3^b \cdot 5^c$, where a, b, c $\in \mathbb{B}$. Since there are two choices for each of a, b, and c, by the counting principle there are $2^3 = 8$ possible products. The resulting set with the operations given in Exercise 3 forms a Boolean algebra. **a.** *Generate D for each of the sets of primes in Exercises 4–6.* **b.** *Give the $+$ and $\cdot$ identities for each corresponding Boolean algebra.* **c.** *Give the rule for forming the complement.*

4. 3, 7	5. 5, 7, 11	6. 2, 3, 5, 7

7. Let $C = \{0, 1, a, b\}$, where 0 and 1 are the additive and multiplicative identities, respectively. Suppose a and b are complements of each other as are 0 and 1. If $[C, +, \cdot, ', 0, 1]$ is a Boolean algebra, fill in the tables below for the operations indicating for each entry the Boolean algebra property used.

x	x'
0	—
1	—
a	—
b	—

$+$	0	1	a	b
0	—	—	—	—
1	—	1	1	1
a	—	—	a	—
b	—	—	—	b

$\cdot$	0	1	a	b
0	0	—	—	—
1	—	—	—	—
a	0	—	a	—
b	0	—	—	b

8. Let $E = \mathbb{B}^2$.
   a. List the elements of E.
   To avoid writing so many parentheses and commas, let's write E equivalently as $\{00, 01, 10, 11\}$. Define the operation of $+$ by adding corresponding digits using the addition of Example 1. Thus, $11 + 01$ is 11, the string containing $1 + 0$ followed by $1 + 1$. Similarly, define the operation $\cdot$ using multiplication from Example 1 of corresponding digits. For example, $11 \cdot 01 = 01$. Complement each element individually: $(01)' = 10$. E, with these operations and the appropriate identities, is a Boolean algebra.
   b. Construct the tables for $+$, $\cdot$, and $'$.
   c. Evaluate $01 + 01 + 00 + 01$.
   d. Evaluate $11 \cdot (01 + 00) + 10$.
   e. Find the additive and multiplicative identities.

9. Actually, the set U_n of n-tuples with elements from $\mathbb{B}$ and the operations defined in Exercise 8 along with appropriate identities always form a Boolean algebra. As we see later in this chapter, we consider computer words as elements of a U_n.
   a. How many elements are in U_n?
   b. Find the identities.
   c. For $n = 3$, evaluate $(101 + 111) \cdot 010$.
   d. For $n = 5$, evaluate $10110 + 01100 + 11100 \cdot 01010$.

10. Let F be the set of all functions from $\mathbb{B}$ to $\mathbb{B}$. One such function is f_0 defined as follows:

    $$f_0(0) = 0 \qquad \text{or}$$
    $$f_0(1) = 0$$

x	$f_0(x)$
0	0
1	0

    a. How many functions are in F?
    Suppose $F = \{f_0, f_1, f_2, f_3\}$ with the functions defined as follows:

x	$f_0(x)$	x	$f_1(x)$	x	$f_2(x)$	x	$f_3(x)$
0	0	0	1	0	0	0	1
1	0	1	1	1	1	1	0

    Define the operations of $+$, $\cdot$, and $'$ as follows:
    $$f(x) + g(x) = \text{MAX}(f(x), g(x))$$
    $$f(x) \cdot g(x) = \text{MIN}(f(x), g(x))$$
    $$f'(x) = \begin{cases} 1 & \text{if } f(x) = 0 \\ 0 & \text{if } f(x) = 1 \end{cases}$$

    In fact, these operations consist of applying the corresponding operation from Example 1 to the individual range elements. F with these operations and the appropriate identities is a Boolean algebra.
    b. Evaluate $f_2 + f_3$:

x	$(f_2 + f_3)(x)$
0	
1	

    c. $f_2 + f_3$ is which function from F?

d. $f_2 \cdot f_3$ is which function from F?

e. Find the additive and multiplicative identities.

f. Find $f_2'(x)$.

g. Fill in the $+$ and $'$ tables for F.

$+$	f_0 f_1 f_2 f_3
f_0	
f_1	
f_2	
f_3	

f	f'
f_0	
f_1	
f_2	
f_3	

11. Actually the set G of functions from $\mathbb{B}^n$ to $\mathbb{B}$ with the operations determined in Exercise 10 along with appropriate identities always forms a Boolean algebra. Elements of G are called **switching** or **combinational functions.** We will see the importance of these functions to computers later in this chapter.

    a. How many functions are in G?

    b. Find the identities f_0 and f_1.

    c. Given the functions f_3 and $f_4 \colon \mathbb{B}^2 \to \mathbb{B}$ below, evaluate $f_3 + f_4, f_3 \cdot f_4,$ and f_4'. (For simplicity we are writing elements from $\mathbb{B}^2$ as ordered strings of two bits.)

x	$f_3(x)$
00	1
01	0
10	1
11	0

x	$f_4(x)$
00	1
01	1
10	1
11	0

x	$(f_3 + f_4)(x)$

x	$(f_3 \cdot f_4)(x)$

x	$f_4'(x)$

12. In Example 1 verify that $(x \cdot y)' = x' + y'$, for all $x, y \in \mathbb{B}$.

13. One interesting application of Boolean algebra and matrix algebra is **Hamming codes,** which can be used to detect and correct errors in transmission of data. First developed by Richard Hamming of Bell Laboratories in 1950, these codes are used in supercomputers for error control. ☐ The (7, 4) Hamming code attaches 3 extra bits for error detection to every 4 bits of information. From the 4 bits of information, x_0, x_1, x_2, x_3, the additional bits, called **check bits,** are determined. One possibility for the check bits follows:

$$x_4 = x_0 + x_1 + x_3 \quad (\text{mod } 2)$$
$$x_5 = x_0 + x_2 + x_3 \quad (\text{mod } 2)$$
$$x_6 = x_0 + x_1 + x_2 \quad (\text{mod } 2)$$

where all arithmetic is done modulo 2. Thus, $1 + 1 = 0$. The resulting 7 bits $(x_0, x_1, \ldots, x_6)$, called a **code word,** is transmitted.

    a. Find the code word for the information (1, 1, 0, 1).

    b. Find the code word for the information (0, 0, 1, 1).

c. How many possible 7-bit words can be received?

d. How many of these 7-bit words are actually code words?

Each 7-tuple, X, received is multiplied by an appropriate parity check matrix, $H_{3 \times 7}$. If

$$HX = \begin{bmatrix} 0 \\ 0 \\ 0 \end{bmatrix}$$

X is a legitimate code word and transmission is assumed to be correct. If $HX \neq \mathbf{0}_{3 \times 1}$, a change in one bit will result in a code word; thus, the correct transmission is interpreted as that code word. For

$$H = \begin{bmatrix} 0 & 1 & 1 & 0 & 1 & 1 & 0 \\ 1 & 0 & 1 & 1 & 0 & 1 & 0 \\ 1 & 0 & 0 & 0 & 1 & 1 & 1 \end{bmatrix}$$

find HX for each of the following X's. Is X a code word? If not, find the best choice for the code word:

e. $\begin{bmatrix} 1 \\ 0 \\ 0 \\ 0 \\ 1 \\ 1 \\ 1 \end{bmatrix}$ **f.** $\begin{bmatrix} 0 \\ 1 \\ 1 \\ 0 \\ 0 \\ 0 \\ 1 \end{bmatrix}$ **g.** $\begin{bmatrix} 0 \\ 1 \\ 1 \\ 0 \\ 0 \\ 1 \\ 1 \end{bmatrix}$

Note: The study of Hamming codes involves much theory of matrix, abstract, and Boolean algebra.

LANGUAGE EXERCISE

14. Consider the Pascal sets D := [2, 4], E := [5, 6, 7], F := [1, 2, 3, 4], each declared as a SET OF 1..7. In the computer implementation of Pascal sets discussed in this section, give the bit string representation for

a. ϕ **b.** The universal set   **c.** D * F

d. E + F   **e.** [5] * D   **f.** [5] * E   ☐

9.2

Properties

In Sections 2.3 we considered a number of properties for $[\mathcal{P}(U), \cup, \cap, ', \phi, U]$ besides those in the definition of an algebra of sets. Corresponding properties were proved in Section 2.4 for the algebra of propositions, $[S, \vee, \wedge, \sim, F, T]$. Armed with the definition from the last section, we are now prepared to prove those properties for Boolean algebras in general. These properties are useful in such applications as symbolic logic and the design of computer circuitry.

One observation that will virtually double our output of properties of a Boolean algebra is that the **dual** of a true statement holds. In the definition of a Boolean algebra, there are two parts to every property. Notice that one part can be derived from the other by switching the operations of $+$ and $\cdot$ and the elements 0 and 1 everywhere they occur. Thus, 4b of the definition, $x \cdot 1 = x$, is the dual of 4a, $x + 0 = x$. We also used duality with set identities in Section 2.3 and with the logic identities in Section 2.4. As another example, in any Boolean algebra the dual of the equality $(x + 0) \cdot (x + y) = x$ is $(x \cdot 1) + (x \cdot y) = x$.

EXAMPLE 1

Prove the **idempotent** properties are true in a Boolean algebra:

$$x + x = x \quad \text{and} \quad x \cdot x = x$$

Solution

You may be startled by these equalities. Everyone knows $2 + 2 \neq 2$ and $2 \cdot 2 \neq 2$. We are not, however, considering the set of real numbers with ordinary addition and multiplication, but rather a Boolean algebra. For instance, in $\mathscr{P}(U)$ the identities $A \cup A = A$ and $A \cap A = A$ seem quite natural. Or in the two-element Boolean algebra $[\mathbb{B}, +, \cdot, ', 0, 1]$ of Example 1 in the last section, $0 + 0 = 0$, $1 + 1 = 1$, as well as $0 \cdot 0 = 0$ and $1 \cdot 1 = 1$. Also, for a condition X that can be TRUE (T) or FALSE (F), X OR X and X AND X have the same value as X. If X were T(F), both compound statements would be T(F). Think of making the statement in English, "I like you, and I like you." The statement is redundant and certainly equivalent to "I like you."

To proceed with the proof of $x + x = x$, we start with the most involved side, $x + x$. A quick glance at the definition of a Boolean algebra in Section 9.1 doesn't give us any obvious clues for simplification, so let's introduce the number 1 using 4b and substitute for it using 5a. The trick of putting in a 1 (or 0) and then substituting can be handy when you are at a loss as to what to do next:

$$
\begin{aligned}
x + x &= (x + x) \cdot 1 && \text{4b, identity} \\
&= (x + x) \cdot (x + x') && \text{5a, complement} \\
&= x + (x \cdot x') && \text{3a, distributive} \\
&= x + 0 && \text{5b, complement} \\
&= x && \text{4a, identity.}
\end{aligned}
$$

The equality $x \cdot x = x$ is true because it is the dual of $x + x = x$. A dual can be justified by making the switches $+ \longleftrightarrow \cdot$ and $0 \longleftrightarrow 1$ in the original proof. Then the dual of each reason is used by exchanging parts a and b of the Boolean algebra definition. Compare the above proof of $x + x = x$ with the following proof of the dual, $x \cdot x = x$:

$$x \cdot x = (x \cdot x) + 0 \qquad \text{4a, identity}$$
$$= (x \cdot x) + (x \cdot x') \qquad \text{5b, complement}$$
$$= x \cdot (x + x') \qquad \text{3b, distributive}$$
$$= x \cdot 1 \qquad \text{5a, complement}$$
$$= x \qquad \text{4b, identity.} \quad \blacksquare$$

EXAMPLE 2

Prove $x + 1 = 1$, and state its dual.

Solution

The dual, $x \cdot 0 = 0$, seems at first glance more obviously true than the given statement. But in $\mathcal{P}(U)$, where the multiplicative identity is U, $A \cup U = U$ seems clear for any set A. And in $[\mathbb{B}, +, \cdot, ', 0, 1]$, $0 + 1 = 1$ and $1 + 1 = 1$ also make sense. In logic, x could be a statement that is either true or false; but when you take the disjunction of x with a statement that is always true, the compound statement must be true. The condition in

    IF (X OR TRUE) THEN...

LANGUAGE EXAMPLE

always evaluates to TRUE in Pascal. $\square$
Starting with the more complicated side you can again substitute for 1.

$$x + 1 = x + (x + x') \qquad \text{5a, complement}$$
$$= (x + x) + x' \qquad \text{2a, associative}$$
$$= x + x' \qquad \text{Example 1, idempotent}$$
$$= 1 \qquad \text{5a, complement.} \quad \blacksquare$$

When convenient, as in the next example, the multiplication symbol will be dropped. Thus, $x' \cdot y'$ can be written as $x'y'$.

The notation x' has been used without formally defining complement. This definition, which follows, is essential to proving the equality in Example 3.

DEFINITION

> Let a be an element of a Boolean algebra, A. $a' \in A$ is the **complement** of a if
>
> $$a + a' = 1 \qquad \text{and} \qquad a \cdot a' = 0.$$

EXAMPLE 3

Solution

Prove $(x + y)' = x'y'$, and state its dual.

This equality and its dual, $(xy)' = x' + y'$, are called **DeMorgan's laws.** These laws probably look more familiar in the context of sets A and B

$$(A \cup B)' = A' \cap B' \quad \text{and} \quad (A \cap B)' = A' \cup B'$$

or in the context of statements p and q,

$$\sim(p \vee q) = \sim p \wedge \sim q \quad \text{and} \quad \sim(p \wedge q) = \sim p \vee \sim q.$$

You also showed in Exercise 12 of the last section that these properties hold in the two-element Boolean algebra on set $\mathbb{B}$.

Before launching into a proof, let's see what $(x + y)' = x'y'$ actually says. Since $(x + y)'$ means "the complement of $(x + y)$," the equality says, "the complement of $(x + y)$ is equal to $x'y'$." To show $x'y'$ is the complement of $(x + y)$, we need to verify the two equalities in the complement definition. Substituting $(x + y)$ for a and $x'y'$ for a', we must verify that

1. $(x + y) + x'y' = 1$ and   2. $(x + y) \cdot x'y' = 0$.

1. Show $(x + y) + x'y' = 1$:

$(x + y) + x'y'$

$\quad = ((x + y) + x')((x + y) + y')$ 3a, distribute $(x + y)$ through

$\quad = (y + (x + x'))(x + (y + y'))$ 1a, commutative; 2a, associative

$\quad = (y + 1)(x + 1)$ 5a, complement

$\quad = 1 \cdot 1$ Example 2, add 1

$\quad = 1.$ 4b, identity

2. Show $(x + y) \cdot x'y' = 0$:

Note: This equality is not the dual of the first equality because the dual of $(x + y) + x'y' = 1$ is $(xy)(x' + y') = 0$.

$(x + y) \cdot x'y'$

$\quad = x \cdot (x'y') + y \cdot (x'y')$ 5b, distribute $x'y'$ through

$\quad = (xx')y' + (yy')x'$ 1b, commutative; 2b, associative

$\quad = 0y' + 0x'$ 5b, complement

$\quad = 0 + 0$ Example 2, multiplication by 0

$\quad = 0$ 4a, identity. ■

EXAMPLE 4

Prove the double complement of x is x, that is,

$$x'' = x.$$

Solution

Recall that $A'' = A$ is true when considering sets, just as $\sim\sim p = p$ is true in logic. In $[\mathbb{B}, +, \cdot, ', 0, 1]$, $0'' = 1' = 0$ and $1'' = 0' = 1$, as well. x'' is rather like a double negative in English; the result is a positive statement.

We want to show the complement of x' is x. Again we use the two parts of the definition and need to prove

1. $x' + x = 1$ and

2. $x'x = 0$.

But these equalities are clearly true by applying the commutative properties to the complement properties 5a and 5b in the Boolean algebra definition. ■

We can use the identities we have proved and the definition of a Boolean algebra to prove other identities and to simplify Boolean expressions and computer circuits.

EXAMPLE 5

Simplify $(x' + y)' + xxy$

Solution

$$
\begin{array}{lll}
(x' + y)' + xxy & = (x' + y)' + xy & \text{idempotent} \\
& = x''y' + xy & \text{DeMorgan} \\
& = xy' + xy & \text{double complement} \\
& = x(y' + y) & \text{distributive} \\
& = x \cdot 1 & \text{complement} \\
& = x & \text{identity}
\end{array}
$$

As we see in Sections 9.3 and 9.4, a circuit could be designed to take input values of 0 and/or 1 for x and y and produce the output $(x' + y)' + xxy$. An easier way to design the circuit is just by using x as the output.

Or you could write a compound condition in a program from Boolean expressions X and Y (each has a value of TRUE or FALSE) that mimics the above:

IF NOT (NOT X OR Y) OR (X AND X AND Y) THEN. . . .

Wouldn't it be simpler to write the following?

IF X THEN. . . . □ ■

LANGUAGE
EXAMPLE

EXERCISES 9.2

1. Fill in the blanks for the steps and reasons in the following proof that $(x + y') = xy$ holds in any Boolean algebra.

$$(x + y')y = y(x + y') \qquad \underline{\quad a \quad}$$
$$= \underline{\quad b \quad} \qquad \text{distributive}$$
$$= yx + 0 \qquad \underline{\quad c \quad}$$
$$= \underline{\quad d \quad} \qquad \underline{\quad e \quad}$$
$$= xy \qquad \underline{\quad f \quad}.$$

2. Prove the property $x \cdot 0 = 0$ from Example 2, giving all the steps, not just resorting to duality.

3. Prove the property $(xy)' = x' + y'$ from Example 3, giving all the steps, not just resorting to duality.

4. Prove $(x'y)'(x + x + y) = x$, the dual of the property from Examples 5, giving all the steps, not just resorting to duality.

5. Prove $x + xy = x$ in a Boolean algebra, and give its dual. *Note:* This equality is called the **absorption property.**

6. Prove $pq' = (p' + q)'$ in a Boolean algebra, and give its dual.

7. Prove the additive identity is unique in a Boolean algebra. *Note:* See Exercise 30, Section 2.3, for a parallel problem in the algebra of sets.

8. Prove the multiplicative identity in a Boolean algebra is unique. *Note:* See Exercise 31, Section 2.3.

9. Using a particular Boolean algebra, give a counterexample to the cancellation property
$$x \neq 0 \quad \text{and} \quad xy = xz \nRightarrow y = z.$$
Note: You are showing that the cancellation property does not hold in a Boolean algebra.

10. Justify each step in the proof that the complement of x is unique: Suppose x' and c are two different complements of x. We will obtain a contradiction by showing $x' = c$.
   a. We know for x that
       1. $x + x' = 1$ **2.** $xx' = 0$ **3.** $x + c = 1$ **4.** $xc = 0$
   Thus,
   b. $x' = 1 \cdot x'$
   c. $\quad = (x + c)x'$
   d. $\quad = xx' + cx'$
   e. $\quad = 0 + cx'$
   f. $\quad = xc + cx'$
   g. $\quad = xc + x'c$
   h. $\quad = (x + x')c$
   i. $\quad = 1 \cdot c$
   j. $\quad = c.$
   Consequently, we have a contradiction and cannot have two distinct complements of x.

11. **a.** Prove $0' = 1$ is always true in a Boolean algebra. *Note:* You are proving the complement of 0 is 1. Use a similar technique to that found in Examples 3 and 4.

b. Prove $1' = 0$.   **c.** In $\mathcal{P}(U)$, find ϕ' and U'

d. In the algebra of propositions, find F' and T'.

a. Simplify the expressions in Exercises 12–18, giving your reasons for each step. b. Afterward, state the dual of each equality of the given expression and its simplified form.

12. $xy + x'y$ **13.** $x + y + yx'$ **14.** $(x' + y')'$

15. $(x + y) + y$ **16.** $x + y'z + u'(x + y'z)$

17. $x + (xy)'$ **18.** $(x + 0)(x + y)$

19. Prove $x + x'y = x + y$, and state its dual.

20. Prove if $ab = a$ then $a' + b = 1$.

21. In a Boolean algebra A define the relation $\leq$ as follows:

$$x \leq y \quad \text{if and only if } xy = x \quad \text{and} \quad x + y = y.$$

Thus, in the two-element Boolean algebra on $\mathbb{B}$, $0 \leq 1$ since $0 \cdot 1 = 0$ and $0 + 1 = 1$. Prove that $(A, \leq)$ is a partially ordered set for any Boolean algebra A. *Note:* $\leq$ is the partial order relation given in Example 1 of Section 4.2 with $\cdot$ interpreted as intersection and $+$ as union.

Prove each of the equalities in Exercises 22–28 is true in a Boolean algebra.

22. $(x + y)'(x' + y')' = 0$

23. $xz + xyz' = x(y + z)$

24. $x + x'y + x'y' = 1$

25. $xy'z' + xyz + x'yz + xyz' = yz + xz'$

26. $(x + y)(x + y') = x$

27. $x'y'z' + xy'z' + x'yz' + xyz' = z'$

28. $xy + x'z + yz = xy + x'z$

9.3

Logic Gates

In 1938 a paper by Claude E. Shannon, "A Symbolic Analysis of Relay and Switching Circuits," was published that gave an electronic circuitry interpretation of Boolean algebra. In one of the most important masters theses ever, Shannon showed how Boolean algebra can be used to create, study, and simplify relay circuits. His paper changed circuit design from an art to a science and prepared the way for the use of electronic circuits in computers, calculators, and telephone systems. □

Figure 9.1

Three basic gates with
tables of inputs and outputs

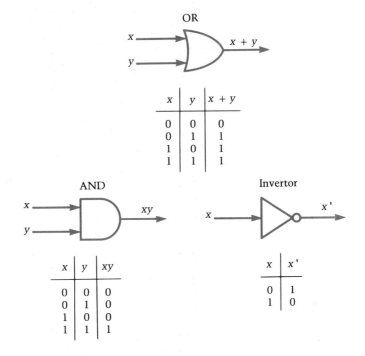

x	y	$x + y$
0	0	0
0	1	1
1	0	1
1	1	1

x	y	xy
0	0	0
0	1	0
1	0	0
1	1	1

x	x'
0	1
1	0

We have seen that there are three basic operations in a Boolean algebra, $+$, $\cdot$, and $'$. Corresponding to these are three basic parts to a logic circuit, **OR gate, AND gate,** and **inverter.** In a **logic circuit** only two possible values, 0 and 1, are present. The gates are the building blocks of this electronic circuitry. Given two binary inputs, x and y, the OR gate yields an output of $x + y$, while the AND gate produces an output of $x \cdot y$. The inverter complements an input of x giving x'. Thus, these logic gates are the physical realization of the operations in a two-element Boolean algebra. The symbols used in engineering for these gates are given in Figure 9.1 along with tables of inputs and outputs. These tables correspond to those for $[\mathbb{B}, +, \cdot, ', 0, 1]$ and $[S,$ OR, AND, NOT, F, T$]$, where S is a set of conditions in a program, in Section 9.1.

An **integrated circuit chip** has a logic circuit on a small piece of semiconductor material such as silicon (see Figure 9.2). Chip technology has advanced from small-scale integration (SSI) with 10 or fewer gates on a chip to very-large-scale integration (VLSI) with hundreds of thousands of gates per chip. The **microprocessor** or CPU on a chip has enabled the development of microcomputers and fueled a computer revolution. The first microprocessor was developed in 1970 by Intel for a Japanese calculator manufacturer. But the calculator company was not sure the chip met its needs adequately or cheaply enough, so Intel was stuck with the product. Not real-

HISTORICAL NOTE

Figure 9.2

UNIVAC I computer; chip compared to the eye of a needle inset in upper, left corner. The UNIVAC I, built in the early 1950s using vacuum tubes, was the first commercial computer. Within 15 years the chip had even greater computing power. Photo courtesy of James E. Stoots, Jr., Lawrence Livermore National Laboratory, University of California. Work performed under the auspices of the U.S. Department of Energy. *Note:* This very computer was used in the November 4, 1952, CBS News Broadcast with Walter Cronkite, "Eisenhower vs. Stevenson," to predict the outcome of the Presidential election. Both the computer and television were in their infancy. With but a few million votes counted, at 8:30 P.M. the computer correctly predicted a landslide for Eisenhower. But polls had indicated a close race, so nervous computer officials changed the values of some of the variables and had the computer recalculate to predict a closer election. Only when the landslide was evident was the original prediction revealed to the television audience. ☐

izing its potential, the young Intel Company hesitantly introduced the chip to the public in 1971. When engineers began to grasp the importance of the invention, sales soared as did microprocessor technology. ☐

The logic gates can be combined to create any number of logic circuits on a chip. Consider the logic network in Figure 9.3. With three Boolean or binary inputs of x, y, and z, the circuit produces one output of u. By ANDing x and y, NOTing z, and ORing the two results, u is obtained. We can use Boolean algebra notation to represent the logic circuit:

$$u = xy + z'.$$

Figure 9.3

Circuit for $u = xy + z'$

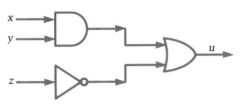

A truth table in Table 9.1 indicates exactly when u is 1. We have a column for values for the inputs x, y, and z and the output u. To make computation of u easier, intermediate results, xy and z', also have columns. Since there are three inputs and each input has two possible values, 0 or 1, there are $2^3 = 8$ rows in the truth table. For consistency, we start with 000 and list all possible input values in columns x, y, and z having the value of z change the fastest and the value of x change the slowest. Observation of the table tells us that the circuit has an output of 1 provided x and y are both 1 or z is 0.

TABLE 9.1
Truth Table for $u = xy + z'$

x	y	z	xy	z'	$u = xy + z'$
0	0	0	0	1	1
0	0	1	0	0	0
0	1	0	0	1	1
0	1	1	0	0	0
1	0	0	0	1	1
1	0	1	0	0	0
1	1	0	1	1	1
1	1	1	1	0	1

The expression $xy + z'$ is an example of a Boolean expression. The variables x, y, and z are Boolean variables with values of 0 or 1 which are combined using the basic Boolean operations. Notice that the definition of Boolean expression given below is recursive.

DEFINITION

> A variable x that can only have a value of 0 or 1 in the two-element Boolean algebra on $\mathbb{B}$ is called a **Boolean variable.**

Let $x_1, x_2, \ldots, x_n$ be Boolean variables. A **Boolean expression** is

1. $0, 1, x_1, x_2, \ldots,$ or x_n
2. $A + B, A \cdot B,$ or A' if A and B are Boolean expressions.

In $u = xy + z'$, output u is actually a function of inputs x, y, and z so that we can write

$$u = f(x, y, z) = xy + z'.$$

The function f is called a switching or combinational function. In this case, $f: \mathbb{B}^3 \rightarrow \mathbb{B}$.

Function f is a **switching** or **combinational function** provided $f: \mathbb{B}^n \rightarrow \mathbb{B}$ for some positive integer n.

EXAMPLE 1

Find another combinational function g equivalent to $f(x, y, z) = xy + z'$ whose circuit is in Figure 9.3. Draw the combinational circuit that corresponds to g.

Solution

By a property in Boolean algebra we can distribute z' through the product:

$$xy + z' = (x + z')(y + z').$$

Thus, one answer for g is

$$g(x, y, z) = (x + z')(y + z').$$

By Boolean algebra properties we can find other such functions. The circuit corresponding to g can be found in Figure 9.4. Though this circuit is equivalent to that in Figure 9.3, Figure 9.4 is more complicated, using 4 gates instead of 3. Functions f and g are equivalent, having the same output for the same input; but the circuit for f is more efficient than that for g. ∎

Figure 9.4

Circuit for $g(x, y, z) =$
$(x + z')(y + z')$

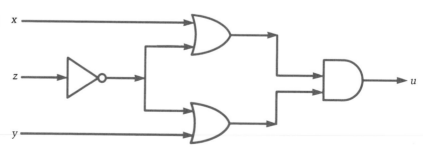

DEFINITION

Two combinational functions $f: \mathbb{B}^n \to \mathbb{B}$ and $g: \mathbb{B}^n \to \mathbb{B}$ are **equivalent** (or **equal**), written $f = g$, if $f(x_1, x_2, \ldots, x_n) = g(x_1, x_2, \ldots, x_n)$ for all $(x_1, x_2, \ldots, x_n) \in \mathbb{B}^n$.

EXAMPLE 2

Write the combinational function f corresponding to the circuit in Figure 9.5; then using the techniques of Boolean algebra, find a simpler, equivalent function g. Draw the truth tables for both f and g and observe the equality of output.

Figure 9.5

Circuit for Example 2

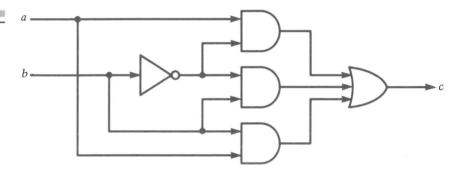

Solution

Notice that to use the value of a in two places we connect another line from the input line. A connection is indicated by a solid dot. When lines cross in the diagram without a solid dot, there is no connection. Also, notice that three lines enter the OR gate. Since the associative property for $+$ holds in a Boolean algebra, grouping is irrelevant, that is

$$(x + y) + z = x + (y + z) = x + y + z.$$

The gate with three inputs could be drawn as in Figure 9.6.

Figure 9.6

(a) Symbol for OR gate with three inputs. (b, c) Two implementations for this OR gate.

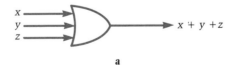

a

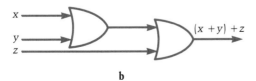

b

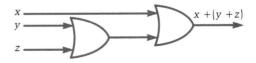

c

To find the Boolean expression let's start at the top left of the diagram in Figure 9.5 where we have inputs a and b' to the top AND gate, giving ab'. The middle AND gate operates on b' and b to produce $b'b$; while the bottom gate ANDs a and b, ab. The output from these three gates are ORed together to give c:

$$c = f(a, b) = ab' + b'b + ab.$$

Now using Boolean algebra, we simplify $f(a, b)$:

$$
\begin{aligned}
ab' + b'b + ab &= ab' + 0 + ab & \text{complement}\\
&= ab' + ab & \text{identity}\\
&= a(b' + b) & \text{distributive}\\
&= a1 & \text{complement}\\
&= a & \text{identity.}
\end{aligned}
$$

An equivalent function is the projection function onto the first coordinate, $g(a, b) = a$, and the circuit doesn't need any gates at all!

$$a \longrightarrow a$$

$$b \longrightarrow$$

The truth table for f is Table 9.2; that of g is Table 9.3. With two inputs we need $2^2 = 4$ rows. Notice the last column of the two tables are equal, reaffirming that f and g are equivalent. ■

TABLE 9.2
Truth Table for $f(a, b) = ab' + b'b + ab$

a	b	b'	ab'	$b'b$	ab	$ab' + b'b + ab$
0	0	1	0	0	0	0
0	1	0	0	0	0	0
1	0	1	1	0	0	1
1	1	0	0	0	1	1

TABLE 9.3
Truth Table for
$g(a, b) = a$

a	b	a
0	0	0
0	1	0
1	0	1
1	1	1

EXERCISES 9.3

*For each of the circuits in Exercises 1–4, give **a.** The combinational function. **b.** The truth table. **c.** Also, for each circuit describe in English the conditions that cause an output of 1. For instance, in Example 1, "xy + z' is 1 if both x and y are 1 or if z is 0."*

1.

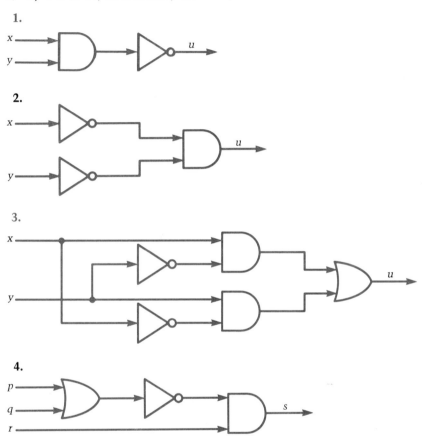

2.

3.

4.

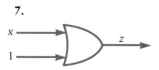

5. Using the properties of Boolean algebra find an equivalent function to that of Exercise 1.

6. Using the properties of Boolean algebra find an equivalent function to that of Exercise 2.

Simplify the combinational circuits in Exercises 7–9 using Boolean algebra. Note: A label of 1 or 0 on a line means that constant value is on the line.

7.

8.

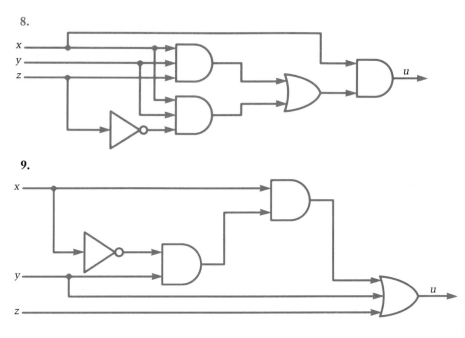

9.

For each of the combinational functions in Exercises 10–13, sketch the corresponding combinational circuit.

10. $h(x, y) = (x + y)'$

11. $f(p, q) = pq'$

12. $f(x, y, z) = x' + y' + z$

13. $g(a, b, c) = ab'c + a'bc$

14. Consider F, the set of all combinational functions from $\mathbb{B}^n$ to $\mathbb{B}$. For f, $g \in F$, define a relation α as follows:

$$f \, \alpha \, g \text{ provided } f \text{ and } g \text{ are equivalent.}$$

a. Prove that α is an equivalence relation.
b. What are its equivalence classes?

9.4

Logic Circuit

In Section 9.3 we saw how to generate a circuit from a Boolean expression and vice versa. But how do you start with an application, generate the truth table, and from there find the Boolean expression and logic circuit?

We will start with a particular example. A task that certainly needs to be performed by a computer is to add two bits. The binary number system will be covered in detail in Chapter 10, but for the time being we only need the following sums:

$$0 + 0 = \quad 0$$
$$0 + 1 = \quad 1$$
$$1 + 0 = \quad 1$$
$$1 + 1 = 10.$$

(As we will see in the next chapter, 10 in the binary number system is equal to 2 in the decimal number system.) For two inputs, x and y, we really desire two outputs: the least significant bit, which we'll call the sum s, and the carry c. The truth table in Table 9.4 follows the above addition with two inputs and two outputs.

Consider the circuits for c and s separately. In some situations you might be able to look at the truth table and figure out the Boolean expression. For instance, c is 1 only in the case where both x and y are 1. Do any of the basic gates perform this function? Yes, sending x and y through the AND gate will produce c; thus, $c = xy$.

Generating output s is more challenging. This output is 1 when exactly one of x and y is 1. The method we will use for generating the Boolean expression for x is called **canonical sum-of-products.** From the table, we see that either the second *or* the third row generate an output of 1. Consequently, because of the "or" in the last sentence, the expression will have the overall structure of

$$(\ ) + (\ ).$$

TABLE 9.4

Truth Table for Computing the Sum s and Carry c of Two Bits $x + y$

x	y	c	s
0	0	0	0
0	1	0	1
1	0	0	1
1	1	1	0

TABLE 9.5

Truth Table for xy, xy', $x'y$, and $x'y'$

x	y	x'	y'	xy	xy'	$x'y$	$x'y'$
0	0	1	1	0	0	0	1
0	1	1	0	0	0	1	0
1	0	0	1	0	1	0	0
1	1	0	0	1	0	0	0

Recall that multiplication, ab, in the two-element Boolean algebra or ANDing will yield 1 in exactly one situation, where both a and b are 1. Consequently, given values for x and y, exactly one of the following products is 1: xy, xy', $x'y$, $x'y'$. Table 9.5 shows all of these products in truth table form with exactly one 1 in each product row. Observe the cases

1. If $x = 0$ and $y = 0$, then only $x'y'$ is 1.
2. If $x = 0$ and $y = 1$, then only $x'y$ is 1.
3. If $x = 1$ and $y = 0$, then only xy' is 1.
4. If $x = 1$ and $y = 1$, then only xy is 1.

Applying this information, we see for inputs of $x = 0$ and $y = 1$ to generate an output of $s = 1$ on the second row of the Table 9.4, we need the product $x'y$. Also, the product xy' gives an output of 1 for the third row. Thus, we have

$$s = x'y + xy'.$$

Notice for the other inputs, s is 0:

$$x = 0 \text{ and } y = 0 \Rightarrow s = 0'0 + 0 \cdot 0' = 1 \cdot 0 + 0 \cdot 1 = 0 + 0 = 0$$
$$x = 1 \text{ and } y = 1 \Rightarrow s = 1'1 + 1 \cdot 1' = 0 \cdot 1 + 1 \cdot 0 = 0 + 0 = 0.$$

We can get each term of the expression without resorting to Table 9.5. If an input is 1, use the variable as is; if the input is 0, complement the variable. Thus, for the second row with $x = 0$ and $y = 1$, we use $x'y$; and for the third row with $x = 1$ and $y = 0$, xy'. The complete circuit is called a **half-adder** and is shown in Figure 9.7.

Figure 9.7

Circuit for a half-adder

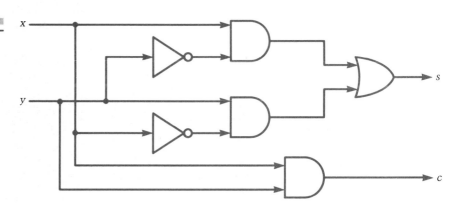

Algorithm for Creating Canonical Sum-of-Products Form of Boolean Expression from Truth Table

1. There will be a term added for each row of the truth table with an output of 1.
2. For each term corresponding to a row, write down the input variables in order, complementing any variable whose value on that row is 0.

EXAMPLE 1

Exercise 12 of Section 5.1 contained a combinational function op: $\mathbb{B}^7 \to \mathbb{B}$ that took a 7-bit string (or ordered 7-tuple) and returned 1 when there were an even number of 1's in the string. The 7-bit string s with op(s) appended was then assured of having an odd number of 1's. For example, op(1001000) = 1 so that 1001000 1 has three 1's; and op(1001010) = 0 so that 1001010 0 also has an odd number of 1's. The value of op(s) is called the **parity bit** in the situation where we are generating **odd parity.** A computer transmitting information will often attach a parity bit to every n bits of information. Here, $n = 7$. At the receiving end, if the 8 bits, 7 bits of information with the parity bit, contain an even number of 1's instead of an odd number, there has been an error in transmission.

What is the Boolean expression to generate odd parity? From such an expression we can create a circuit. The first step is to build the truth table. With an input string of 7 bits we need $2^7 = 128$ rows. To make the problem more manageable for this example, let us suppose that input strings are only 3 bits long. The same procedure would be used for 7 bits.

Solution

Write down all possible selections of input bits. For any row with an even number of 1's, put a 1 in the output column so that the completed row now has an odd number of 1's. The final truth table is in Table 9.6.

TABLE 9.6
Truth Table for Function op of Example 1

x	y	z	$op\,(x,\,y,\,z)$
0	0	0	1
0	0	1	0
0	1	0	0
0	1	1	1
1	0	0	0
1	0	1	1
1	1	0	1
1	1	1	0

We are now ready to apply the canonical sum-of-products method. Since there are four rows with an output of 1, there are four terms in the expression.

$$(\quad) + (\quad) + (\quad) + (\quad).$$

In the first row all input values are 0, so each variable is complemented before multiplication: $x'y'z'$. The fourth row indicates that only x is complemented: $x'yz$. Continuing in the same fashion for the sixth and seventh rows, we generate the canonical sum-of-products form of the Boolean expression as

$$x'y'z' + x'yz + xy'z + xyz'. \quad \blacksquare$$

The canonical sum-of-products form is not necessarily the most efficient Boolean expression for a truth table. We saw in the last section how properties of Boolean algebra can be used to simplify such expressions. In fact, there are other methods based on Boolean algebra that can be used to minimize a circuit. These methods will be left for you to explore in future courses.

EXERCISES 9.4

1. The mapping f in Exercise 28 of Section 5.1 took a 6-tuple of bits and returned 0 if any of the input bits was 0.
   a. What situation(s) will cause an output of 1?
   b. Without drawing the truth table, give the Boolean expression and circuit for this function.

Give the truth table, canonical sum-of-products form, simplified Boolean expression, and resulting circuit diagram for each of the combinational functions in Exercises 2–5.

2. $f(x, y) = \begin{cases} 1 & \text{if } y = 0 \\ 0 & \text{otherwise} \end{cases}$

3. $f(x, y, z) = \begin{cases} 1 & \text{if exactly two of } x, y, \text{ and } z \text{ are } 1 \\ 0 & \text{otherwise} \end{cases}$

4. $f(x, y, z) = \begin{cases} 1 & \text{if } x = z \\ 0 & \text{otherwise} \end{cases}$

5. $f(x, y, z) = \begin{cases} 1 & \text{if } x = 1 \text{ or } y \neq z \\ 0 & \text{otherwise} \end{cases}$

6. A committee has three members, two regular members and a chairperson. Members are to vote electronically on motions with yes $\equiv 1$ and no $\equiv 0$. The chairperson has veto power over any motion. The single output of the circuit indicates if the motion passed (yes) or not (no).
   a. Give the truth table.
   b. Find the canonical sum-of-products form of the Boolean expression.
   c. Simplify the Boolean expression.
   d. Draw the simplified circuit.

7. The circuit for a half-adder given in this section can be illustrated in a block diagram as in Figure 9.8.

Figure 9.8

Block diagram of a half-adder

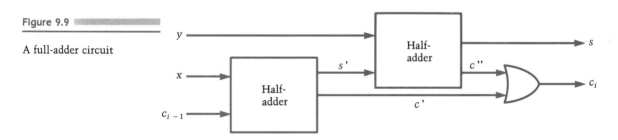

A **full-adder** adds three inputs, x, y, and a previous carry, and produces a sum (s) and carry (c_i). Therefore, a full-adder can be used to add two n-bit numbers, two bits and a carry at a time. We can use the half-adders to design the full-adder. Besides the sums listed at the first of this section, we need the following equality in the binary number system:

$$1 + 1 + 1 = 11.$$

As we will see in Chapter 10, 11 in the binary number system equals 3 in the decimal number system.

To construct a full-adder, we first add x and the previous carry, c_{i-1}, with a half-adder, and generate a sum s' and carry c'. Now add s' and y with another half-adder, getting the final sum s and a carry c''. We have a final sum s but two carries, c' and c''. If either of those are 1 the carry out of the full-adder should be 1. Thus, OR c' and c'' together to get the carry out at the ith step c_i. (c' and c'' could not both be 1 since we are adding at most three 1's for the three inputs x, y, and c_{i-1}.) The block diagram for the full-adder is given in Figure 9.9.

Figure 9.9

A full-adder circuit

a. Draw the truth table for inputs x, y, c_{i-1} and outputs c_i and s.
b. Give values of s', c', s, c'', and c_i with inputs of $c_{i-1} = x = 1$ and $y = 0$.
c. Repeat part b for all inputs equal to 1.
d. Apply the full-adder to add 1011 and 0011. In other words, the least significant bits are processed as x and y. What should c_0 be on this first step? Give x, y, c_{i-1}, s, and c_i at every step by filling in the table below. Notice the carry out at step 1, $c_1 = 1$, is the carry in for step 2, $c_{2-1} = c_1 = 1$.

Step	x	y	c_{i-1}	c_i	s
1	1	1		1	0
2	1	1	1		
3	0	0			
4	1	0			

8. The n-to-2^n decoder with n input lines and 2^n output lines was used in Exercise 31 of Section 1.1. The device decodes an n-bit binary input as one of 2^n numbers. The $2^3 = 8$ integers from 0 to 7 are represented by 3-bit binary numbers as follows:

Binary number	Decimal number	Binary number	Decimal number
000	0	100	4
001	1	101	5
010	2	110	6
011	3	111	7

If, for instance, the binary number 011 ($x = 0$, $y = 1$, $z = 1$) is input, then the fourth output line u_3 and no other is activated (see Figure 9.10).

Figure 9.10

Block diagram of an $n \times 2^n$ decoder

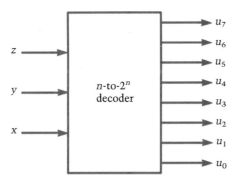

a. Give the Boolean expressions for each of u_0, u_1, . . . , and u_7.
b. Draw the circuit diagram for u_0.
c. Draw the circuit diagram for u_3.

9. Use a block diagram of a full-adder (Exercise 7) to design an incrementer which will generate an output of $x + 1$ upon input of x.

10. Gray codes were covered in Section 8.4. This type of code encodes integers so that in going from one integer to the next only one bit is changed. One example of a Gray code, along with the corresponding numbers in the decimal and binary number systems, is given below.

Decimal	Binary	Gray	Decimal	Binary	Gray
N	wxyz	abcd	N	wxyz	abcd
0	0000	0000	8	1000	1100
1	0001	0001	9	1001	1101
2	0010	0011	10	1010	1111
3	0011	0010	11	1011	1110
4	0100	0110	12	1100	1010
5	0101	0111	13	1101	1011
6	0110	0101	14	1110	1001
7	0111	0100	15	1111	1000

For four input lines, w, x, y, z, there are four output lines, a, b, c, d.

a. Find and simplify the Boolean expressions for a, b, c, and d.

b. Suppose a, b, c, and d are the input lines and w, x, y, and z are the output lines. Give the simplified Boolean expressions for w, x, y, and z to convert a number from a Gray code representation to the corresponding binary number.

11. Consider a combinational circuit that accepts a 2-bit number and returns the square of that number. (See Exercise 10 for the binary representations of the decimal numbers 0, 1, 2, 3, 4, and 9.) How many

   a. Input variables are needed?   **b.** Output variables?

   c. Give the truth table.

   d. Find the Boolean expression for each output variable.

   e. Design the simplified circuit.

12. Gates can be combined to form other gates. Some common ones with their symbols are given in Figure 9.11. Give the truth table for each.

Figure 9.11

Symbols for XOR, NAND, and NOR gates

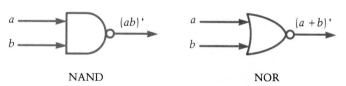

XOR

NAND

NOR

13. Redesign the half-adder circuit using an XOR gate as defined in Exercise 12.

14. Redesign each of the circuits of Exercises 1–4 of Section 9.3 to use the gates from Exercise 12.

15. Redesign each of the circuits of Exercises 10–13, Section 9.3, to use at least one of the gates from Exercise 12 above.

Chapter 9   REVIEW

Terminology/Algorithms	Pages	Comments
Boolean algebra	327	Set with $+$, $\cdot$, $'$, 0, 1 and next 5 properties
Commutative properties	327	$x + y = y + x$, $x \cdot y = y \cdot x$
Associative properties	327	$x + (y + z) = (x + y) + z$ $x \cdot (y \cdot z) = (x \cdot y) \cdot z$
Distributive properties	327	$x + (y \cdot z) = (x + y) \cdot (x + z)$ $x \cdot (y + z) = (x \cdot y) + (x \cdot z)$
Identities, 0 and 1	327	$x + 0 = x$, $x \cdot 1 = x$
Complement	327	$x + x' = 1$, $x \cdot x' = 0$
$[\mathbb{B}, +, \cdot, ', 0, 1]$	327	2-element Boolean algebra
Dual	334	$+ \longleftrightarrow \cdot$, $0 \longleftrightarrow 1$
Idempotent properties	335	$x + x = x$, $x \cdot x = x$
$x + 1$, $x \cdot 0$	336	$x + 1 = 1$, $x \cdot 0 = 0$
DeMorgan's laws	337	$(x + y)' = x'y'$, $(xy)' = x' + y'$
x''	338	x
OR gate	341	$x + y$
AND gate	341	xy
Inverter	341	x'
Boolean variable	343	Has value of 0 or 1
Boolean expression	344	Uses Boolean variable(s), 0, 1, $+$, $\cdot$, $'$
Logic circuit $\rightarrow$ Boolean expression	343	Substitute Boolean operators for gates
Switching or combinational function	344	$f: \mathbb{B}^n \rightarrow \mathbb{B}$
Boolean expression $\rightarrow$ logic circuit	344	Use gates
Equivalent or equal combinational functions	345	Same truth values
Canonical sum-of-products	349	Truth table $\rightarrow$ Boolean expression

REFERENCES

Augarten, S. *Bit by Bit, An Illustrated History of Computers.* New York: Tickner and Fields, 1984.

Computer Museum, Boston.

Fagen, M. D., ed. *A History of Engineering and Science in the Bell System,* 165. New York: Bell Telephone Laboratories, Incorporated, 1978.

Gardner, M. "Boolean Algebra," In *Mathematics: An Introduction to Its Spirit and Use,* 174. San Francisco: Freeman, 1979.

Levine, R. D. "Supercomputers," *Scientific American* 246 (1): 118–180, January 1982.

Pfeiffer, J. E. "Symbolic Logic." In *Mathematics: An Introduction to Its Spirit and Use,* 175–177. San Francisco: Freeman, 1979.

Rice, B. F., and C. O. Wilde. "Error Correcting Codes I," *UMAP,* Unit 346, 1979.

Zientara, M. *The History of Computing.* Framingham, MA: CW Communications, 1981.

10 BINARY AND HEXADECIMAL NUMBER SYSTEMS

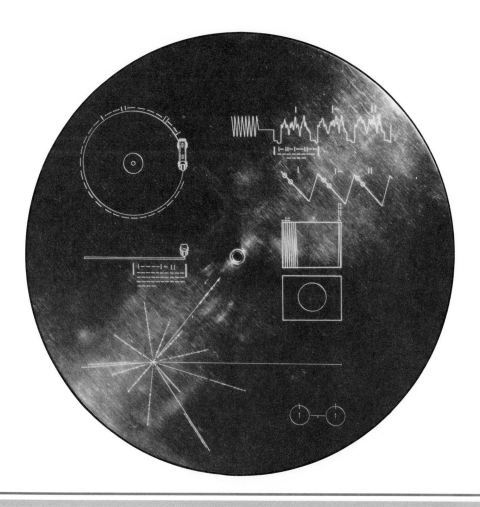

Cover of Voyager Record. A record of our civilization for extraterrestrial beings is attached to the exterior of the Voyager spacecraft. How to play the record and the solar system's location are displayed using the binary number system with "−" meaning "0" and "I" indicating "1".

Binary Number System

Modern digital computers represent everything with 0's and 1's. What seems to be a foregone conclusion today was actually brilliant insight in the formative days. John Vincent Atanasoff had many frustrations while inventing the first specialized electronic digital computer, the ABC or Atanasoff-Berry computer. Exasperated after grappling with many difficult problems, including how numbers and characters should be represented, he got into his car late one night and raced to nowhere on the lonely highways of Iowa. After several hours of dazed driving, he stopped at a roadside cafe on the Iowa-Illinois border. As he relaxed there, insight came to him about several things, including the idea to use the binary system. Two things are much easier to represent electronically than ten; 0 can be represented by a voltage of, say, 0 to 2.3, and 1 by voltage of 2.3 or greater. As often happens after intense, lengthy concentration, the leap to understanding occurred. □

To understand binary representation, let's first consider the decimal number 5863.7, five thousand eight hundred sixty-three and seven-tenths. The number consists of 5 thousands, 8 hundreds, 6 tens, 3 ones, and 7 tenths. In expanded form we have

$$5 \times 1000 + 8 \times 100 + 6 \times 10 + 3 \times 1 + 7 \times 1/10.$$

Notice that $1/10$, 1, 10, 100, and 1000 are all powers of 10: 10^{-1}, 10^0, 10^1, 10^2, 10^3, respectively. Thus, we could write

$$5 \times 10^3 + 8 \times 10^2 + 6 \times 10^1 + 3 \times 10^0 + 7 \times 10^{-1}.$$

Numbers in base 2 are represented similarly, but there are only 2 digits, and the expansion is in powers of 2, not 10. Hence, in the binary number 1101.1, the number immediately to the left of the binary point indicates the number of 1's, the 0 to the left indicates the number of 2's, the next 1 represents the number of 2^2's or 4's, and the leftmost 1 gives the number of 2^3's or 8's. The 1 to the right of the binary point represents 2^{-1} or $1/2$. The expansion reveals what the number is in base 10:

$$
\begin{array}{cccccc}
1 \times 2^3 & + & 1 \times 2^2 & + & 0 \times 2^1 & + 1 \times 2^0 + 1 \times 2^{-1} = \\
1 \times 8 & + & 1 \times 4 & + & 0 \times 2 & + 1 \times 1 + 1/2 \quad = \\
8 & + & 4 & + & 0 & + \quad 1 + 0.5 \quad = 13.5
\end{array}
$$

Thus, 1101.1 in binary is 13.5 in decimal notation. Where there is a possibility of confusion we will write the base as a subscript to the number such as 1101.1_2 and 13.5_{10}.

EXAMPLE 1   What decimal number does 110110.01 represent?

Solution $1 \times 2^5 + 1 \times 2^4 + 0 \times 2^3 + 1 \times 2^2 + 1 \times 2^1 + 0 \times 2^0 + 0 \times 2^{-1} + 1 \times 2^{-2} =$

$32 + 16 + 4 + 2 + 1/4 = 54.25.$ ▪

Notice with each position advanced to the left, the value is doubled: 1/4, 1/2, 1, 2, 4, 8, 16, 32, etc.

EXAMPLE 2   What is 1001100111 in base 10?

Solution   In going from right to left, double the numbers in your head, writing down any number that has a 1 in the corresponding position. The number sequence is 1, 2, 4, 8, 16, 32, 64, 128, 256, 512, 1024, . . . , and we write

$1 + 2 + 4 + 32 + 64 + 512 = 615.$ ▪

In Section 10.3 we will examine a method for converting numbers from base 10 to base 2, but for small numbers we can accomplish the conversion by inspection. Split up each decimal number into a sum of powers of 2 and place a 1 in each corresponding position for the binary number. The first few positions are

$\overline{32}\ \overline{16}\ \overline{8}\ \overline{4}\ \overline{2}\ \overline{1}.$

So, for example,

$19_{10} = 16 + 2 + 1 = \underline{1}\ \underline{0}\ \underline{0}\ \underline{1}\ \underline{1}_2.$
$\phantom{19_{10} = 16 + 2 + 1 = }16\ 8\ 4\ 2\ 1$

EXAMPLE 3   Express the following decimal numbers in binary notation: 5, 8, 30, and 48.

Solution $5\ \ = 4 + 1 = 101_2$
$8\ \ = 1000_2$
$30 = 16 + 8 + 4 + 2 = 11110_2$
$48 = 32 + 16 = 110000_2.$ ▪

We count in the decimal system without thinking, but analysis of the process can help us count in the binary system. There are 10 digits, 0 through 9. For incrementing a position when the largest digit 9 has been reached, 0 is placed in that position and 1 is carried to the left. Thus, 10 follows 9. The number after 29 is 30:

$$29 + 1 = (2 \times 10 + 9) + 1$$
$$= 2 \times 10 + (9 + 1)$$
$$= 2 \times 10 + 1 \times 10 = 3 \times 10 = 30.$$

Similarly, the number after 5999 is 6000, and 10,000 follows 9999.

$$5999 + 1 \rightarrow 5999 \rightarrow 5999 \rightarrow 5999 = 6000$$

$$9999 + 1 \rightarrow 9999 = 10,000.$$

In the binary system we run out of digits in a position quickly because there are only two, 0 and 1. So in this number system $0 + 1 = 1$, but $1 + 1$ cannot be 2; there is no digit 2 in base 2! We make the rightmost or least significant position 0 and carry a 1 to the left to obtain $10_2 = 2_{10}$. Now, $2 + 1 = 10_2 + 1_2 = 11_2 = 3_{10}$. What comes after 11_2?

$$11 + 1 \rightarrow 11 \rightarrow 11 = 100.$$

EXAMPLE 4

Count in the binary number system from 0 to 20.

Solution

	Binary		Binary		Binary
0	0	7	111	14	1110
1	1	8	1000	15	1111
2	10	9	1001	16	10000
3	11	10	1010	17	10001
4	100	11	1011	18	10010
5	101	12	1100	19	10011
6	110	13	1101	20	10100 ■

EXAMPLE 5

Give the next binary number in each case: 11111000, 100011, 1111111.

Solution

We group each number from the right into 4-bit strings for ease of reading.

$$1111\ 1000 + 1 = 1111\ 1001$$
$$10\ 0011 + 1 = 10\ 0100$$
$$111\ 1111 + 1 = 1000\ 0000. \ \blacksquare$$

Finding the next smaller integer follows the reverse algorithm to addition.

Algorithm to Subtract 1 from an Integer

> If the least significant (rightmost) digit is greater than 0,
>   then subtract 1 from the least significant digit
> else
>   a. replace it with the largest digit in that base (9 in base 10, 1 in base 2) and
>   b. subtract 1 from the number to the left of the digit.

Thus, $623_{10} - 1 = 622$, but $630 - 1 = 629$. Similarly, $10111_2 - 1 = 10110$, while $10110_2 - 1 = 10101$. This recursive process yields, $3000 - 1 = 2999$ in the decimal system and $1000 - 1 = 111$ in the binary system.

EXAMPLE 6

Decrement by 1 the following binary numbers: 100010011, 1011100, 1000000.

Solution

$$1\ 0001\ 0011 - 1 = 1\ 0001\ 0010$$
$$101\ 1100 - 1 = \quad 101\ 1011$$
$$100\ 0000 - 1 = \quad 11\ 1111. \ \blacksquare$$

Subtraction can help us in figuring out the largest numbers that a particular computer can represent. An **unsigned number** does not have a plus or minus. What is the largest unsigned integer that can be represented in 8 bits? It is easy enough to write in binary notation, $1111\ 1111_2$, but what is the equivalent decimal value? Instead of adding a long string of powers of 2, notice that this binary number is 1 less than $1\ 0000\ 0000_2$. Remembering that the least significant position indicates the number of 1's or 2^0's, we see that the 1 of this number is in the position for the number of 2^8's. Since $2^8 = 256$, $1111\ 1111_2 = 256_{10} - 1 = 255_{10}$. The smallest unsigned integer that can be represented with 8 bits is $0000\ 0000 = 0$. Thus, in an 8-bit byte, we can represent 256 integers, from 0 to 255.

By the way, the term "byte" was coined in the late 1950s from "bite," but the spelling was changed to avoid an accidental drop of an "e," which would convert the word to "bit." Some had resisted use of the term "bit" ten years earlier, branding the acronym for "binary digit" as an "irresponsible vulgarity." □

EXAMPLE 7

In some microcomputers integers are stored in 16 bits. How many different unsigned integers can be represented in these machines?

Solution

The largest possible unsigned integer is $1111\ 1111\ 1111\ 1111_2$ and the smallest is 0. Thus, in all there are $1\ 0000\ 0000\ 0000\ 0000_2 = 2^{16} = 65,536$ different unsigned integers than can be stored. ■

EXAMPLE 8

In the memory of a computer, bytes are numbered with addresses starting with 0. Suppose an address can be represented with 14 bits. How many different locations are possible? How many K does the computer have?

Solution

The largest address is $11\ 1111\ 1111\ 1111_2$, which is $2^{14} - 1$. Since there are $2^{14} = 16,384$ bytes, memory is of size $2^{14} = 2^4 \times 2^{10} = 16K$. ■

EXERCISES 10.1

Express the binary numbers in Exercises 1–20 in the decimal number system.

1. 111	**2.** 1011	**3.** 1000
4. 1111	**5.** 10010	**6.** 110001
7. 111101	**8.** 1000000	**9.** 10011010
10. 11110000	**11.** 11001001001	**12.** 10 0000 0000
13. 111 1111 1111 1111		**14.** 0.11
15. 0.001	**16.** 0.1001	**17.** 101.101
18. 1001.1001	**19.** 11101.011	**20.** 11111.11111

Express the base 10 numbers of Exercises 21–36 in base 2.

21. 4   **22.** 7   **23.** 11   **24.** 15

25. 3   **26.** 19   27. 25   **28.** 10

29. 29   **30.** 6   **31.** 35   32. 40

33. 9   **34.** 39   **35.** 43   **36.** 24

37. Count from 21 to 45 in the binary number system.

Increment and decrement by 1 each of the binary numbers in Exercises 38–45.

38. 11001   **39.** 1011   40. 1000   **41.** 11100

42. 11111   **43.** 1001111   **44.** 1010011   **45.** 11000

Find the smallest and largest unsigned integers that can be expressed in the number of bits in Exercises 46–51.

46. 4   **47.** 10   **48.** 32

49. 7   **50.** 12   **51.** 6

If a memory address is n bits long, how many bytes and how many K are addressable for n equal to each of the numbers in Exercises 52–55?

52. 9   **53.** 8   54. 22   **55.** 29

If memory contains n bytes, how many bits must be in the address for n to equal each of the numbers in Exercises 56–62?

56. 128   57. 8K   **58.** 6K   **59.** 32K

60. 48K   **61.** 64K   **62.** 256K

63. An n-to-2^n decoder reads in an n-bit number, decodes it or figures out the corresponding decimal number from 0 to $2^n - 1$, and sends a pulse across the corresponding output line. For example, if in a 3-to-8 decoder the input is 101, output line 5 becomes active. Which line is activated by an input of

     a. 100?   **b.** 010?

64. An **encoder** has the opposite task of the decoder. On an 8-to-3 encoder, if line 6 is active, what is the binary output?

65. Write an algorithm to increment by 1 an integer in base 10 or base 2.

66. Consider the following sequence of binary numbers.

         0.01, 0.1, 1, 10, 100, 1000, 10000, 100000.

     a. What happens to the 1 in reading the sequence from left to right, from one number to the next?

     b. What happens to the decimal values of these numbers in reading the sequence?

    c. Do your observations still hold for the following sequence? 0.101, 1.01, 10.1, 101, 1010, 10100, 101000

    d. Based on your observation, double 101110.01.

    e. Check your work for part d by converting the binary numbers to base 10.

    f. Reading the sequences from parts a and c from right to left, what happens to the values?

67. Why might a microcomputer manufacturer equip the product with 131,072 bytes of memory instead of some round number like 130,000?

10.2

Hexadecimal Number System

In dealing with the binary number system in the last section, we saw that long strings of bits are difficult to read. Imagine in the days before high-level languages, writing programs with only 0's and 1's. To keep insanity from setting in, other number systems, the octal and hexadecimal, are used as abbreviations.

In Example 4 of Section 10.1 the binary representations for decimal numbers 0 through 20 were given. Notice that 16 numbers can be represented in 4 bits, from 0 to $1111_2 = 2^4 - 1 = 15_{10}$. The hexadecimal set has 16 digits to represent these numbers,

$$\mathbb{H} = \{0, 1, 2, 3, 4, 5, 6, 7, 8, 9, A, B, C, D, E, F\}.$$

In the binary or base 2 system we have 2 digits; in the decimal or base 10 system there are 10; and in the **hexadecimal** or **base 16 system** there are 16. Conversion from hexadecimal to decimal notation involves the same procedure as conversion from binary to decimal, except expansion is in powers of 16 instead of powers of 2.

EXAMPLE 1

Express F5C.4A with the equivalent decimal number.

Solution

We must convert the hexadecimal digits to decimal numbers as well as expand: A → 10, B → 11, C → 12, D → 13, E → 14, F → 15. Thus,

$$F5C.4A = 15{\times}16^2 + 5{\times}16^1 + 12{\times}16^0 + 4{\times}16^{-1} + 10{\times}16^{-2}$$

$$= 15{\times}256 + 5{\times}16 + 12{\times}1 + 4/16 + 10/256$$

$$= 3932.2890625. \ \blacksquare$$

For expressing small integers in hexadecimal we can use inspection, breaking the number into a sum of powers of 16: 1, 16, 256, 4096, ...

EXAMPLE 2

Express 45_{10} in hexadecimal.

Solution

$45 = 2 \times 16 + 13 = 2D_{16}$. ■

Counting or incrementing also proceeds as we indicated in the decimal and binary number systems. When you run out of digits in the rightmost position, put a 0 there and increment the number to the left. Thus, the number after F is $10_{16} = 16_{10}$. Decrementing in base 16 also proceeds by the same algorithm presented in Section 10.1.

EXAMPLE 3

Count from 14_{10} to 33_{10} in base 16.

Solution

Dec.	Hex.	Dec.	Hex.	Dec.	Hex.	Dec.	Hex.
14	E	19	13	24	18	29	1D
15	F	20	14	25	19	30	1E
16	10	21	15	26	1A	31	1F
17	11	22	16	27	1B	32	20
18	12	23	17	28	1C	33	21

■

EXAMPLE 4

Increment and decrement by 1 each of the following numbers: $X = $ 3C, 99, 9F, FF, D0, 600.

Solution

X	X + 1	X − 1
3C	3D	3B
99	9A	98
9F	A0	9E
FF	100	FE
D0	D1	CF
600	601	5FF

■

EXAMPLE 5

What is the largest unsigned integer that can be expressed in two hexadecimal digits?

Solution

$FF_{16} = 100_{16} - 1 = 16_{10}^{2} - 1 = 255_{10}$. ■

LANGUAGE EXAMPLE

As mentioned at the first of this section, hexadecimal numbers are used extensively as an abbreviation for binary numbers. A **memory dump,** such as that in Figure 10.1 from an aborted COBOL program, is displayed in hexadecimal as opposed to binary. A table indicating how characters are encoded in a computer, such as in ASCII, is generally written in hexadecimal (Appendix A). A knowledge of the hexadecimal system is also essential for an understanding of programming and debugging in assembler language. □

Figure 10.1

Partial memory dump resulting from an aborted COBOL program

```
D2018016 E04A47F0 C20E5840 B04C9620 40015880 B04858E0 80349180 E00E4770
C20E5880 8008D23B 8000C4A1 D202B0F4 4018F363 8027B0F4 D200802E B0F7DC07
8027A070 D5075000 C4304780 C1D8D207 80345000 47F0C1F0 D5075008 C4304780
C1F058E0 B0485840 E008D207 40345008 5880B048 41E0C414 94F78007 5080B050
50E0B054 58F08040 4110B050 05EF5840 B04CD501 4001C4F2 4770C318 91404000
4780C23E 5880B048 5080B050 41E0C41C 50E0B054 58F08040 4110B050 05EF5840
B04C9180 40004780 C2BE9500 501C4780 C2BE4820 501A5E20 40101F33 4330501C
06304430 C4044780 C2BE9680 40015880 B04858E0 80349180 E00E4770 C2BE58E0
8008D236 E000C438 4430C40A D203B0FA 4018F363 E01AB0F4 D200E021 B0F7DC07
E01AA070 94F78007 5080B050 41E0C414 50E0B054 58F08040 4110B050 05EF5880
B04CD501 8001C4F2 4770C318 58E0B048 50E0B050 5080B054 58F0C42C 4110B050
05EF12FF 4780C318 55F0C418 4770C2FC 5880B04C 96088001 47F0C318 55F0C420
4770C310 58E0B04C 9620E002 47F0C318 5880B04C 96108001 9180B0F3 4780C336
5800B04C 18880410 80040A09 5880B04C 1FEE50E0 800C5880 B04CD501 8001CAF2
4780C3A0 91408002 4780C388 5880B048 58E08034 9180E00E 41600010 4770C3A0
58E08008 4110B050 D20CE000 C4DD94F7 80075080 B05041E0 C41450E0 B05458F0
804005EF 47F0C3A0 5860B04C 91206002 5860B04C 41600008 47F0C3A0 41600004
5880B04C 1F00D201 B0F08001 58208010 48308016 D23F8000 B0B0D201 8001B0F0
50208010 40308016 5880B048 1FFFD200 8078B0F2 4070804A 58D0D004 43F0C428
BF07C429 181B5810 D0185810 10005810 105858F0 106405EF 18F658E0 D00C980C
D01407FE D5005000 2000D200 E00E5000 00000001 00000002 00000004 00000005
00000008 80000000 000000F8 00E68578 40404040 40404040 C3D6D5E3 D9D6D340
C2D3D6C3 D9D6D5E8 D440C3C8 C5C3D2C2 D3E2F1F7 F0F0F2C9 40D3D6C1 C440C6C1
C440C1C3 40C6D6D9 40D4D6C4 C5D361C6 D6D9D4C1 E3E3C5D9 40404040 40404040
C9D3C5CA D9C1C7C5 40C1C3C3 C5E2E240 CED6E4D3 C440D5D6 E340D9C5 C1C440C2
40E2E3D6 40C1E340 40404040 40404040 40C6D6D9 40404040 40404040 40C9D5E5
D3D6C3D2 40D4D6C3 C5D30000 00000000 00000000 00000000 E9C1D7C1 F8404040
C1D3C9C4 D1C2C2F2 F1F2F5C2 D3E2D8C9 C6D4E340 D1C2C2F2 D1C2C2F2 F1F2F5C2
F8F3F2F7 D1C2C2F2 F1F2F5C2 D3E2D8C9 C6D4E340 D1C2C2F2 D1C2C2F2 F1F2F5C2
D3E2D8C9 D1C2C2F2 F1F2F5C2 D3E2D8C9 C6D4E340 D1C2C2F2 D1C2F4F0 15C2D3E2
D3E2D8C9 D5075000 C4304780 C1D8D207 80345000 47F0C1F0 D5075008 C4304780
D8C6D6D9 B0485840 E008D207 40345008 5880B048 41E0C414 94F78007 5080B050
BF07CB51 58F08040 4110B050 05EF5840 B04CD501 4001C4F2 4770C318 91404000
18DBD207 5880B048 5080B050 41E0C41C 50E0B054 58F08040 4110B050 05EF5840
C2D440C3 40004780 C2BE9500 501C4780 C2BE4820 501A5E20 40101F33 4330501C
D7D9D6C4 C4044780 C2BE9680 40015880 B04858E0 80349180 E00E4770 C2BE58E0
D9C9C1D3 E000C438 4430C40A D203B0FA 4018F363 E01AB0F4 D200E021 B0F7DC07
4060B110 94F78007 5080B050 41E0C414 50E0B054 58F08040 4110B050 05EF5880
43E0A021 8001C4F2 4770C318 58E0B048 50E0B050 5080B054 58F0C42C 4110B050
C0F85880 4780C318 55F0C418 4770C2FC 5880B04C 96088001 47F0C318 55F0C420
B04C58A0 58E0B04C 9620E002 47F0C318 5880B04C 96108001 9180B0F3 4780C336
C13E58E0 18880410 80040A09 5880B04C 1FEE50E0 800C5880 B04CD501 8001CAF2
E0204870 91408002 4780C388 5880B048 58E08034 9180E00E 41600010 4770C3A0
000B19AE 4110B050 D20CE000 C4DD94F7 80075080 B05041E0 C41450E0 B05458F0
C18E920A 47F0C3A0 5860B04C 91206002 5860B04C 41600008 47F0C3A0 41600004
186A4770 1F00D201 B0F08001 58208010 48308016 D23F8000 B0B0D201 8001B0F0
B120E000 40308016 5880B048 1FFFD200 8078B0F2 4070804A 58D0D004 43F0C428
B12E4770 181B5810 D0185810 10005810 105858F0 106405EF 18F658E0 D00C980C
88B00004 58F08040 4110B050 05EF5840 B04CD501 4001C4F2 4770C318 91404000
B8200004 5880B048 5080B050 41E0C41C 50E0B054 58F08040 4110B050 05EF5840
45E0C928 40004780 C2BE9500 501C4780 C2BE4820 501A5E20 40101F33 4330501C
B116CB12 C4044780 C2BE9680 40015880 B04858E0 80349180 E00E4770 C2BE58E0
4A30CB12 E000C438 4430C40A D203B0FA 4018F363 E01AB0F4 D200E021 B0F7DC07
47F0C2C2 94F78007 5080B050 41E0C414 50E0B054 58F08040 4110B050 05EF5880
B1144780 8001C4F2 4770C318 58E0B048 50E0B050 5080B054 58F0C42C 4110B050
B12E4020 4780C318 55F0C418 4770C2FC 5880B04C 96088001 47F0C318 55F0C420
```

We have seen that a group of 4 binary digits can be represented by a hexadecimal digit. A complete conversion to hexadecimal consists of splitting the binary number into groups of 4 bits, starting from the binary point, padding with zeros if necessary, and expressing each set in hexadecimal. For example, $1011100.101 = 0101\ 1100.1010 = 5C.A$ since $0101_2 = 4 + 1 = 5_{10} = 5_{16}$; $1100_2 = 8 + 4 = 12_{10} = C_{16}$; and $1010_2 = 8 + 2 = 10_{10} = A_{16}$. Let's examine the expansion of this number more closely to see why the conversion works:

$0101\ 1100.1010$

$$= (1 \times 2^6 + 1 \times 2^4) + (1 \times 2^3 + 1 \times 2^2) + (1 \times 2^{-1} + 1 \times 2^{-3})$$
$$= (2^2 \times 2^4 + 1 \times 2^4) + (8 + 4) + (2^{-1} \times 2^4 \times 2^{-4} + 2^{-3} \times 2^4 \times 2^{-4})$$
$$= (2^2 \quad + 1)2^4 \quad + \quad 12 \quad + (2^3 + 2^1)2^{-4}$$
$$= (4 \quad + 1)16 \quad + \quad 12 \quad + (8 \quad + 2)1/16$$
$$= \quad 5 \times 16 \quad + \quad 12 \quad + 10 \times 16^{-1} = 5C.A_{16}.$$

EXAMPLE 6

Express each binary number as a hexadecimal:

$110111,\ 1101110.1,\ 1010100101.100011.$

Solution

Remembering to group from the binary point, we have

$$11\ 0111 = 37_{16}$$
$$110\ 1110.1000 = 6E.8_{16}$$

since
$$1110 = 8 + 4 + 2$$
$$= 14_{10} = E_{16}.$$

$$10\ 1010\ 0101.1000\ 1100 = 2A5.8C$$

since
$$1010 = 8 + 2$$
$$= 10_{10} = A_{16}$$

and
$$1100 = 8 + 4$$
$$= 12_{10} = C_{16}.$$

Conversion from hexadecimal to binary notation is equally as easy. Just express each hexadecimal digit in 4 bits.

EXAMPLE 7

Convert the hexadecimal numbers B3 and F8D47.C to equivalent binary numbers.

Solution

Remembering to express $3_{10} = 11_2$ in 4 bits we have

$$B3 = 1011\ 0011 \quad \text{since } B_{16} = 11_{10} = 8 + 2 + 1.$$
$$F8D47.C = 1111\ 1000\ 1101\ 0100\ 0111.1100 \quad \text{since } F = 15_{10},$$
$$D = 13_{10}, \quad \text{and} \quad C = 12_{10}. \quad \blacksquare$$

EXAMPLE 8

Using the ASCII conversion table in Appendix A, write in hexadecimal and binary notation the name "UNIVAC 1" stored in memory.

Solution

Looking up each individual character we have

    55 4E 49 56 41 43 20 31

Notice that blank is a character as well as "1" in this case. The actual storage in the computer is in base 2 and is much more difficult for people to read and interpret:

    0101 0101 0100 1110 0100 1001 0101 0110 0100 0001 0100
    0011 0010 0000 0011 0001.   ■

The base 8 or **octal number system** with digits 0 through 7 is used for abbreviation of binary numbers in the assembler language of the CRAY supercomputer. In the octal system one octal digit represents three bits instead of four bits as with a hexadecimal digit. Emanuel Swedberg proposed in the early 1700s that Sweden change from the decimal to the octal system for arithmetic, weights, and measures. King Charles XII was won over to the concept of the number system being based on some power of 2. They even considered base 64, but finally rejected the idea because this system requires 64 different digits. The king was killed by a cannon ball, however, and Sweden was never converted to the octal system.   □

In the UNIX operating system one can assign read (r), write (w), and execute (x) permission on each file for the owner, for his or her group, and for all others. Thus, there are 9 on-off switches with a 1 bit indicating that permission to use the corresponding resource is granted. With the format

      rwx   *rwx*   *rwx*
    owner   group   public

111101100_2 or 754_8 indicates rwxr-xr--. Everyone is allowed to read the file; both the owner and group can execute; but only the owner can write to the file. To change your file (myfile) to have this protection you could write the change mode statement

```
chmod 754 myfile  □
```

EXERCISES 10.2

Express the hexadecimal numbers of Exercises 1–12 in decimal notation.

1. D	**2.** E2	**3.** C00	**4.** 987
5. A0FF	**6.** FFF	**7.** FFFF	**8.** 2B835
9. A.4	**10.** CD.D3	**11.** 4E.8	**12.** 100.01

13. Express the hexadecimal numbers from Exercises 1–12 in binary notation.

Express the decimal numbers of Exercises 14–21 in hexadecimal notation.

14. 12	**15.** 24	**16.** 33	**17.** 27
18. 125	**19.** 62	**20.** 160	**21.** 176

Start from each of the hexadecimal numbers in Exercises 22–27 and give the next four hexadecimal integers.

22. B8 **23.** 5C0 **24.** 2D **25.** 9D **26.** FD **27.** FFF

Increment by 1 each hexadecimal number in Exercises 28–33.

28. 9999 **29.** F0AC **30.** 9F **31.** 60BF **32.** EEF **33.** FFFFF

Decrement by 1 each hexadecimal number in Exercises 34–39.

34. 9999 **35.** F0AC **36.** F0 **37.** CD0 **38.** A000 **39.** E0500

40. Give the largest hexadecimal number, written in hexadecimal and decimal notation, that can be expressed in
   a. three hexadecimal digits.   **b.** four hexadecimal digits.

Express each binary number of Exercises 41–46 as a hexadecimal number.

41. 1001.0011	**42.** 1001010.1	**43.** 11010010.00001
44. 1100.11	**45.** 11111.1111111	**46.** 100000.00111

Encode each character string of Exercises 47–50 into ASCII.

47. LISP   **48.** COBOL   **49.** 3:25 P.M.   **50.** 6 E05

51. The contents of a part of memory are given in hexadecimal representation as 52 45 41 44 20 41 5B 36 5D. Assuming we have character
strings, decode this part of memory using the ASCII encoding scheme.

52. Repeat Exercise 51 using 58 20 2B 20 31 2E 35 3B.

*Several 6502 assembler statements are in Exercises 53–55. In each case a
hexadecimal number is given after a dollar sign, which indicates how
far forward to branch based on the results of the last arithmetic. Convert each hexadecimal number to its decimal equivalent. A comment
explaining each command is given after the semicolon.*

53. BNE $20;   BRANCH IF NOT EQUAL 0

54. BEQ $7F;   BRANCH IF EQUAL 0

55. BMI $3D;   BRANCH IF MINUS

56. Suppose you know an unsigned integer is contained in 4 bytes (8 bits
each) in memory. Convert this number, 01 00 A5 F3, to base 10.

57. Suppose in a computer there are 8 bits in a byte and 4 bytes in a word.
   a. How many hexadecimal digits are there in a byte?
   b. How many hexadecimal digits are there in a word?
   c. 6 hexadecimal digits can be expressed in how many bytes?
   d. 13 hexadecimal digits can be expressed in how many bytes?

58. Suppose in the representation of a real number in a particular computer
the mantissa or fractional part is given as a binary number in 24 bits
while an exponent of 16 is given in 7 bits.
   a. What is the largest possible power of 16?
   b. What is the largest possible number?

*Express the octal numbers in Exercises 59–62 with decimal represent-
ation.*

59. 73   **60.** 25   **61.** 604   **62.** 7136

*Express the decimal numbers of Exercises 63–66 with octal represent-
ation.*

63. 13   **64.** 27   **65.** 33   **66.** 47

*Start from each octal number in Exercises 67–70 and give the next four
octal integers.*

67. 6   **68.** 37   **69.** 75   **70.** 7704

Increment by 1 each octal number in Exercises 71–73.

71. 777   **72.** 77777   **73.** 7577

74. Decrement by 1 each octal number.
    a. 7000   **b.** 6030

75. Give the largest octal number, written in octal and decimal notation, that can be expressed in
    a. 3 octal digits.   **b.** 4 octal digits.

76. Write a change mode UNIX command to provide the following permission.
    a. rwx--x---   **b.** r-x--xr-- ☐

LANGUAGE
EXERCISE

10.3

Conversion from Decimal

We have seen how to convert a base 2, 8, or 16 number to base 10. But how do we go the other way around? Example 2 of the last section gives us a hint. In that problem we converted 45_{10} to hexadecimal by finding how many 16's were in the number and what was left over. 45 divided by 16 gives a quotient of 2 and remainder of 13. Since we have two 16's and thirteen 1's in 45, the hexadecimal number is 2D. To make it easier to see in larger problems, write the quotient below instead of above the dividend, and write the remainder off to the side as follows:

$$16\underline{|45}$$
$$2 \quad 13$$

EXAMPLE 1

Convert the decimal number 1069 to the equivalent hexadecimal number.

Solution

We begin as above

$$16\underline{|1069}$$
$$66 \quad 13$$

Thus, $1069 = 66 \times 16 + 13$. We know the number of units (13), but we are not through because 66 is greater than 15. Divide the quotient, 66, by 16 to obtain the following:

$$16\underline{|66}$$
$$4 \quad 2$$

$66 = 4 \times 16 + 2$. Consequently,

$$1069 = 66 \times 16 + 13$$
$$= (4 \times 16 + 2) \times 16 + 13$$
$$= (4 \times 16 \times 16 + 2 \times 16) + 13$$
$$= 4 \times 16^2 + 2 \times 16^1 + 13$$
$$= 42D_{16}$$

Instead of writing the problem as two separate divisions, continue down the page as follows:

$$16\underline{\mid 1069}$$
$$16\underline{\mid \quad 66 \quad 13}$$
$$4 \quad\;\; 2$$

We stop when the quotient is less than 16. Then read off the last quotient and remainders in reverse order as indicated by the arrow, being careful to convert the decimal numbers between 10 and 15 to the hexadecimal digits.

To check our answer, use the method of Section 10.2 to convert 42D to decimal notation:

$$42D_{16} = 4 \times 16^2 + 2 \times 16^1 + 13$$
$$= 4 \times 256 + 2 \times 16 + 13 = 1069. \quad \blacksquare$$

EXAMPLE 2

Convert the decimal number 0.15625 to hexadecimal notation.

Solution

Here, we are trying to find the number of 16^{-1}'s, 16^{-2}'s, 16^{-3}'s, etc., in 0.15626. To find the number of each nonnegative power of 16 in a number, we divided by 16. To find the number of each negative power of 16 in a number, we multiply by 16.

$$0.15625 \times 16 = 2.5 = 2 + 0.5 \tag{1}$$

2 is the first hexadecimal number after the hexadecimal point. Repeat the process with 0.5:

$$0.5 \times 16 = 8 \tag{2}$$

Thus, $0.15625_{10} = 0.28_{16}$. Checking, we have

$$2 \times 16^{-1} + 8 \times 16^{-2} = 2/16 + 8/256$$
$$= 0.125 + 0.03125 = 0.15625.$$

To justify the work in obtaining the answer 0.28_{16}, let us rearrange equation (2).

$$0.5 = 8 \times 16^{-1}.$$

Then substituting 8×16^{-1} for 0.5 in equation (1), we have

$$0.15625 \times 16 = 2 + 8 \times 16^{-1}.$$

Multiplying by 16^{-1} we have

$$\begin{aligned}
0.15625 &= (2 + 8 \times 16^{-1}) \times 16^{-1} \\
&= 2 \times 16^{-1} + 8 \times 16^{-2} \\
&= 0.28_{16}. \quad \blacksquare
\end{aligned}$$

EXAMPLE 3

Convert the decimal number 64,462.75 to hexadecimal.

Solution

Separate the number into the integer and fractional parts. Proceed in the same way as Example 1 for the integer part.

```
16 | 64462
16 |  4028    14
16 |   251    12
        15    11
```

Hence, since $15_{10} = F$, $11_{10} = B$, $12_{10} = C$, and $14_{10} = E$, $64{,}462_{10} = FBCE_{16}$. Then proceed as in Example 2 to convert 0.75 to hexadecimal: $0.75 \times 16 = 12$, so $0.75_{10} = 0.C_{16}$. The final answer is $FBCE.C_{16}$.

Checking, we find

$$\begin{aligned}
FBCE.C &= 15 \times 16^3 + 11 \times 16^2 + 12 \times 16 + 14 + 12 \times 16^{-1} \\
&= 15 \times 4096 + 11 \times 256 + 12 \times 16 + 14 + 12/16 \\
&= 64{,}462.75. \quad \blacksquare
\end{aligned}$$

Calculators are certainly marvelous tools to use with these problems. Let's repeat the integer part of the problem showing some hints to make life easier. First, if you have a memory in your calculator, store 16 there since 16 will be used repeatedly.

64462 ÷ 16 (memory) gives 4028.875. Write down the integer part, 4028, as the quotient. Subtract 4028 from the value in the calculator (4028.875), giving the fractional part (0.875). Multiply 16 (memory) by the value in the calculator (0.875), giving the remainder (14). Repeat the process with the new quotient, 4028, and continue. To reiterate the steps with 16 in memory: Repeat until the quotient (Q) is less than 16:

   1. Q ÷ memory

   2. Q ← integer part

   3. Write down Q

   4. − Q

   5. × memory

   6. Write down remainder, the result of Step 5.

Some calculators will do these conversions and hexadecimal arithmetic, but it is recommended that you not use these features until you understand the procedures yourself.

Another process is to convert from a decimal integer to any base. Proceed as in Example 1 but divide by the base to which you are converting.

Algorithm for Converting from Base 10 Integer Q to a Base b Number

1. Repeat until $Q < b$:

   a. Calculate $Q ÷ b$.

   b. Q ← quotient from Step a. Write this new Q below the old Q.

   c. Write the remainder from Step a to the right of the new Q.

2. Convert each remainder and Q to a base b digit.

3. The base b number has the digits of Step 2 read from bottom to top.

To convert from a decimal fraction to any base proceed as in Example 2 but multiply by the appropriate base.

Algorithm for Converting from Base 10 Fractional Number $F < 1$ to a Base b Number

1. Write down a point to start recording the answer.

2. Repeat until you have the desired number of decimal places:

   a. Calculate $F × b$.

   b. F ← fractional part of product from Step a.

   c. Convert each integer part from Step a to a base b digit and record it on the right of the developing answer.

EXAMPLE 4

Solution

Convert 221.6_{10} to binary.

We use the first algorithm with repeated division by 2 to convert 221. Since 2 is the divisor each time, it is unnecessary (and tedious!) to write it repeatedly, so we will record it only once.

$$
\begin{array}{r|l}
2\,|\,221 & \\
\hline
110 & 1 \\
55 & 0 \\
27 & 1 \\
13 & 1 \\
6 & 1 \\
3 & 0 \\
1 & 1
\end{array}
$$

Since 1 and 0 in decimal translate unchanged to binary, we can write down the answer immediately as $1101\,1101_2$. To convert the fractional part we multiply 0.6 times 2:

	Answer
$0.6 \times 2 = \mathbf{1}.2$	$0.\mathbf{1}$
$0.2 \times 2 = \mathbf{0}.4$	$0.1\mathbf{0}$
$0.4 \times 2 = \mathbf{0}.8$	$0.10\mathbf{0}$
$0.8 \times 2 = \mathbf{1}.6$	$0.100\mathbf{1}.$

Notice that we are now back to 0.6. The process will go on forever in this repeating binary expansion. Thus,

$$0.6_{10} = 0.\overline{1001}_2.$$

Hence, the final answer is $11011101.\overline{1001}$. Though 221.6 has a terminating decimal expansion, in binary it has a repeating expansion.

It is always a good idea to check your work since careless mistakes are so easy to make:

$$11011101_2 = 1 \times 128 + 1 \times 64 + 1 \times 16 + 1 \times 8 + 1 \times 4 + 1 = 221$$

$$0.10011_2 = 1 \times 1/2 + 1 \times 1/16 + 1 \times 1/32$$
$$= 0.5 + 0.0625 + 0.03125 = 0.59375,$$

which rounds to 0.6. The result is not exactly 0.6 since the answer is $0.1001\overline{1001}$, not 0.10011.

Since hexadecimal is an abbreviation for binary, another way to solve this problem is to convert 221 to hexadecimal and then to binary. Thus,

$$221_{10} = DD_{16} = 1101\ 1101_2.$$

For the fractional part we have

$$0.6 \times 16 = 9.6.$$

9 is the first hexadecimal digit after the point, and we use 0.6 for multiplication again. Thus,

$$0.6_{10} = 0.\overline{9}_{16} = 0.\overline{1001}_2.\ \blacksquare$$

The repeating binary expansion for the answer points out a problem of expressing numbers exactly in the computer. Suppose our computer stores floating point numbers as normalized numbers with the fractional part in 12 bits. Let 221.6 be read and converted to normalized binary notation:

$$221.6 = DD.\overline{9}_{16} = 1101\ 1101.\overline{1001} = 0.1101\ 1101\ \overline{1001} \times 2^8.$$

The fractional part or **mantissa** is stored as $1101\ 1101\ 1001 = DD9_{16}$; the exponent is also held in memory. If the number is eventually printed, it must be converted back to decimal:

$$0.1101\ 1101\ 1001 \times 2^8 = 1101\ 1101.1001 = 221.5625_{10}.$$

The absolute error is

$$221.6 - 221.5625 = 0.0375$$

and the relative error is

$$0.0375/221.6 = 0.000169 = 0.0169\%.$$

This conversion between bases with a finite amount of storage is the very reason we should not make a test like

IF X = Y THEN. . .

when X and Y are real numbers. Suppose for instance that Y had been read to be 2216.0, and X had been previously assigned as 10 times a variable with value 221.6. First of all, by the methods we have covered, the value of Y is not truncated:

$$2216.0 = 8A8.0_{16} = 1000\ 1010\ 1000.0$$
$$= 0.1000\ 1010\ 1000 \times 2^{12} = 0.8A8 \times 2^{12}.$$

We saw in Example 4 that 221.6 is stored as $0.DD9 \times 2^8 = 221.5625$. This number times 10 is $2215.625 = 8A7.A = 0.8A7A \times 2^{12}$ which would be truncated and stored as $0.8A7 \times 2^{12} = 2215.0$. Though we know that $221.6 \times 10 = 2216.0$, because of previous truncation, the computer finds that the numbers stored under the names X and Y are not equal. The problem has been exaggerated here for ease of calculation by saying the fractional part is stored in only 12 bits. For example, the Apple II microcomputer uses 32 bits to store the mantissa. The concept, however, is the same. Therefore, to test equality of real numbers, see if they are equal within a certain number of decimal places. The check of condition

IF ABS(X − Y) < 1.0E−8 THEN...

**LANGUAGE
EXAMPLE**

asks if X and Y are equal to within 8 decimal digits. Written mathematically, the statement is testing the validity of

$$|X - Y| < 10^{-8} \Longleftrightarrow -10^{-8} < X - Y < 10^{-8}.$$

Similarly, in working with matrices, where you expect a number of 0 entries, it is a good idea to have a statement similar to the following in a loop:

IF ABS (A[I,J]) < 1.0E−8 THEN
    A[I,J] := 0;

If the value is within 10^{-8} either way of 0, the number probably should be exactly 0. ☐

EXERCISES 10.3

Convert the decimal numbers in Exercises 1–8 to the equivalent hexadecimal numbers.

1. 121   **2.** 0.28125   **3.** 186.4   **4.** 0.93

5. 2626.625   **6.** 3335.7   **7.** 24,577   **8.** 454,660

Convert the decimal numbers in Exercises 9–13 to equivalent binary numbers without converting to hexadecimal first.

9. 23   **10.** 0.8125   **11.** 56.75   **12.** 87.7   **13.** 291

Convert the following decimal numbers in Exercises 14–18 to binary representation by converting to base 16 first.

14. 63   **15.** 0.45   **16.** 224.9

17. 28.015625   **18.** 554

Suppose the decimal numbers in Exercises 19–23 are stored normalized in base 2 with 8 bits for the fractional part. **a.** *Write the fractional part in hexadecimal after truncation, along with the exponent of 2. Find the* **b.** *absolute error and* **c.** *relative error for each.*

19. 0.82   **20.** 0.23   **21.** 0.0129   **22.** 7.1   **23.** 501.0

Convert the decimal numbers in Exercises 24–27 to hexadecimal notation, then normalize in base 16 by moving the hexadecimal point to the left of the first nonzero hexadecimal digit and multiplying by an appropriate power of 16.

24. 0.0024   **25.** 6.1   **26.** 20.25   **27.** 239.9

28. Suppose 0.6 is read into a variable X, and suppose the computer holds 4 hexadecimal digits in the normalized fraction.
    a. What will these digits be?
    b. If the value of X is printed with 5 decimal digits, what is printed?

Convert the decimal numbers in Exercises 29–31 to octal representation.

29. 57   **30.** 132   **31.** 3038

Convert the octal numbers in Exercises 32–34 to hexadecimal numbers without using the decimal system.

32. 644   **33.** 377   **34.** 7125

Convert the hexadecimal numbers in Exercises 35–37 to base 8 without using the decimal system.

35. DE   **36.** 9C   **37.** 6B8

38. Suppose an audio system can be controlled by a microcomputer. The status of the audio box is held in a location, X. There are 8 bits in this byte with a 1 indicating that a status is active. Suppose the status byte follows:

    NOT_USED   POWER   STOP   RUN   FAST_FWD   REWIND   SEARCH   RECORD
          7   6   5   4   3   2   1   0

    Thus, the binary number 01010001 indicates the audio deck power is on and the machine is recording. The BASIC command PEEK(X) retrieves this value in decimal.
    a. What value is retrieved?
    b. If the value 97 is retrieved, what is the binary number and the status?
    c. What decimal number is obtained to indicate the machine is on and in the process of rewinding? ☐

    LANGUAGE EXERCISE

39. The Burks-Goldstine-von Neumann report in 1947 recommended use of the binary instead of the decimal system in computers for the following reasons: (1) logic is fundamentally binary; (2) speed; (3) greater simplicity; (4) in decimal 40 bits are needed for accuracy of 10 decimal digits (precision of 10^{10}) since each digit is encoded in 4 bits, but in binary only about _____ digits are needed. For example, in decimal, 29 is encoded

HISTORICAL NOTE

as 0010 1001, where each digit is converted to 4 bits separately; but in binary 29 is encoded as 1 1101. ☐

a. Fill in the blank above.

How many bits are needed for 10^{12}

b. In decimal?   c. In binary?

40. Write an IF-clause in BASIC to test if two variables X and Y are equal to within 5 decimal digits. ☐

10.4

Two's Complement

HISTORICAL
NOTE

Several methods have been used for expressing negative numbers in the computer. The most obvious way is to convert the number to binary and stick on another bit to indicate sign, 0 for positive and 1 for negative. Suppose integers are stored using this **signed-magnitude** technique in 8 bits so that the leftmost or most significant bit holds the sign while the remaining bits represent the magnitude. Consequently, $+41_{10}$ is represented by 0010 1001 = 29_{16}, while -41_{10} is stored as 1010 1001 = $A9_{16}$. Early computers used signed-magnitude, but the circuitry for arithmetic was complicated, and the mapping from the signed-magnitude representation to the decimal integers was not one-to-one. ☐ Consider how you would represent 0. As all zeros, 0000 0000, seems easy enough, but that represents $+0$. The representation 1000 0000 indicates -0. With two ways to store 0, programming to check if a variable had a value of 0 meant asking if its value was $+0$ or -0.

A method of representing numbers that avoids this problem and simplifies arithmetic and circuitry is called **two's complement.** You can still tell the sign by looking at the most significant bit, but for negative numbers the rest of the number looks different. Considering -41 again, first, convert the absolute value of magnitude to binary.

$41_{10} = 0010\,1001.$

Then take the complement of each bit in 𝔹. Consequently, 1's become 0's, and 0's become 1's. This number is called the **one's complement** of 41.

$1101\ 0110.$

To complete the procedure, increment by 1 the one's complement.

$-41_{10} = 1101\,0111_2 = D7_{16}.$

Conversion of $41 = +41$ to two's complement consists merely of expressing the number in binary.

$41 = +41 = 0010\,1001 = 29_{16}.$

Another way to take the one's complement is to subtract the number from all 1's. After all, $1 - 0 = 1$ and $1 - 1 = 0$. Thus,

$$
\begin{array}{r}
1111\ 1111 \\
-\ 0010\ 1001 \\
\hline
1101\ 0110.
\end{array}
$$

Algorithm for Computing the Two's Complement of a Binary Integer

If the number is nonnegative then
  Stop
else
   a. Find the one's complement by complementing each
      bit, $0 \longleftrightarrow 1$, or by subtracting the number from all 1's.
   b. Increment the result by 1.

EXAMPLE 1

Express 104 and -104 in two's complement representation with 8 bits.

Solution

Convert 104 to binary. For faster arithmetic, let's convert to hexadecimal, then to binary.

$$16 \underline{|104}$$

$$6\quad 8$$

$$104_{10} = 68_{16} = 0110\ 1000_2.$$

Since 104 is positive, we are through for the first number.
    With -104, we complement and increment

$$1001\ 0111$$

$$-104 = 1001\ 1000_2 = 98_{16}.$$

A trick you may choose to use in taking the two's complement of a binary number like 0110 1000 is to start at the left, complementing every bit down to, but not including, the rightmost 1. Thus, the bits of 0110 are each complemented, while those of 1000 are not. ∎

As we have seen before, it is often faster to work with hexadecimal than binary numbers, particularly if integers are expressed in, say, 32 bits instead of the 8 of our examples. Moreover, in a dump of memory used for debugging (see Figure 10.1 at Section 10.2) or in assembler language, hexadecimal notation is used instead of binary. Let's see how we would express -62 in 8 bits with two's complement representation, performing all calculations in hexadecimal.

EXAMPLE 2

Express −62 in two's complement representation using the hexadecimal number system.

Solution

Find 62 in base 16:

$$16\lfloor\underline{62}$$
$$3\quad 14.$$

Thus, $62 = 3E = 0011\,1110_2$. Since −62 is less than 0, find the one's complement by subtracting from all 1's:

$$\begin{array}{rr} 1111\,1111 = & FF \\ -0011\,1110 & -\,3E \\ \hline 1100\,0001 & C1. \end{array}$$

Eight 1's form FF. Also, $F - E = 15_{10} - 14_{10} = 1$ and $F - 3 = 15_{10} - 3 = 12_{10} = C$. Now, incrementing, we have

$$1100\,0010 \quad \text{or} \quad C2.$$

In hexadecimal, we can combine taking the one's complement and incrementing just as we can in binary: $(FF - 3E) + 1 = (FF + 1) - 3E$. Combining the 1 with the right $F = 15$, we have

$$\begin{array}{rr} F\,(16) \\ -3\quad E \\ \hline C\quad 2. \end{array}$$

Actually, this procedure is called taking the **16's complement** of the number. ■

**Algorithm for Expressing in Hexadecimal Notation
the Two's Complement of a Decimal Integer**

1. Convert the absolute value to hexadecimal notation.
2. If the number is nonnegative then
       Stop
     else
       a. Subtract the least significant hexadecimal digit
          from 16 and
       b. Subtract each of the other digits from F.

EXAMPLE 3

Express -425 and -48 in two's complement notation using four hexadecimal digits to represent 16 bits.

Solution

a. Convert 425 to hexadecimal.

$$16\underline{|425}$$
$$\underline{|\ 26\ \ \ 9}$$
$$1\ \ 10$$

Thus, $425_{10} = 1A9_{16}$, so the two's complement is

$$
\begin{array}{r}
F\ F\ \ F(16) \\
-\ 0\ 1\ A\ \ \ 9 \\
\hline
-425 = \quad F\ E\ 5\ \ \ 7.
\end{array}
$$

b. Converting 48 to hexadecimal, we have

$$16\underline{|48}$$
$$3\ \ \ 0.$$

Completing the problem with $48_{10} = 30_{16}$, we subtract

$$
\begin{array}{r}
F F\ F(16) \\
-\ 0 0 3\ \ \ 0 \\
\hline
F F C(16).
\end{array}
$$

Since, $F = 15$ is the largest hexadecimal digit, we carry the one 16 to the 16's digit giving

$$-48 = FFD0. \quad ∎$$

Notice in the representation of $-104_{10} = 1001\,1000_2 = 98_{16}$ of Example 1 and $-425 = 1111\,1110\,0101\,0111 = FE57$ of Example 3, the most significant bit is 1, indicating a negative number. 104 and 425 have representations, respectively, of

$$0110\,1000 = 68_{16}$$

and

$$0000\,0001\,1010\,1001 = 01A9_{16}.$$

For the number to be negative, the most significant (leftmost) hexadecimal digit must be $8 = 1000_2$ or greater so that the most significant bit is set or equal to 1. Numbers with a most significant hexadecimal digit of $0 = 0000_2$ through $7 = 0111_2$ are positive. Thus, suppose we are looking at a hexadecimal dump of memory and know we have an 8-bit number at a position. We discern that the number 3C is positive, while A6 is negative. Looking at these hexa-decimal numbers, what decimal numbers do they represent? The positive number has a simple conversion back to decimal as was done with unsigned integers: $3C = 3 \times 16 + 12 = 60$. The beauty of two's complement representation is that to convert to decimal we just take the two's complement of the negative number and expand in the decimal number system:

$$
\begin{array}{r}
F(16) \\
-\ A\quad 6 \\
\hline
5\quad A = 5 \times 16 + 10 = 90.
\end{array}
$$

Thus, $A6 = -90_{10}$.

Algorithm for Converting a Two's Complement Hexadecimal Number to a Decimal Number

1. If the most significant hexadecimal digit is 8 or greater then
   a. The final answer is negative.
   b. Take the two's complement of the number: subtract the least significant digit from 16 and the others from F.
   else
   a. The final answer is positive.
2. Convert the result of Step 1 to a decimal number, combining the base 10 number with the sign.

EXAMPLE 4

If a computer uses 8-bit, two's complement representation for integers, what decimal numbers do the following represent?   **a.** $7F_{16}$ **b.** 80_{16} **c.** FF_{16}

Solution

a. 7F is a positive integer, so

$$7F = 7 \times 16 + 15 = 127.$$

Moreover, $7F = 0111\,1111$ is the largest positive number that can be expressed in 8 bits.

b. $80_{16} = 1000\,0000$ is a negative number. Taking the two's complement, we have

$$
\begin{array}{r}
F(16) \\
-\ 8\quad 0 \\
\hline
7(16) \rightarrow 80.
\end{array}
$$

Interestingly, 80_{16} is its own two's complement:

$$80_{16} = -80_{16} = -(8 \times 16 + 0) = -128_{10}.$$

This is the smallest negative number that can be expressed in 8-bit, two's complement representation. Notice, if we have a number whose unsigned representation is greater than 80_{16}, we are just subtracting more from F(16), resulting in a smaller absolute value:

c.

$$
\begin{array}{r}
F(16) \\
-\ F\ \ F \\
\hline
0\ \ 1.
\end{array}
$$

Thus, $FF = -1_{10}$. The range of integers is from 80_{16} to $7F_{16}$ or -128_{10} to 127_{10}. Notice we have one more negative integer than positive. ■

EXAMPLE 5

Find the range of the decimal numbers that can be stored in a computer in 16-bit, two's complement representation.

Solution

The largest positive number is

$$7FFF = 7 \times 4096 + 15 \times 256 + 15 \times 16 + 15 = 32,767.$$

The smallest negative number is 8000_{16}. Taking the two's complement, we have

$$
\begin{array}{r}
F\ F\ F(16) \\
-\ 8\ 0\ 0\ \ 0 \\
\hline
7\ F\ F(16) \rightarrow 8000.
\end{array}
$$

Therefore, $-8000_{16} = -8 \times 4096 = -32,768$ is the smallest negative integer. The range of integers is from $-32,768$ to $32,767$. This is precisely the range of integers in an Apple II microcomputer. ■

EXERCISES 10.4

Express each of the decimal numbers of Exercises 1–8 in 8-bit, two's complement representation. Write your answer in hexadecimal notation.

1. 91	**2.** −91	**3.** 115	**4.** −37
5. −64	**6.** −151	**7.** 128	**8.** −6

Express each of the decimal numbers of Exercises 9–16 in 16-bit, two's complement representation. Write your answer in hexadecimal notation.

9. 91   **10.** −91   **11.** −151   **12.** −8435

13. 30,583   **14.** −30,583   **15.** 40,000   **16.** −22,559

The numbers in Exercises 17–24 are stored as 8-bit, two's complement numbers, expressed here in hexadecimal notation. Find the decimal equivalent of each.

17. E   **18.** D4   **19.** 63   **20.** 9D

21. CC   **22.** E1   **23.** B0   **24.** 4E

The numbers in Exercises 25–31 are stored as 16-bit, two's complement numbers, expressed here in hexadecimal notation. Find the decimal equivalent of each.

25. 8C0C   **26.** 7000   **27.** A000   **28.** E1A7

29. FFFF   **30.** FFDE   **31.** 43C0

32. Find the range of decimal integers that can be expressed with two's complement representation
a. With 32 bits.   **b.** With 64 bits.

The numbers in Exercises 33–36 are expressed in two's complement representation. In each case, give the larger of the two signed integers.

33. 7C and 2A   **34.** 3F and D0

35. FF and 93   **36.** 0D32 and A432

37. Suppose we know a block of eight 8-bit, two's complement numbers are stored in the computer starting at a particular location. Seeing this hexadecimal dump, how many data values are positive, and how many are negative?

7E   AF   00   26   88   9F   03   D6

10.5

Arithmetic

We did a limited amount of addition using a carry as we counted in various bases. Recall in incrementing that if the digit to which we are adding 1 is the largest in that base, we write down 0 and carry 1 to be added to the next position, such as in the following examples:

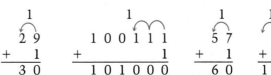
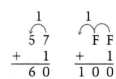

	Base 10	Base 2		Base 8	Base 16
	1	1		1	1
	2 9	1 0 0 1 1 1		5 7	F F
	+ 1	+ 1		+ 1	+ 1
	3 0	1 0 1 0 0 0		6 0	1 0 0.

Notice how the carry occurs in each problem:

$29 + 1 = (2 \times 10 + 9) + 1 = 2 \times 10 + 1 \times 10 = 3 \times 10.$

$100111 + 1 = (2^5 + 2^2 + 2 + 1)_{10} + 1$
$$= 2^5 + 2^2 + 2 + 2 = 2^5 + 2^2 + 2 \times 2$$
$$= 2^5 + 2^2 + 2^2$$
$$= 2^5 + 2 \times 2^2 = 2^5 + 2^3 = 101000_2.$$

$FF + 1 = (15 \times 16 + 15)_{10} + 1 = 15 \times 16 + 16$
$$= 16 \times 16 = 1 \times 16^2 = 100_{16}.$$

$57_8 + 1 = (5 \times 8 + 7)_{10} + 1 = 5 \times 8 + 8 = 6 \times 8 = 60_8.$

Adding numbers other than 1 is similar. Consider the binary sum $1011 + 11$:

$$
\begin{array}{r}
1\,\curvearrowright \\
1\ 0\ 1\ 1 \\
+\qquad 1\ 1 \\
\hline
1\ 1\ 1\ 0
\end{array}
$$

$1 + 1 = 2_{10} = 10_2$ -------- $\left\{\begin{array}{l}\text{Addition of the two least}\\ \text{significant bits gives 0}\\ \text{with a carry of 1.}\end{array}\right.$

$1 + 1 + 1 = 3_{10} = 11_2$ ------ $\left\{\begin{array}{l}\text{3 is bigger than the largest}\\ \text{digit in base 2; 2 divides into}\\ \text{3 with a quotient of 1 (carry)}\\ \text{and a remainder of 1 (write}\\ \text{down).}\end{array}\right.$

$1 + 0 = 1$

EXAMPLE 1

Perform these addition problems with *unsigned,* base 16 integers and check your work in the decimal system.

Solution

a. $\begin{array}{r} B\ E \\ +\ 2\ C \\ \hline E\ A \end{array}$

$14 + 12 = 26 = 16 + 10$

$1 + 11 + 2 = 14$

Check

$BE = 190$
$2C = \underline{\ \ 44}$
$EA = 234_{10}$

Solution

b.

$$
\begin{array}{r}
2.C\ 7 \\
5.F \\
+\ 4\ E.D\ 3 \\
\hline
5\ 7.8\ \ A
\end{array}
$$

$$12 + 15 + 13 = 40 = 2 \times 16 + 8$$
$$2 + 2 + 5 + 14 = 23 = 1 \times 16 + 7$$

Check

$$
\begin{array}{rl}
2.C7 = & 2.77734375 \\
5.F = & 5.9375 \\
4E.D3 = & 78.82421875 \\
\hline
57.8\ A = & 87.53906250_{10}. \quad \blacksquare
\end{array}
$$

One of the big advantages of two's complement arithmetic, is that we can use the same circuitry for adding positive or negative numbers. Reconsider Example 1a as 8-bit, two's complement numbers. While 2C still represents 44_{10}; BE represents -66_{10} (F(16) − BE = 42_{16} = $4 \times 16_{10} + 2 = 66$); and EA is -22_{10} (F(16) − EA = $16_{16} = 16_{10} + 6 = 22$). Notice, however, that even considered as *signed* numbers, the answer is still correct:

$$
\begin{array}{rl}
BE = & -66 \\
+\ 2C = +\ & 44 \\
\hline
EA = & -22_{10}.
\end{array}
$$

EXAMPLE 2

Perform the addition of the following 16-bit, two's complement numbers and check your answer in base 10.

Solution

Addition		Check in base 10

$$
\begin{array}{rcrcr}
F\ D\ C\ F & = & -0231_{16} & = & -561 \\
+\ 9\ 0\ \ C\ 3 & = & -6F3D & = & -28477 \\
\hline
8\ E\ 9\ 2 & = & -716E & = & -29038
\end{array}
$$

$$15 + 3 = 18 = 16 + 2$$
$$1 + 12 + 12 = 25 = 16 + 9$$
$$15 + 9 = 24 = 16 + 8.$$

Since there is not room for the last carry in the 16 bits, it is thrown away. Sometimes this carry out causes a problem, called **overflow,** where the final answer is incorrect because it has the wrong sign. For example, if we had added **F**DCF and 80C3 in 16 bits, the result would have been 7E92; but you cannot add two negatives and get a positive. There simply are not enough bits to express the answer. Overflow can also occur when you add two positive integers and get a negative result. ■

Subtraction in different bases also parallels subtraction in the decimal number system. To subtract y from x in base 10, you change the sign of y and add. For example, $10 - (-7) = 10 + +7 = 17$ or $35 - 6 = 35 + (-6) = 29$. Similarly, in binary to subtract y from x, you take the two's complement of y and add.

EXAMPLE 3

Perform the subtraction of the following 8-bit, two's complement numbers.

Solution

Check in base 10

a. $\begin{array}{r} 0011\ 1011_2 \\ -\ 1110\ 1001 \end{array} \rightarrow \begin{array}{r} 0011\ 1011 \\ +\ 0001\ 0111 \\ \hline 0101\ 0010_2 \end{array} \begin{array}{l} =\ \ \ \ 59 \\ =\ -(-23) \\ =\ \ \ 82_{10} \end{array}$

b. $\begin{array}{r} \text{D3} \\ -\text{C2} \end{array} \rightarrow \begin{array}{r} \text{D3} \\ +\ \text{3E} \\ \hline 11_{16} \end{array} \begin{array}{l} =\ \ -45 \\ =\ -(-62) \\ =\ \ \ 17_{10}. \end{array}$ ■

EXAMPLE 4

Suppose X is an 8-bit **register** or high-speed location in the arithmetic/logic unit of a computer.

a. If X initially contains 0 and is decremented by 1, what value in hexadecimal will there be in X?

b. If we interpret FF as an unsigned integer, what decimal number does it represent?

c. If we interpret FF as a signed integer, what decimal number does it represent?

d. Suppose the value of X, which was initialized to 0, is an index to a loop. Each time through the loop, the contents of X are decremented. Suppose at the end of the loop is the instruction in assembler language

BNE LOOP

which means branch to the top of the loop as long as the contents of X are "not equal to zero." How many times will the loop be executed?

LANGUAGE
EXAMPLE

Solution

a.
$$\begin{array}{r} 00 \\ - \ 01 \end{array} \rightarrow \begin{array}{r} 00 \\ + \ \mathrm{FF} \\ \hline \mathrm{FF} \end{array}$$

b. As an unsigned integer, $\mathrm{FF} = 15 \times 16 + 15 = 255$.

c. As a signed integer, $\mathrm{FF} = -(\mathrm{F}(16) - \mathrm{FF}) = -1$.

d. The loop is executed 256 times since at the top of the loop, X has the values 0, FF, FE, FD, . . . , 3, 2, 1. □ ■

We will not deal with multiplication or division here except by powers of 2. When an integer is stored in a register, the operation of (logical) shifting the number to the left or right often can be accomplished readily. One bit falls off the end into oblivion and a 0 is used to pad at the other end. Notice what happens to the value stored in an 8-bit register, as the contents are shifted several times to the left. Suppose the starting value is 9_{10}:

Binary number	Decimal number
0000 1001 ← 0	9
0 ← 0001 0010 ← 0	18
0 ← 0010 0100 ← 0	36
0 ← 0100 1000	72 .

If the number is a signed integer, shifting again will result in a negative number and thus cause overflow. If it is interpreted as an unsigned integer, we can get in one more shift before that most significant 1 falls off the end. The reason shifting to the left causes doubling goes back to our representation of numbers:

$$\begin{aligned} 9 \quad &= 1001_2 = 2^3 + 1 \\ 2 \times 9 &= 2(2^3 + 1) \\ &= 2^4 + 2 \\ &= 10010_2 = 18. \end{aligned}$$

Shifting to the right results in an integral halving of the unsigned integer:

Binary number	Decimal number
0 → 0000 1001	9
0 → 0000 0100 → 1	4
0 → 0000 0010 → 0	2
0 → 0000 0001 → 0	1 .

Of course, division by 2 must be integer division so that 9/2 results in 4 instead of 4.5.

There is a type of shifting (arithmetic shifting) that will double and halve signed numbers, but it will not be covered in this book. Arithmetic with floating point numbers has also been omitted. However, this introduction to other number systems should provide a solid background for further study.

EXERCISES 10.5

Find the sums of the unsigned numbers in the indicated bases in Exercises 1–6.

	Base 2		Base 8		Base 16
1.	11 1110 1111	**2.**	34	**3.**	3F0
	+ 10 1101		+ 562		+ DF

4.	111	**5.**	2.7	**6.**	AA.
	1001		73.0		2DB.B
	1 1011		+ 4.4		+ C.C3
	+ 101				

For Exercises 7–10 perform the indicated operations on the 8-bit, signed integers in base 2. Convert to base 10 to check your answer.

7.	1001 1100	**8.**	0010 1011	
	+ 0111 0100		+ 0101 1101	

9.	1110 0000	**10.**	0011 0110	
	+ 1001 1100		− 1111 0110	

For Exercises 11–14 perform the indicated operations on the 8-bit, signed integers represented in hexadecimal. Convert to base 10 to check your answer.

11.	8A	**12.**	BC	**13.**	DA	**14.**	7C
	+ 2F		− 14		− E3		+ C7

For Exercises 15–18 perform the indicated operations on the 16-bit, signed integers represented in hexadecimal. Convert to base 10 to check your answer.

15.	624D	**16.**	FFFF	**17.**	D4E5	**18.**	5AE2
	+ 1591		+ FFFF		− 882F		− E350

19. Express −3
    a. In 2 hexadecimal digits.
    b. In 4.
    c. In 8.

20. Suppose in a certain assembler language BNE LOOP says if the result of the last computation is not 0, add the value of LOOP to the contents of PC, the program counter register, to compute the address of the location of the instruction to execute next. Thus, LOOP is an offset from the address in the PC. If LOOP is positive, you branch forward in the program. If LOOP is negative, you branch back. Suppose the value of LOOP is an 8-bit signed integer and PC is a 16-bit nonnegative integer. For the following values of LOOP and PC, indicate the address of the next instruction and whether you are branching forward or backward. Be careful to convert the value of LOOP to an equivalent 16-bit number before addition.

$$\text{LOOP} = 2C \qquad \text{PC} = A035_{16} \quad \square$$

21. Repeat Exercises 20 using
$$\text{LOOP} = FC \qquad \text{PC} = 3821_{16}$$

22. Suppose you are writing the BNE instruction and want to branch 39_{10} bytes beyond the address in the PC. What is the offset in hexadecimal? (See Exercise 20.)

23. Answer the question in Exercise 22 if you want to branch 25_{10} bytes back from the address in the PC. $\square$

24. The memory of the Apple II microcomputer is divided into sections called **pages.** There are 256 pages, each containing 256 bytes. Addresses have the page number followed by the byte number on the page, both in hexadecimal. Thus, D635 stands for byte $35_{16} = 53_{10}$ on page $D6_{16} = 214_{10}$.

    a. How many bytes are in the memory?

    b. If pages D0 through FF are in read-only-memory, ROM, how many pages, expressed in hexadecimal and decimal notation, are in ROM?

    c. How many bytes, expressed in hexadecimal and decimal notation, are in ROM?

    d. How many bytes, expressed in hexadecimal notation, are in pages 0 through 7?

    e. Suppose pages 0 through 7 are reserved for the system except for bytes 00 to CF on page 3. How many bytes, expressed in hexadecimal and decimal notation, are reserved for the system on these pages?

Suppose the numbers in Exercises 25–27 are stored as unsigned integers in 8-bit registers. Double and halve each, expressing the answers in the indicated base.

25. $0001\ 1101_2$ **26.** CF_{16} **27.** 129_{10}

Fill in the power of 2 or base 2 number to complete each equality in base 2 for Exercises 28–31.

28. $110.1111_2 = 0.110111_2 \times 2^{-}$ **29.** $0.001001_2 = 0.1001_2 \times 2^{-}$

30. $\underline{\hspace{1.5cm}} = 0.11_2 \times 2^{-4}$ **31.** $\underline{\hspace{1.5cm}} = 0.11_2 \times 2^4$

Chapter 10   REVIEW

Terminology/Algorithms	Pages	Comments
Binary or base 2 number system	360	Digits from $\mathbb{B}$
Base 2 → base 10	360	Expand in powers of 2
Counting in base 2	362	Algorithm
Decrementing in base 2	363	Algorithm
Largest unsigned integer in n bits	363	$2^n - 1$
Hexadecimal or base 16 number system	366	Digits from $\mathbb{H}$
Base 16 → base 10	366	Expand in powers of 16
Counting in base 16	367	Algorithm
Decrementing in base 16	367	Algorithm
Largest unsigned integer in n hexadecimal digits	367	$16^n - 1$
Base 2 → base 16	369	Split into groups of 4 bits from binary point, convert each group to hexadecimal
Base 16 → base 2	369	Each hexadecimal digit → binary
Octal or base 8 number system	370	Digits from $\{0, 1, 2, 3, 4, 5, 6, 7\}$
Base 10 integer → base b	376	Algorithm
Base 10 fractional number → base b	376	Algorithm
Signed-magnitude	381	$0 \equiv +,\ 1 \equiv -,$ magnitude → base 2
→ Two's complement representation in base 2	382	Algorithm
→ Two's complement representation in base 16	383	Algorithm
Two's complement, base 16 number → base 10	385	Algorithm

Terminology/Algorithms	Pages	Comments
Addition in bases	388	Algorithm
Overflow	389	Wrong sign because of carry out
Subtraction in bases	390	Algorithm

REFERENCES

Buchholz, W. "Origin of the Word Byte." *Byte* 144, February 1977.

Gardner, W. D. "John V. Atanasoff: He and Mauchley Are 'Pretty Good Friends'." *Datamation* 20: 88–89, February 2, 1974.

Glaser, A. *History of Binary and Other Nondecimal Numeration.* Los Angeles: Tomash, 1971.

Hinkel, J. S. *Apple Machine/Assembly Language Programming.* Scottsdale, AZ: Gorsuch Scarisbrick, 1986.

Hopcroft, J. E., and J. D. Ullman. *Introduction to Automata Theory, Languages, and Computation.* Reading, MA: Addison-Wesley 1979.

Interactive Audio, "Designing an Interface Card." *Byte* 10 (6): 113–117, June 1985.

Sagan, C. *et al. Murmurs of Earth, the Voyager Interstellar Record.* New York: Random House, 1978.

Tanenbaum, A. S. *Structured Computer Organization,* 2nd ed. Englewood Cliffs, NJ: Prentice-Hall, 1984.

Tropp, H. S., ed. "Origin of the Term Bit." *Annals of the History of Computing* 6 (2): 152–155, April 1984.

11 *ANALYSIS OF ALGORITHMS*

Foggy Chessmen. This computer graphics scene is composed of three separate images: the chessmen, their reflections, and the chess board.

Exponentials and Logarithms

The function $h(n) = 2^n$ has arisen in many contexts in this book such as computing the number of subsets of a set, the number of rows in a truth table, the number of n-bit numbers, and the maximum number of leaves at level n of a binary tree. This function is one example of an exponential function. Another such function, $g(x) = 10^x$, has special significance because of the decimal number system. And $\exp(x) = e^x$, which is built into many computer languages, is employed in a variety of applications involving continuous exponential growth. The logarithmic functions, which are inverses of the corresponding exponential functions, are used in computer science in such areas as the analysis of the efficiency of algorithms. For example, as discussed in Section 11.2, a binary search of a sorted list of n items for a particular element takes an average of $\lceil \log_2(n + 1) \rceil$ inquiries into the list.

We have been most concerned with the exponential function $h(n) = 2^n$ with the domain restricted to the set of integers. Under such a restriction, the graph is as in Figure 11.1. The graph of the continuous, increasing function $h(x) = 2^x$ with domain $\mathbb{R}$ and range $\mathbb{R}^+$, the set of positive real numbers, is given in Figure 11.2. Notice that the graphs of $g(n) = (1/2)^n$ and $g(x) = (1/2)^x$ in Figure 11.3 are

Figure 11.1

Graph of $h(n) = 2^n$, $n \in \mathbb{I}$

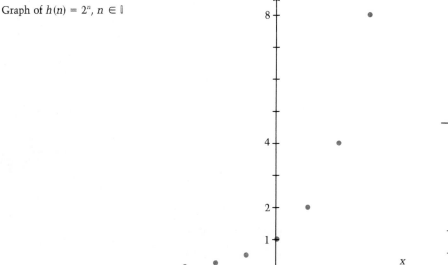

n	2^n
0	1
1	2
2	4
3	8
-1	1/2
-2	1/4
-3	1/8

Figure 11.2

Graph of $h(x) = 2^x$, $x \in \mathbb{R}$

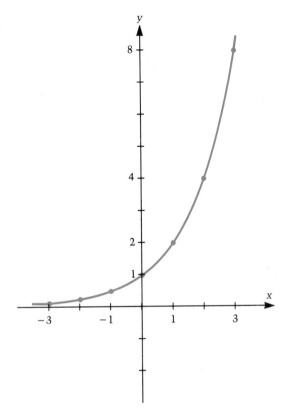

x	2^x
0	1
1	2
2	4
3	8
-1	1/2
-2	1/4
-3	1/8

Figure 11.3

Graphs of (**a**) $g(n) = (1/2)^n$, $n \in \mathbb{I}$ and (**b**) $g(x) = (1/2)^x$, $x \in \mathbb{R}$

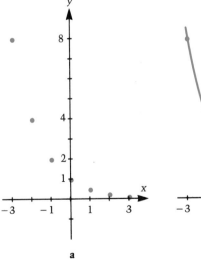

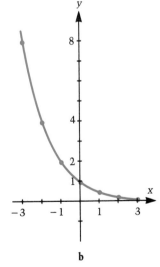

a

b

n	$(1/2)^n$
0	1
1	1/2
2	1/4
3	1/8
-1	2
-2	4
-3	8

decreasing; that is, observing the graph from left to right, the graph goes down. Exponential functions have the general form $f(x) = b^x$ where b is positive and not equal to 1. The exception for 1 is because $g(x) = 1^x = 1$, whose graph is a horizontal line; and this type of graph is not typical of the family of exponential functions. Graphs of exponential functions $f(x) = b^x$ may be more or less steep, but have the general shape of $h(x) = 2^x$ for $b > 1$ or $g(x) = (1/2)^x$ for $0 < b < 1$.

DEFINITION

An **exponential function** is of the form $f(x) = b^x$, where b is a positive real number not equal to 1.

Another important exponential function is $\exp(x) = e^x$, where e is the irrational real number $e = 2.71828. . .$, which is the value that $(1 + 1/x)^x$ approaches as x gets larger and larger without bound. For a bank account with an annual rate of growth r, a principal P, and n compounding periods per year, the amount present after t years is

$$A = P(1 + r/n)^{nt}.$$

Letting $x = n/r$ so that $xr = n$, we also have

$$A = P[(1 + 1/x)^x]^{rt}.$$

If, however, there is continuous compounding, the number of compounding periods per year, n, goes to infinity as does x, and the formula becomes

$$A = Pe^{rt}.$$

This formula is used not only for the growth of your continuously compounded bank account, but also for radioactive decay and population growth.

EXAMPLE 1

Suppose you invest $1000 in two accounts. One account pays 8.1% compounded quarterly while the other pays 7.5% compounded continuously. How much is in each account after 5 years?

Solution

First, as always, we assign the variables:

$$P = 1000 \quad \text{and} \quad t = 5$$

for both accounts.

For the quarterly compounded account

$n = 4$

$r = 0.081$

$A = P(1 + r/n)^{nt} = 1000(1 + 0.081/4)^{4 \times 5}$

$\quad = 1000(1.02025)^{20} \approx \$1493.25.$

(Be sure to use your calculator, which hopefully has keys to evaluate x^y and e^x.)

For the continuously compounded account

$r = 0.075$

$t = 5$

$A = Pe^{rt} = 1000e^{0.075 \times 5} = 1000e^{0.375} \approx \$1454.99.$

Here, the first account is a better investment. ■

Many problems involve being given a base and its value to some power for which we need to calculate the exponent. For example, suppose we want to know how long it will take to double the amount in the 7.5% continuously compounded account. We have $A = 2P$, $r = 0.075$, and want to calculate t in $2P = Pe^{0.075t}$, or after canceling P, t in $2 = e^{0.075t}$. We want to get the t "down" to solve for it. What is needed is the inverse of the exponential function, the natural logarithm or logarithm to the base e function, $\exp^{-1}(x) = \ln(x) = \log_e(x)$.

Before completing this problem, let us look at the logarithm to the base 2 function, $h^{-1}(x) = \log_2(x)$, the inverse function of $h(x) = 2^x$. Recall from Section 5.4 that (x, y) is on the graph of h^{-1} if and only if (y, x) is a point on the graph of h. Thus, the domain and range values are switched as follows:

1. $x = 2^y \iff \log_2 x = y.$

Figure 11.4 illustrates this exchange of elements in the ordered pairs. The domain of $h^{-1}(x) = \log_2 x$ is $\mathbb{R}^+$ (the range of $h(x) = 2^x$), while the range of $\log_2$ is $\mathbb{R}$ (the domain of h). Moreover, when given the value of a logarithm, you are given the exponent, and the base of the logarithm is the base in the exponential form. Since these functions are the inverses of each other, their composition always gives the domain element:

2. $\log_2(2^x) = x = {_2}\log_2 x.$

Figure 11.4

Graphs of $h(x) = 2^x$ and $h^{-1}(x) = \log_2(x)$

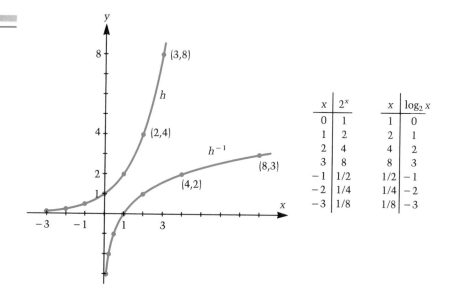

x	2^x		x	$\log_2 x$
0	1		1	0
1	2		2	1
2	4		4	2
3	8		8	3
-1	1/2		1/2	-1
-2	1/4		1/4	-2
-3	1/8		1/8	-3

DEFINITION

The **logarithm to the base b function,** $g(x) = \log_b(x)$, for $b \neq 1$, is the inverse of the exponential function $f(x) = b^x$.

There are several properties of logarithms which can be more easily understood by looking at the exponential function and then using the equivalence with property 1 above.

3. Since $1 = 2^0$, **$\log_2 1 = 0$.**
4. Because $2 = 2^1$, **$\log_2 2 = 1$.**
5. Since when multiplying with the same base we add exponents as in $2^c \cdot 2^d = 2^{c+d}$ and since logarithms are the exponents, we have **$\log_2 uv = \log_2 u + \log_2 v$.** For example, since

$$2^5 = 32 \iff \log_2 32 = 5$$
$$2^2 = 4 \iff \log_2 4 = 2$$
and $2^3 = 8 \iff \log_2 8 = 3$

we have

$$\log_2 (4 \times 8) = \log_2 32 = 5 = 2 + 3 = \log_2 4 + \log_2 8.$$

6. Similarly, since division with a common base results in subtraction of exponents, $2^c/2^d = 2^{c-d}$, we have **$\log_2 u/v = \log_2 u - \log_2 v$.** We see that

$$\log_2 32/4 = \log_2 8 = 3 = 5 - 2 = \log_2 32 - \log_2 4.$$

7. Because the raising of a power to a power is equivalent to multiplying the exponents, $(2^c)^d = 2^{cd}$, we have $\log_2 u^v = v \log_2 u$. For example, $2^6 = 64$, and

$$\log_2 (8^2) = \log_2 64 = 6 = 2 \times 3 = 2 \log_2 8.$$

These properties hold for all logarithms, regardless of the base. Restated in their general form, the properties are as follows:

1. $x = b^y \Longleftrightarrow \log_b x = y$
2. $\log_b (b^x) = x = b^{\log_b x}$
3. $\log_b 1 = 0$
4. $\log_b b = 1$
5. $\log_b uv = \log_b u + \log_b v$
6. $\log_b u/v = \log_b u - \log_b v$
7. $\log_b u^v = v \log_b u$

HISTORICAL NOTES

John Napier, a Scottish baron who considered mathematics a hobby, published his invention of logarithms in 1614. Unlike most other scientific achievements, his work was not built upon that of others. This highly original invention was welcomed enthusiastically by the population which had great difficulty with even elementary arithmetic, a subject not usually taught in the schools. Suddenly it was found that problems of multiplication and division could be reduced to much simpler problems of addition and subtraction (see properties 5 and 6 above). Napier also contributed to mathematics by providing a convenient way of writing real numbers by using a decimal point. It wasn't until the 18th century when the notation e was introduced and evaluated to 23 places by Euler. ☐ We now return to the length of time needed to double our money.

EXAMPLE 2

Suppose you invest money in an account that pays 7.5% compounded continuously. How long will it take you to double your investment?

Solution

The values of the variables are $A = 2P$ with P a constant, $r = 0.075$, $t = ?$.

$$2P = Pe^{0.075t}$$
$$2 = e^{0.075t}.$$

The natural logarithm of both sides will give us the exponent.

$$\ln 2 = \ln e^{0.075t} = 0.075t.$$

Now, using your calculator, we find

$$t = (\ln 2)/0.075 \approx 9.242 \text{ years.} \quad \blacksquare$$

EXAMPLE 3

How many bits are needed to express the integers from 0 to 337,000 in binary?

Solution

We want to find the integer n such that

$$337,000 \le 2^n - 1 \quad \text{or} \quad 337,001 \le 2^n$$

because the maximum number to be expressed in n bits is $2^n - 1$. For the time being let us look at the equality:

$$337,001 = 2^n.$$

To solve for the exponent, take the logarithm to the base 2 of both sides:

$$\log_2 337,001 = \log_2 2^n = n.$$

$\log_2 337,001$ is probably not an integer, and we want to make sure n is large enough to represent all the numbers from 0 to 337,000, so we take the ceiling to find the exact value of n:

$$\lceil \log_2 337,001 \rceil = n.$$

We run into a problem, however, in evaluating $\log_2 337,001$ since your calculator probably only has keys for ln and $\log_{10}$. An additional formula can be helpful:

$$\log_2 337,001 = (\ln 337,001)/(\ln 2) \approx 18.4.$$

Thus, $n = 19$ bits are needed to represent 0 through 337,000. You might want to verify that the binary representation for 337,000 is 101 0010 0100 0110 1000. $\quad \blacksquare$

In general, the **change of base formula** is

$$\log_b x = \frac{\log_a x}{\log_a b}$$

for $a, b \in \mathbb{R}^+$ and not equal to 1. Consequently, the same result is obtained for Example 3 if we used base 10 (or any other base) instead of base e:

$$\log_2 337,001 = (\log_{10} 337,001)/(\log_{10} 2) \approx 8.4.$$

The logarithmic and exponential functions are used to implement the real number powers such as $x = b^y$ in the computer. For example, given an assignment statement X = 7.2 ** 3.9 in FOR-TRAN, the exponent is "brought down" by taking the logarithm of $X = 7.2^{3.9}$.

$$\ln X = \ln 7.2^{3.9} = 3.9 \ln 7.2$$
$$\approx 3.9 \times 1.974081026 \approx 7.698916001.$$

Then the exponential function of both sides gives X:

$$e^{\ln X} = X \approx e^{7.698916} \approx 2205.95544. \quad \square$$

Checking if our result seems reasonable, we can evaluate $7^4 = 2401$ to approximate the actual answer.

This method of using the fact that the logarithmic and exponential functions are inverses of each other works very nicely, but presents problems with negative bases. Suppose X = (−5)**2.0. With an integer exponent, (−5)**2 is computed as (−5)*(−5) to yield $(−5)^2 = 25$. Using logarithms with the real exponent of 2.0, we have

$$\ln X = \ln(−5)^{2.0} = 2.0 \ln(−5).$$

But now we have problems. Try taking $\ln(−5)$ with your calculator. You get an error, and well you should because the domain of the ln function is $\mathbb{R}^+$. You cannot take the logarithm of a negative number. Thus, in programming, write integer exponents without a decimal point. Otherwise, the power will be evaluated using logarithms or not at all.

EXERCISES 11.1

1. Graph the functions $f(x) = 10^x$ and $f^{-1}(x) = \log_{10} x$ on the same coordinate system.

2. Graph the functions $f(x) = 0.1^x$ and $f^{-1}(x) = \log_{0.1} x$ on the same coordinate system.

3. Graph the functions $f(x) = e^x$ and $f^{-1}(x) = \ln x$ on the same coordinate system.

Evaluate the numbers in Exercises 4–29 without use of a calculator.

4. $\log_3 3$ 5. $\log_3 3^7$ 6. $(\log_3 3)^7$

7. $7(\log_3 3)$ 8. $\log_3 81$ 9. $\log_3(3^7 \cdot 81)$

10. $\log_3 3^7 \cdot \log_3 81$ 11. $\log_3(81/3^7)$ 12. $(\log_3 81)/(\log_3 3^7)$

13. $\log_4 1$ 14. $\log_4 0$ 15. $10^{\log_{10} 7}$

16. $100^{\log_{10} 7}$ 17. $10,000,000^{\log_{10} 7}$ 18. $\log_{10} 10^6$

19. $\log_{10} 100^6$ **20.** $\log_{10} 10,000^6$ **21.** $\log_2(4 + 4)$

22. $\log_2 8$ **23.** $\log_2 5/\log_2 125$ **24.** $\log_2(5/125)$

25. $\log_2(1/32)$ **26.** $\log_2(1/1024)$ **27.** $\ln e^{7x}$

28. $e^{\ln 9.3}$ **29.** $\ln(-3)$

Using a calculator evaluate the logarithms in Exercises 30–32.

30. $\log_2 17.3$ **31.** $\log_5 10.2$ **32.** $\log_3 143.8$

Write the exponential equations in Exercises 33–35 as equivalent logarithmic equations.

33. $2^9 = 512$ **34.** $\sqrt[3]{64} = 4$ **35.** $10^{-4} = 0.0001$

Write the logarithmic equations in Exercises 36–38 as equivalent exponential equations.

36. $\log_4 8 = 3/2$ **37.** $\log_{10} 2 = 0.30103$ **38.** $\ln 16.445 = 2.8$

39. Consider the fact that $2^{10} = 1024 \approx 10^3$. 2^{70} is approximately what power of 10?

40. a. 10^{15} is approximately what power of 2?
   b. Thus, approximately how many binary digits will be needed to represent 10^{15} as an integer?
   c. Solve $10^{15} = 2^x$ more precisely using $\log_{10}$.

41. Repeat Exercise 40 using 10^{19}.

42. Avogadro's number is 6.022045×10^{23}.
   a. Represent 6.022045 as 10 to a real number exponent.
   b. Represent Avogadro's number as 10 to a real number exponent.
   c. How many bits are necessary to hold Avogadro's number as an integer?

43. Suppose you invest $500 in a money market account that pays a 9.2% annual interest rate compounded quarterly. How much will be in the account at the end of
   a. 3 years?   **b.** 10 years?
   c. How long will it take to triple the original investment?

44. Repeat Exercise 43 assuming continuous compounding.

45. Suppose the population of a human cell culture grows continuously. If there are 5×10^4 cells present initially growing at a rate of 150% per day, how many are there at the end of 3 days?

46. In a culture growing at a continuous rate, if there are 8.2×10^6 cells present after 3 days of growth at a rate of 60% per day, how many were there initially?

47. Suppose a cell culture grows continuously. If 7.56×10^5 cells were in the culture initially and 2.85×10^8 after 3 days, what was the growth rate?

48. Suppose you place a certain amount in an account that pays a guaranteed rate of 6% compounded monthly. How long does it take for your money to
   a. Double?   b. Quadruple?
   c. Be 8 times the original amount?
   d. Be 16 times the original amount?
   e. Compare the length of time it took to get twice as much money as it took to get 16 times as much money.

49. Suppose researchers have found it takes 2 hours for a certain bacteria count to double. How long will it take for a colony to increase from a count of 7000 bacteria to 9.3×10^8 bacteria?

LANGUAGE
EXERCISE

50. Write a Pascal (or FORTRAN) function to calculate the amount A when compounding continuously, given P, r, and t. *Note:* EXP(X) evaluates e^x. ☐

LANGUAGE
EXERCISE

51. Suppose your Pascal compiler does not implement exponents but does have the EXP and LN functions. Write your own Pascal function, POWER, to evaluate B^U as POWER(B, U). ☐

52. A radioactive substance diminishes with time. Since it is decaying, the rate r has a negative value. Suppose the half-life for the substance is 5730 years; i.e., after 5730 years the amount present A is half the original amount P.
   a. Find r.
   b. If one kilogram was present originally, how much will remain after 10,000 years?

53. The Richter scale, $\log_{10}(I/I_0)$, indicates the intensity I of an earthquake relative to the intensity I_0 of a particular small quake.
   a. An earthquake that is 1000 times the intensity of I_0 measures what on the Richter scale?
   b. The San Francisco earthquake in 1906 measured 8.6 on the Richter scale. How many times more intense was it than I_0?

54. a. A binary tree of height n and with the maximum number of nodes has how many leaves at level n?
   b. Suppose a binary tree has all its leaf nodes at level n. If the tree has 512 leaves, what is its height?
   c. What is the length of the longest path in the tree of part b?

55. Repeat Exercise 54 for a 3-ary tree, a tree with at most 3 children for each node. For part b use 729 leaves.

Prove each of the properties in Exercises 56–61 using properties 1 and/or 2: (1.) $x = b^y \Longleftrightarrow \log_b x = y$ (2.) $\log_b(b^x) = x = b^{\log_b x}$.

56. Prove $\log_b b = 1$.   57. Prove $\log_b 1 = 0$.

58. Prove $\log_b uv = \log_b u + \log_b v$.   59. Prove $\log_b u/v = \log_b u - \log_b v$.

60. Prove $\log_b u^v = v \log_b u$.

61. Prove $\log_b x = \dfrac{\log_a x}{\log_a b}$.

Complexity

In Section 3.3 on loops we touched upon the subject of analysis of algorithms which is the study of the efficiency of algorithms, how much time and/or how much space they consume. Since that section, we have learned about some formulas involving sigma notation, about binary trees, and about the logarithmic and exponential functions, all of which will help us now.

Exercise 12 in Section 3.3 involved analysis of an algorithm to search through an array, item by item, looking for X. Perhaps we are looking through a class roll for the name of a student or an inventory list for a particular item. Either the index of the array item or 0 was to be returned, depending on whether X was in the array or not. The algorithm to perform this **sequential search** of the array A of N elements is as follows:

**Algorithm for Sequential Search for X
in Array A of N Elements**

```
1. [Loop]        For I from 1 to N do:
   A. [Found?]      If X = A[I], then print I and stop.
2. [Not Found]  Print 0.
```

If X is found immediately, statement A is executed once. We use the notation, $O(1)$, to indicate in this **best case** where the algorithm goes through the loop one time. This notation is called "big oh" notation for obvious reasons. In the **worst case** X is not found or is the Nth element, and we say the run time is $O(N)$. To calculate the average number of times step A is executed assuming X is in the array, we must consider every possibility. If X is found as an Ith item, for $I = 1, 2, \ldots,$ or N, step A is executed I times. With N situations we have an average of

$$\frac{(1 + 2 + \cdots + N)}{N} = \frac{\displaystyle\sum_{i=1}^{N} i}{N}$$

$$= \frac{\dfrac{N(N + 1)}{2}}{N} \qquad \text{formula}$$

$$= \frac{N(N + 1)}{2N}$$

$$= \frac{N+1}{2} \qquad \text{cancel}$$

$$= \frac{N}{2} + \frac{1}{2} \le N.$$

This last inequality indicates that the average number of times step A is executed is less than or equal to N. We say that, assuming X is in the array, this algorithm has an average run time of $O(N)$. We don't worry about the constants since the actual average run time, which depends on external factors such as the machine and compiler, is still a multiple of N. We just look for a simple function of N, disregarding constant multiplies or sums.

DEFINITION

Let f and g be functions with domain $\mathbb{I}^+$. $f(n)$ is of **complexity** or **order at most $g(n)$**, written with **big oh** notation as $f = O(g)$, if there exist a positive constant c and a positive integer n_0 such that

$$|f(n)| \le c|g(n)|$$

for all $n > n_0$.

Should $f(n)$ and $g(n)$ always be nonnegative, the absolute values, of course, are unnecessary. In a certain sense this definition is saying that f is dominated by g.

The following theorem tells us how we can strip away unnecessary parts to arrive at a simplified $O(f)$. The statement in part a of the theorem says that for any positive constant a, $a \cdot f$ and f are equally efficient. This statement allows us to say the average run time for the sequential search is $O(N)$ as well as $O(2N)$. The statement in part b says that an algorithm that has a run time of $O(n)$ is more efficient than one with a run time of $O(n^2)$; a $O(n^2)$ algorithm is more efficient than one that is $O(n^3)$, etc. As the integer exponent increases, the efficiency decreases. And the statement of part c says when we have a polynomial, the degree of the polynomial indicates the efficiency. For example, $3x^5 + 12x^3 + x^2 - 3x + 4 = O(x^5)$, or $3x^5 + 12x^3 + x^2 - 3x + 4$ is of complexity at most the polynomial x^5. Also, as we have noted, $1/2 \, N + 1/2 = O(N)$. An algorithm which is of an order at most $f(n)$, a polynomial, is called a **polynomial-time algorithm**.

THEOREM

a. For $a > 0$, $af = O(f)$ and $f = O(af)$.
b. n^{m-1} is $O(n^m)$, but n^m is not $O(n^{m-1})$ for $m, n \in \mathbb{I}^+$.
c. If $f(n) = \sum_{i=0}^{m} a_i n^i$ is a polynomial of degree m, then f is $O(n^m)$.

Proof

a. af is $O(f)$ since for constant $c = a$, $af \leq cf = af$. f is $O(af)$ since for $c = 1/a$, $f \leq c(af) = (1/a)(af) = f$.
b. Suppose $m, n \in \mathbb{I}^+$. $f(n) = n^{m-1}$ is $O(n^m)$ since for $c = 1$, $n^{m-1} \leq n^m$. But $g(n) = n^m$ is not $0(n^{m-1})$: There is no constant c with $n^m \leq cn^{m-1}$ for all n since solution of this inequality yields $n \leq c$ which makes c depend on n and hence be nonconstant.
c. $n^i \leq n^m$ for $i = 0, 1, 2, \ldots, m$. Therefore,

$$|f(n)| = |a_0 n^0 + a_1 n^1 + \cdots + a_m n^m|$$
$$\leq |a_0|n^0 + |a_1|n^1 + \cdots + |a_m|n^m$$
$$\leq |a_0|n^m + |a_1|n^m + \cdots + |a_m|n^m$$
$$= (|a_0| + |a_1| + \cdots + |a_m|)n^m.$$

Since $(|a_0| + |a_1| + \cdots + |a_m|)$ is a positive constant, $f(n)$ is $O(n^m)$.

EXAMPLE 1

Indicate why each statement is true.

a. $17N = O(N)$ and $N = O(17N)$.

Solution

This statement is true by part a of the theorem.

b. $f(n) = 17n + 5$ is $O(n)$

Solution

Since $f(n)$ is a polynomial, part c applies.

c. $f(n) = n$ is $O(n^2)$, but $g(n) = n^2$ is not $O(n)$.

Solution

The statement is true by part b of the theorem. Hence, an algorithm whose run time is of order at most n is more efficient than one that is $O(n^2)$.

d. $f(N) = \log_2 N$ is $O(N)$.

Figure 11.5

Graphs of $f(N) = \log_2 N$ and $g(N) = N$

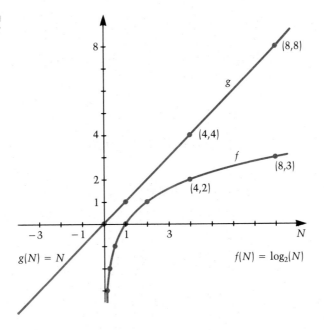

$g(N) = N$ $f(N) = \log_2(N)$

Solution

As can be observed in Figure 11.5, for all N, $\log_2 N < N$, so $\log_2 N$ is $O(N)$. An algorithm that runs in $\log_2 N$ time is more efficient than one that runs in N time. ∎

For a list or array A of n items, we have seen that the sequential search method to look for X has complexity $O(n)$ in the worst case. For an array arranged in ascending order, the binary search algorithm is an improvement with complexity $O(\log_2 n)$. When n is about 20 or less, however, a sequential search is usually better because, among some other overhead constraints, the array must be sorted before using the latter method.

The binary search method is similar to looking through the phone book for a name. You may open to the middle of the book and decide if you need to proceed to the front half or back half. You take the appropriate half and cut it in half, deciding the section at which to look next. With each cut, you halve the number of names you are searching, a process which narrows down the field rapidly. You would never think of searching through the phone book sequentially from page one for a name; the binary search algorithm is much more efficient for a large list.

**Algorithm for Binary Search for X
in an Ordered Array A of N Items**

> Variables L and R are the left and right boundaries, re-
> spectively, of indices to be searched. The index I is always
> $L \leq I \leq R$.
> 1. [Initialize] $L \leftarrow 0$ and $R \leftarrow N + 1$
> 2. Repeat:
>   a. [Midpoint] $I \leftarrow \lfloor (L + R)/2 \rfloor$.
>   b. [Not Found] If $I = L$, print 0 and stop.
>   c. [Found]   If $X = A[I]$, print I and stop.
>   d. [Continue]   If $X < A[I]$, then $R \leftarrow I$
>   else $L \leftarrow I$.

EXAMPLE 2

Consider the array A of $N = 7$ elements in Figure 11.6. Show the steps of a binary search for $X = 'P'$ in the array.

Solution

Following the algorithm we have:

   1. $L = 0$ and $R = 8$
 2a. $I = \lfloor (0 + 8)/2 \rfloor = 4$
 2d. $X = 'P' > A[4] = 'H'$. Therefore, $L = 4$ (We now are con-
       sidering the right part of the table.)
 2a. $I = \lfloor (4 + 8)/2 \rfloor = 6$
 2c. $X = 'P' = A[6]$, so $I = 6$ is returned. ■

Figure 11.6

Array A of 7 characters in
alphabetic order

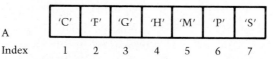

A

'C'	'F'	'G'	'H'	'M'	'P'	'S'

Index   1   2   3   4   5   6   7

It took two comparisons with the binary search, instead of six for a linear search, to find $X = 'P'$. The greatest number of comparisons for the search in Example 2 would be three in the cases where $X = 'C', 'G', 'M', 'S'$ or where X is not in the list. For an array with $n = 2^m - 1$ items, we need at most m comparisons because each time we are cutting the list in half. We need to answer at most m true/false questions, "Is X in the left sublist or not?," to pinpoint X at index 0 (not included) through $2^m - 1$. This combinatorics prob-lem is the same as encountered in determining that $2^m - 1$ is the largest integer that can be represented in m bits. Writing $n = 2^m - 1$, we have

$$n + 1 = 2^m$$

or equivalently

$$\log_2(n + 1) = m.$$

An array of n items must be searched at most $\lceil \log_2(n + 1) \rceil$ times to find X. We take the ceiling function for situations where $\log_2(n + 1)$ is not an integer, that is, where n is not one less than an integral power of 2. Since

$$\lceil \log_2(n + 1) \rceil \leq 2 \log_2(n)$$

we can say that in the worst case the binary search algorithm has complexity of $O(\log_2(n))$.

In Section 8.4 we considered the traveling salesman problem of finding the best route to take in making a round trip tour of n cities. One algorithm consists of seeing which of $(n - 1)!$ possible routes is shortest. This algorithm is of complexity $O(n!)$. As mentioned earlier, for an n of only about 20 or 30 the algorithm would take a computer billions of years to complete. Similarly, someone suggested in the early 1900s to automate chess by looking ahead at every possible path through the game. Claude Shannon noted in 1950 that it would take a supercomputer 10^{90} years to figure out the first move! No more efficient algorithms have ever been found to solve such problems. $\square$

Another problem for which a polynomial-time algorithm has not been discovered is factoring a number into primes, and, in fact, the lack of an efficient algorithm to factor large numbers makes public-key cryptography secure. Listing all the subsets of a set of n elements is also known to be of exponential order $O(2^n)$ since there are 2^n subsets. It simply isn't known if there are algorithms of polynomial order, $O(n^m)$ for some constant $m \in \mathbb{I}^+$, to do these tasks. Table 11.1 shows the dramatic difference in values of functions of n as n gets larger. Notice also for the first few values of n that the polynomial function $f(n) = n^3$ is greater than the exponential function $g(n) = 2^n$, but 2^n exceeds n^3 when $n = 10$.

TABLE 11.1
Comparison of Functional Values for Various Values of n

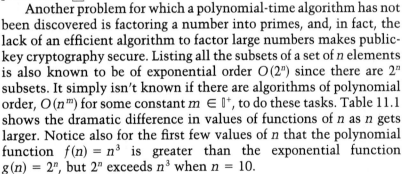

$\log_2 n$	0	1	2	3	4	5
n	1	2	4	8	16	32
$n \log_2 n$	0	2	8	24	64	160
n^2	1	4	16	64	256	1,024
n^3	1	8	64	512	4,096	32,768
2^n	2	4	16	256	65,536	4.29×10^9
10^n	10	100	10,000	10^8	10^{16}	10^{32}
$n!$	1	2	24	40,320	2.09×10^{13}	2.63×10^{35}
n^n	1	4	256	1.68×10^7	1.84×10^{19}	1.46×10^{48}

EXERCISES 11.2

Indicate   ***a.*** *if f is O(g) and*   ***b.*** *if g is O(f) for each of Exercises 1–16.*

	f	g		f	g
1.	$n^{7/6}$	n^2	**2.**	$\log_2(\log_2(n))$	$\log_2(n)$
3.	$n \log_2(n)$	2^n	**4.**	$\log_2(\log_2(n))$	$n \log_2(n)$
5.	n^n	$n!$	**6.**	n^2	$n \log_2(n)$
7.	5^n	7^n	**8.**	$400n^3 + n + 348$	n^3
9.	$\log_2 n$	n	**10.**	e^n	$n!$
11.	n^{35}	2^n	**12.**	$6\log_2 n - 14$	n^2
13.	2^n	n^2	**14.**	$(334n - 18)/117$	n
15.	$85\log_2 n$	n	**16.**	$9n^{200} + 8n^{45} + 5n^{99}$	6^n

Give the best "big oh" for each of Exercises 17–25.

17. $75n^4$ **18.** $32n^{4/3} + n^{5/2}$ **19.** $6\log_2 n + 8n$

Note: See Sections 6.2 and 6.4 for formulas involving the Σ notation.

20. $\displaystyle\sum_{i=1}^{n} i$ **21.** $\displaystyle\sum_{i=0}^{n} 3^i$ **22.** $\displaystyle\sum_{i=1}^{n} i^2$

23. $\dfrac{8n^5 - 15n}{7n^2}$ **24.** $\dfrac{(2n-3)(5n+6)}{n-4}$ **25.** $\log_2 n + (\log_2 n)^2$

What is the run-time complexity based on N for each of the program segments in Exercises 26–28?

26.
```
FOR I := 1 TO N DO
        A[I] := X;
```

27.
```
FOR I := 1 TO N DO
        FOR J := 1 TO N DO
                FOR K := 1 TO N DO
                        A[I,J,K] := X;
```

28.
```
FOR I := 1 TO N DO
        FOR J := 1 TO N DO
                A[I, J] := X;  ☐
```

29. Apply the binary search algorithm to find each of the following characters in the array A, and give the number of times the loop is executed in each case:

    a. 'S'   **b.** 'X'   **c.** 'K'
    d. 'D'   **e.** 'E'   **f.** 'U'

Array A

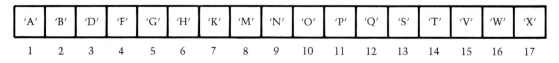

'A'	'B'	'D'	'F'	'G'	'H'	'K'	'M'	'N'	'O'	'P'	'Q'	'S'	'T'	'V'	'W'	'X'
1	2	3	4	5	6	7	8	9	10	11	12	13	14	15	16	17

30. On the same coordinate system graph each of the following functions for $x \geq 0$: $f(x) = 2$, $g(x) = \log_2 x$, $h(x) = x$.

31. Repeat Exercise 30 using $f(x) = x^2$, $g(x) = x^3$, $h(x) = 2^x$.

32. Consider the algorithm of Section 10.1 to subtract 1 from a binary integer $n < 2^m$, where m is the number of bits in the representation of n.
    a. What is the complexity of the algorithm?
    b. Is the order of this algorithm exponential?

33. Suppose our method of finding a prime factor of an integer n is to attempt dividing by primes p until $\mod_p(n) = 0$ or until $\sqrt{n} \leq p$ as discussed in Exercise 24 of Section 1.1. Let $n < 2^m$, where m is the number of bits in the representation of n; and assume we have a list of primes less than 2^m.
    a. What is the method's complexity?
    b. Is the order of this algorithm exponential?

34. Suppose we want to arrange the elements of array A with N items into ascending order. An insertion sort compares every pair of elements, switching the values of those that are out of order, A[I − 1] > A[I].
    a. How many possible pairs are compared?
    b. What is the maximum number of exchanges?
    c. What is the complexity of this algorithm in the worst case?
    d. Is this a polynomial-time algorithm?

Find the complexity of the algorithms in Exercises 35–38.

35. The algorithm in Section 5.2 to compute the SUM of n numbers.

36. The algorithm in Section 5.2 to compute the MEAN of n numbers.

37. Of the algorithm in Section 5.3 to compute the MAX of n numbers.

38. Of Kruskal's algorithm in Section 8.4.

39. Let $n = 32$. Suppose a step of an algorithm is executed every micro-second.
    a. How many hours would it take to execute an algorithm that takes 2^n steps?
    b. How many years for $n!$ steps?
    c. How long for $\log_2 n$ steps?
    d. How many seconds for n^3 steps?

40. a. Suppose you want to write down all 3-character strings (repetitions allowed) you can compose from a set S of 10 characters. How many strings will you have?
    b. How many strings will you write if S has 20 characters?
    c. How many strings will you write if S has n characters?
    d. What is the "big oh" of this problem?
    e. Is the problem of exponential order?
    f. Polynomial order?

41. Consider the QUIK sort algorithm to sort a set S. An element x is picked from S and then S is split into two sets, S_1 and S_2, of elements less than

x and elements greater than x, respectively. The procedure is repeated for each of these sets. We continue until the subsets have only one element each. Gluing the sets back together, we have an ordered set. The algorithm for QUIK sort follows.

QUIK(S):

1. If $n(S) = 1$, return S.
2. Take any element $x \in S$.
3. Partition S into sets $S_1 = \{y \mid y < x\}$, $\{x\}$, and $S_2 = \{y \mid y > x\}$.
4. Concatenate the elements of QUIK(S_1), $\{x\}$, and QUIK(S_2).

a. Perform the procedure for the list 6, 2, 8, 5, 1, 7, 9.
b. Perform the procedure for the list 9, 8, 7, 6, 5, 2, 1.
c. Find the "big oh" for the best case of a QUIK sort.
d. Find the "big oh" for the worst case of a QUIK sort.

Exercises 42–45 are found as sample questions in the 1984 AP Course Description in Computer Science.* For Exercise 42 suppose that L is a list of distinct numbers arranged in increasing order and that a binary search is to be used to determine whether some number x occurs in this list.

42. If L contains 127 entries and x actually occurs as the 80th entry of L, how many entries from L will be examined by the search?
    (A) 1   (B) 3   (C) 12.7   (D) 80   (E) 127

43. Suppose one wishes to determine the second smallest number in a list of four distinct numbers by comparing numbers in that list. What is the maximum number of comparisons that will be made by the most efficient algorithm for this task?
    (A) Fewer than 3   (B) 3   (C) 4   (D) 5   (E) More than 5

LANGUAGE EXERCISES

Exercises 44 and 45 refer to the following Pascal procedure which sorts 15 real numbers from smallest to largest by performing an insertion sort.

```
const
  n = 15;
type
  Collection = array[0..n] of real;
procedure sort (var a: Collection);
  {procedure to sort a[i],...,a[n] into increasing
  order}
```

*AP questions selected from *AP Course Description in Computer Science.* College Entrance Examination Board, 1984. Reprinted by permission of Educational Testing Service, the copyright owner of the sample questions.

```
              var
                i,p : integer;
                Temp : real;
              begin
              for i := 2 to n do
                begin {start pass i}
                  Temp := a[i];
                  a[0] := Temp; {sets sentinel to stop search for}
                                     {insertion point               }
                  p := i;
                  while a[p-1] > Temp do
                    begin
                      a[p] := a[p-1];
                      p := p-1
                    end; {end while}
                  a[p] := Temp
                end {end for}
              end;
```

44. Which of the following assertions is an invariant of the for-loop, i.e., which is true at the beginning of each pass through the loop?
    (A) $a[j] \leq a[k]$ if $1 \leq j < k \leq n$
    (B) $a[j] \leq a[k]$ if $1 \leq j \leq i < k \leq n$
    (C) $a[j] \leq a[k]$ if $1 \leq j < i \leq k \leq n$
    (D) $a[j] \leq a[k]$ if $1 \leq j < k < i$
    (E) $a[j] \leq a[k]$ if $1 \leq j < k \leq i$

45. In the program above, the number of times the condition
    $$a[p-1] > Temp$$
    is evaluated grows proportional to what function of n?
    (A) n (B) $n!$ (C) $n * \log(n)$ (D) 2^n (E) n^2 ☐

Chapter 11   REVIEW

Terminology/Algorithms	Pages	Comments
Exponential function	400	$f(x) = b^x, b > 0, b \neq 1$
Amount, n compounding periods per year	400	$A = P(1 + r/n)^{nt}$
Amount, continuous compounding	400	$A = Pe^{rt}$
Logarithm to base b	402	Inverse of $f(x) = b^x$

Terminology/Algorithms	Pages	Comments
Properties of logarithms	403	$x = b^y \iff \log_b x = y$ $\log_b(b^x) = x = b^{\log_b x}$ $\log_b 1 = 0$ $\log_b b = 1$ $\log_b uv = \log_b u + \log_b v$ $\log_b u/v = \log_b u - \log_b v$ $\log_b u^v = v \log_b u$
Change of base formula	404	$\log_b x = \dfrac{\log_a x}{\log_a b}$
Sequential search	408	Algorithm
$f(n)$ of complexity or order at most $g(n)$	409	$\lvert f(n) \rvert \leq c \lvert g(n) \rvert$ for some c, $n > n_0$
Big oh, $f = O(g)$	409	f of complexity at most g
Polynomial-time algorithm	409	$O(n^m)$, $m \in \mathbb{I}^+$
Properties of big oh	410	For $a > 0$, $af = O(f)$ and $f = O(af)$. n^{m-1} is $O(n^m)$ but n^m is not $O(n^{m-1})$ for $m, n \in \mathbb{I}^+$. Polynomial of degree m is $O(n^m)$.
Binary search	412	Algorithm

REFERENCES

Augarten, S. *Bit by Bit, An Illustrated History of Computers*. New York: Tickner and Fields, 1984.

Cajori, F. *A History of Mathematical Notations, Higher Mathematics*, Vol. 2. Chicago: Open Court, 1929.

Cralle, R. K. "Banking Mathematics." *Tentacle* 3(9): 40–41, September 1983.

Cralle, R. K. "Wilderness Arithmetic." *Tentacle* 3(6): 21–24, June 1983.

Educational Testing Service, *1984 AP Course Description in Computer Science*. Princeton, NJ: College Board Publications, 1984.

Lakshmivarahan, S. "Algorithms for Public Key Cryptosystems." In *Advances in Computers*, M. Yovits, ed., 22: 45–108. New York: Academic Press, 1983.

Michie, D. *Machine Intelligence and Related Topics.* New York: Gordon & Breach Science, 1982.

Roberts, F. S. *Applied Combinatorics.* Englewood Cliffs, NJ: Prentice-Hall, 1984.

Tucker, A. *Applied Combinatorics,* 2nd ed. New York: Wiley, 1984.

Whitted, T. and D. M. Weimer. "A Software Testbed for the Development of 3D Raster Graphics Systems." *ACM Transactions on Graphics* 1(1): 43–58, January 1982.

PARTIAL TABLE OF ASCII DATA REPRESENTATION

Character	Hexadecimal Notation	Character	Hexadecimal Notation	Character	Hexadecimal Notation
space	20	6	36	J	4A
!	21	7	37	K	4B
"	22	8	38	L	4C
#	23	9	39	M	4D
$	24	:	3A	N	4E
'	27	;	3B	O	4F
(	28	<	3C	P	50
)	29	=	3D	Q	51
*	2A	>	3E	R	52
+	2B	?	3F	S	53
−	2D	A	41	T	54
.	2E	B	42	U	55
/	2F	C	43	V	56
0	30	D	44	W	57
1	31	E	45	X	58
2	32	F	46	Y	59
3	33	G	47	Z	5A
4	34	H	48	[	5B
5	35	I	49	]	5D

LARGEST PRIME LESS THAN A GIVEN NUMBER

Number	Prime	Number	Prime
100	97	1000	997
200	199	2000	1999
300	293	3000	2999
400	397	4000	3989
500	499	5000	4999
600	599	6000	5987
700	691	7000	6997
800	797	8000	7993
900	887	9000	8999

GLOSSARY OF
COMPUTER SCIENCE TERMS

ABC computer   First special purpose electronic computer; solved systems of equations; designed by Atanasoff and Berry in the late 1930s.

ABS   Pascal function to return the absolute value of a number.

Ada   Computer language developed under the auspices of the U.S. Department of Defense; named after Lady Ada Augusta Lovelace.

address   Number identifying a memory location.

ALGOL   Computer language defined in the late 1950s; given precise definition using Backus-Normal form.

algorithm   Method for doing something in a finite number of steps.

analytic engine   Machine designed by Charles Babbage in 1800s to solve general arithmetic problems; never completed.

AND gate   Gate with inputs of x and y, output of $x \cdot y$.

APL   Programming language that is functional in nature; all operations have the same priority.

argument   Value passed to a function or procedure.

array   Data structure that contains data items of the same type and uses one variable name.

artificial intelligence   The study of having the computer exhibit human-like intelligence.

ASCII   American Standard Code for Information Interchange; a coding scheme.

assembler language   Low-level programming language that uses mnemonics for the machine language instructions.

assignment   Giving a value to a variable.

Atanasoff-Berry computer   ABC computer.

atom   The fundamental unit in LISP; a string of letters and/or digits.

attribute   In a relational data base, a field in a relation.

automata theory   Study of mathematical models of systems with discrete inputs and discrete outputs, such as computers.

auxiliary storage   Memory device to store data outside the computer; secondary storage.

Backus-Normal form   Used to give a formal definition of a language.

bandwidth   Data-transfer rate.

BASIC   Programming language popular on microcomputers.

baud   Unit for data-transfer rate; usually means bits per second.

best case   In execution of an algorithm, the most advantageous situation, often the situation that takes the shortest amount of time to execute.

binary   Base 2.

binary digit   0 or 1; abbreviated as "bit."

binary search   Method of looking for an item in a list; list cut in half with each inquiry.

binary tree   Rooted tree in which every node has at most two children; data structure with data stored at the nodes of a binary tree structure.

bit   0 or 1; binary digit.

bits per inch   On computer tape the number of bits that can be stored in a one-inch track.

block   On computer tape or disk, a group of records.

BNF   Backus-Normal form.

bpi   Bits per inch on a track of tape.

bug Error in a computer program.

byte Group of bits to encode a character; usually 8 bits.

bytes per inch On computer tape the number of bytes that can be stored in one inch.

call To cause a subroutine to be executed.

CAR LISP function that returns the first element of a list.

cathode ray tube terminal Input/output device with screen and keyboard.

CDC 7600 A type of supercomputer.

CDR LISP function that returns a list with the first element missing.

central processing unit The division that controls the computer and performs arithmetic and logic operations.

CHR Pascal function that returns the character that corresponds to a given decimal number under the coding scheme.

character Symbol, such as a letter, digit, or punctuation mark, that can be stored in a computer.

check bit Bit used in the detection and correction of errors in the transmission of data; used in a Hamming code.

chip A small piece of semiconductor material, usually silicon, etched with a logic circuit.

circuit System of electrical components and lines controlling the flow of electrical current.

clock cycle Basic measurement of time for execution in a computer.

clock rate Frequency; number of cycles per unit of time.

COBOL Computer language widely used in business applications.

code word Sequence of bits in a code that contains information and check bits; used in a Hamming code.

color intensity Degree of brightness of a color in a graphics application.

column-major order Data for a two-dimensional array stored a column at a time.

combinational circuit Circuit where the outputs are uniquely determined by the inputs.

combinational function Function from $\mathbb{B}^n$ to $\mathbb{B}$. Mathematical model of a combinational circuit; switching function.

compiler Program to translate a program written in a high-level language to machine language.

computer integer Integer that can be stored in a particular computer.

concatenate To join two character strings to form one.

CONS LISP function to concatenate an atom onto a list.

control structure Pattern in structured programming.

CPU Central processing unit.

CRT terminal Cathode ray tube terminal.

CRAY computer A brand of supercomputer.

cryptography Study of encoding and decoding secret messages.

cycle time Unit of time for measuring all actions in the central processing unit.

cylinder In a disk pack, the set of all tracks directly above or below one another.

data A collection of facts to be processed into information.

data base A collection of integrated data for a variety of applications, stored to minimize repetition.

data structure A formal skeleton for holding data.

data type A characterization of a data value, such as INTEGER, REAL, CHARACTER.

data-transfer rate Speed with which data is transmitted.

decode Process of determining the meaning of a coded message.

decoder Circuit with n input lines and 2^n output lines.

decrement To subtract an amount from a variable.

delay Execution of statements and/or an empty loop to cause a pause.

density Measure of the number of bits stored per unit length on a tape.

digital computer Computer where data can only be a finite number of discrete values.

digitize To represent a picture with digital data.

dimension An indication of the number of elements in an array.

directory A subset in a partition of files; used with UNIX operating system.

discrete Composed of distinct elements.

disk Secondary storage unit that resembles a phonograph record.

disk drive Device to read data from a disk into the computer and write information from the computer onto a disk.

diskette Flexible disk for microcomputers and minicomputers.

disk pack Collection of disks stored as a unit.

distributed data processing Computing in an environment where various components of the computer system are separated geographically.

dithering Method for reconstructing a digitized picture.

dither matrix Threshold matrix used in dithering.

domain For a function $f: A \rightarrow B$, the set A; in a relational data base, the set of all possible values for an attribute.

doubly linked list A data structure that is a sequence of nodes containing data; each node has a forward and backward pointer.

EBCDIC Extended Binary Coded Decimal Interchange Code; a coding scheme used primarily in IBM mainframes.

encode Process of translating a message into code.

encoder Circuit with 2^n input lines and n output lines.

encryption Process of translating a message into a secret code.

end-of-file Indication that no more data is present in a file.

ENIAC First general purpose electronic digital computer; built by Mauchly and Eckert in the mid 1940s.

even parity Method of detecting errors in transmission; each byte or word has an added bit called a parity bit so that in all there are an even number of one's.

execute To run a program or segment of program on the computer.

field Single data item.

file Collection of related records.

finite-state machine Mathematical model that can be used to simulate many input/output, computational devices; 5-tuple of a set of states, an input alphabet, an output alphabet, a next-state function, and an output function.

floating point number Real number stored as mantissa and exponent.

flops Number of floating point operations per second.

formal grammar Description of the syntax of a language; consists of a vocabulary or set of symbols, a nonempty subset of terminal symbols, start symbol, and a finite set of productions or rules.

FORTH Stack-oriented programming language.

FORTRAN First major high-level programming language; used primarily in scientific applications.

full-adder Logic circuit that adds two bits and the carry out from the previous addition and produces outputs of sum and carry bits.

function Binary relation on the cross product of sets $A \times B$ where for each element in A (domain) there corresponds exactly one element in B (codomain); subprogram that returns one value.

functional programming Programming in a language that implements all operations by using functions.

g Giga.

gate Fundamental logic circuit with one or more inputs and one output.

giga (g) 10^9; prefix meaning billion.

graphics Area of computer science that studies how to display pictures on a computer output device; computer-generated picture.

grammar Formal grammar; description of the syntax of a language.

Gray code Binary code where the representations for successive integers differ by only one bit.

half-adder Logic circuit that adds two bits and produces an output of a sum and a carry bit.

Hamming code Code invented by Richard Hamming, where each code word contains information and check bits.

hard copy Output from the computer displayed on paper.

hard disk Inflexible, fixed disk contained in its own disk drive.

hardware Computer equipment.

hardwired Operation determined by wired connection.

hashing function Function that returns the location of a record in an array or on a disk file given the record's key.

hexadecimal Base 16.

hierarchy chart Diagram of the routines and the subroutines they call in a program.

high-level language Any programming language where one statement corresponds to several machine-language instructions.

IBG Inter block gap.

index Constant or variable indicating a particular element in an array.

indices Plural of "index."

increment To add an amount to a variable.

infix notation Arithmetic expression with each operation written between its operands.

information Processed data.

inorder traversal Process of traveling through a binary tree by traversing the left subtree, visiting the node, then traversing the right subtree.

initialize To assign a value to a variable at the start of computation.

input Data read into the computer.

input alphabet In a finite-state machine, the set of symbols that can be used as input.

input/output (I/O) Pertaining to devices used for reading data into the computer and obtaining information from the computer.

inquiry Request for information from a data base.

INT In FORTRAN, the function that truncates real numbers to their integer part; in BASIC, the floor function.

integrated circuit Complex circuit on a chip.

integrated circuit chip Chip; a small piece of semiconductor material, usually silicon, etched with a logic circuit.

I/O Input/output.

iterate To execute a loop.

inter block gap (IBG) Space between blocks on a tape.

inter record gap (IRG) Space between unblocked records on a tape.

inverter NOT gate; produced output of x' with an input of x.

IRG Inter record gap.

K Not as a prefix, 1024 bytes; as a prefix, kilo or 1000.

key Field(s) that uniquely identifies a record.

kilo 1000; prefix meaning thousand.

linked list Sequence of nodes that can contain information with pointers from each node to the next node.

LISP List-oriented programming language used extensively in artificial intelligence.

list Data structure where each item contains within itself a pointer to the next item; in LISP, a set of parentheses enclosing any number of atoms and lists.

logic circuit Circuit where voltage on each line is interpreted as 0 or 1.

logic gate Gate; fundamental logic circuit with one or more inputs and one output.

logical operator AND, OR, or NOT.

loop control structure Control structure to provide a looping mechanism.

loop, program Repetitive control structure; sequence of statements in a program that are repeated while or until a condition is true.

m Milli.

μ Micro.

M Mega.

machine language Language of the computer; instructions written using the octal or hexadecimal number system.

MACSYMA Programming language that does symbol manipulation.

main memory Storage in the computer used directly by the central processing unit.

mainframe Largest, fastest, most expensive type of computer.

mantissa Fractional part of a real number expressed in normalized exponential notation; for example, 12345 in $0.12345 \times 10^3 = 123.45$.

Mark I First electromechanical computer; built in the early 1940s under the direction of Aiken.

MAX Function to return the maximum of a list of numbers.

MAXINT In Pascal, the built-in constant whose value is the largest integer that can be stored in a particular computer.

mega (M) 10^6; prefix meaning million.

megahertz (MHz) Million cycles per second.

memory Storage area for data.

memory dump Listing of the contents of an area of main memory.

Mflops Million floating point operations per second.

MHz Megahertz.

micro (μ) 10^{-6}; prefix meaning millionth.

microprocessor Central processing unit on a chip.

microcomputer Smallest, slowest, least expensive type of computer.

milli (m) 10^{-3}; prefix meaning thousandth.

MIN Function to return the minimum of a list of numbers.

minicomputer Type of computer intermediate in size, speed, and expense.

MOD Function in Pascal and some other languages to return modulo an integer.

module Logical section of a program.

most significant digit Leftmost nonzero digit in a number.

multiprogramming Several programs in a computer arranged to execute in sequence so that when one cannot run another will.

n Nano.

NAND gate Gate with inputs of x and y and output of $(xy)'$.

nano (n) 10^{-9}; prefix meaning billionth.

nesting of loops One loop contained completely within another.

network In computer science, a system of interconnections of terminals and computer systems.

next-state function In a finite-state machine, a function that returns the next state when given the present state and input; transition function.

node Point.

NOR gate Gate with inputs of x and y and output of $(x + y)'$.

normalized exponential notation Real number written with the decimal point to the immediate left of the first nonzero digit; for example, $123.45 = 0.12345 \times 10^3$.

octal Base 8.

odd parity Method of detecting errors in transmission; each byte or word has an added bit called a parity bit so that in all there are an odd number of one's.

one-dimensional array Array consisting of one column of elements; vector; n-tuple.

one's complement Changing all 1's to 0's and 0's to 1's in the binary representation of a number.

operating system Set of programs that control the operations of the computer.

OR gate Gate with output of $x + y$ given inputs of x and y.

ORD Pascal function that returns the decimal number that corresponds to a given character under the coding scheme.

output Information coming from a computer.

output alphabet In a finite-state machine, a set of symbols that can be used as output.

output function In a finite-state machine, a function to return the output given a state.

overflow The result of an arithmetic operation exceeding the capacity of allotted storage; in n-bit, two's complement arithmetic would have the wrong sign.

p Pico.

package System of programs used by more than one organization.

pad To fill the unused portion of a field.

page Fixed length section of memory.

parallel processing Several processors simultaneously contributing to the execution of one program.

parameter Dummy variable used in the definition of a function or subroutine.

parity Method of detecting errors in transmission; each byte or word of data has an added bit, called a parity bit, so that there are an odd or an even number of one's.

parity bit Extra bit added to a byte or word to create odd or even parity.

parse Process of creating a parse tree.

parse tree Tree used to separate a language element into its syntactic parts.

Pascal Popular structured programming language.

password Private word that must be used by an individual to gain access to a particular computer.

PDP-8 Early minicomputer.

PEEK Function in BASIC to return the contents of a requested memory location.

peripheral equipment Devices external to the computer used for input, output, and secondary storage.

pico (p) 10^{-12}; prefix meaning trillionth.

pixel Picture element; smallest unit in a graphics display, such as a dot on a screen.

PL/I   Programming language for scientific and business applications.

Polish notation   Arithmetic expression written in prefix or postfix notation.

POP   Process of pulling an element off of a stack.

postfix notation   Arithmetic expression written with each operation following its operands; type of Polish notation; suffix notation.

postorder traversal   Process of traveling through a binary tree by traversing the left subtree, traversing the right subtree, then visiting the node.

PRED   Pascal function to return the predecessor of an argument of ordinal type.

prefix notation   Arithmetic expression written with each operation preceding its operands; a type of Polish notation.

preorder traversal   Process of traveling through a binary tree by visiting the node, traversing the left subtree, then traversing the right subtree.

printout   Output printed on paper.

procedure   Subprogram that can return several values.

processor   Central processing unit.

program   Sequence of instructions for the computer.

program verification   Process of proving a program is correct.

pseudo-random number   Number generated by an algorithm so that a sequence of numbers so generated appears to be random.

public-key cryptosystem   Method for encoding and decoding secret messages where some of the important numbers for decoding are made public.

punched card   Cardboard card with holes indicating coded data; used as an input/output medium to the computer.

PUSH   Process of placing an element on a stack.

query   To ask for information from a data base system.

queue   Sequence of elements awaiting processing where the first element placed in the queue is the first element processed.

QUIK sort   Recursive algorithm to sort a set into ascending order: for an element x the set is split into a set of elements less than x and a set of elements greater than x; the procedure is repeated for each subset.

RAM   Random access memory.

random access memory (RAM)   Memory to and from which a user can write and read.

random number   Number produced by chance.

read   To obtain data.

read-only memory (ROM)   Memory from which a user can read but to which a user cannot write.

read/write head   Device to read, write, or erase data on tape or disk.

record   Collection of related fields.

recursive function   Function that calls itself.

register   High-speed memory location in the central processing unit.

relation   Set of ordered n-tuples.

relational data base   Data base where data is organized into relations.

relative file organization   Sequence of possible key values is in one-to-one correspondence with the sequence of locations in a file.

ROM   Read-only memory.

root   Unique vertex with zero in-degree in a rooted tree.

rooted tree   Connected digraph with no cycles and with a unique vertex, the root, which has zero in-degree.

rotation   In graphics, the turning of a picture about a fixed point or line.

ROUND   Function in Pascal to round a real number to the nearest integer.

row-major order   Data for a two-dimensional array stored a row at a time.

run   To execute.

SAS   Statistical Analysis System; system of programs used for statistical analysis of data.

scaling   In graphics, the process of changing the size of a picture.

s-expression   Symbolic expression in LISP; atom or list.

secondary storage   Memory device to store data outside the computer.

seed   Number used to start generation of a sequence of pseudo-random numbers.

selection control structure   Control structure for IF-THEN or IF-THEN-ELSE.

semiconductor   Material that will conduct electricity under the proper conditions; integrated circuits are made with semiconductor material such as silicon.

sequential control structure   Control structure for one statement to be executed after another in sequence.

sequential search   Method of looking for an item in a list by going through the list in order.

set of computer real numbers   Numbers expressed with a terminating decimal expansion that can be stored in the computer.

shift (logical)   Move the bits in a register to the left or to the right, one bit falling off the end, the other side padded with 0.

signed-magnitude   Representation of real numbers where the most significant bit holds the sign ($+ \equiv 0$; $- \equiv 1$) and the remaining bits represent the magnitude.

small-scale integration (SSI)   Type of integrated circuit with the smallest number of gates on a chip.

software   Programs.

sort   To place in ascending (or alphabetic) or descending order.

speed of tape   Rate at which tape is moving.

SPSS   Statistical Package for the Social Sciences; system of programs used for statistical analysis of data.

SQR   Pascal function to return the square.

SQRT   Pascal function to return the square root.

SSI   Small-scale integration.

stack   Data structure of a list of elements where the last item added to the stack is the first item taken from the stack.

start symbol   In a grammar, a symbol used to indicate a beginning point.

state   Condition of a finite-state machine.

state graph   A digraph to picture a finite-state machine.

statement   Instruction in a programming language.

storage   Memory.

storage density   Density; measure of the number of bits stored on a tape.

string   Connected sequence of bits or characters treated as a unit.

structured programming   Programming technique using only three control structures (sequential, selection, and loop), program modules, and top-down design.

subroutine   Subprogram; section of code that is executed only when called by another routine, often the main program. Value(s) may be passed to the routine for the parameter(s) and/or returned from the routine from the parameter(s).

SUCC   Pascal function to return the successor of an argument of ordinal type.

suffix notation   Arithmetic expression written with each operation following its operands; a type of Polish notation; postfix notation.

SUM   MACSYMA function to return a sum.

supercomputer   Largest, fastest, most expensive of the mainframe computers.

switching function   Function from $\mathbb{B}^n$ to $\mathbb{B}$; mathematical model of a combinational circuit; combinational function.

symbolic expression   In LISP, s-expression; an atom or list.

table   Array in COBOL.

tape   Magnetized strip of material that is a secondary storage device.

tape speed   Rate at which tape is moving.

terminal   Input/output unit with keyboard and screen or printer used to send data and programs to a computer and to receive information from a computer.

test data   Data used to test the operation of a program.

threshold matrix   Matrix T used to map each element m_{ij} of a matrix M to 0 or 1 depending if $m_{ij} < t_{ij}$ or not; called a dither matrix when used in dithering.

time-sharing   Method for several people to use the resources of the computer at the same time; computer time is shared so that each person has the impression he or she is the sole user of the system.

top-down programming   Structured programming technique that divides the problem into increasingly more detailed modules.

tracks   On disk, concentric circles of data; on tape, parallel lines of data running the length of the tape.

transition function   In a finite-state machine, a function that returns the next state when given the present state and input; next-state function.

translation   In graphics, moving the picture parallel to the axes.

traversal   Examining each node of a tree.

tree   Connected, acyclic graph; often indicates rooted tree; data structure with data stored at the nodes of a rooted tree structure.

TRUNC Pascal function to truncate a real number to its integer part.

truncate To cut off digits at the end of a number.

truth table Table listing all possible true/false values for inputs along with values for the resulting output in a logic circuit or a logic statement.

Turing machine Model of a procedure that can be performed on a digital computer; hypothetical machine used to understand computability; consists of a "tape" containing a finite string of characters from the alphabet, a read/write head, and a finite set of instructions.

two-dimensional array Array consisting of rows and columns; matrix; table.

two's complement Method of representing signed binary numbers; leave a positive binary number as is, and for a negative binary number take the one's complement and add 1.

unblocked Not blocked; records read from or written to a secondary storage medium one at a time.

UNIVAC First computer sold commercially (early 1950s).

UNIX Type of operating system.

unsigned number Number written without a sign.

user Someone who uses a computer system.

user id String of characters used to identify a user to a computer system.

very large scale integration (VLSI) Type of integrated circuit with the largest number of gates on a chip.

visiting a node In tree traversal, reading the contents of a node.

vocabulary Set of symbols that can be used in a language generated by a formal grammar.

VLSI Very large scale integration.

word Group of bits considered as a unit of information.

worst case In execution of an algorithm, the most detrimental situation, often the situation that takes the longest to execute.

XOR gate Gate with inputs of x and y and output of $xy' + x'y$.

zero out Process of assigning 0 to all elements of an array.

ANSWERS TO SELECTED EXERCISES

CHAPTER 1

SECTION 1.1

3. $I, I_c, N, I^+, Q, R, P, H$ **12.** $\mathbb{R}, \mathbb{R}$ **13.** Q, R, R_c **24. a.** 15   **25. a.** $2^6 = 64$ **b.** $2^4 = 16$ **28. a.** 10^3
e. None (Wishful thinking!)   **30. a.** $2^5 = 32$ **31. a.** 3   **32.** $2^{16} = 65,536$ **35.** $2^{13}K = 8162K$
38. 450,000,000 psec = 450,000 nsec = 450 μsec   **43. a.** 2^{12} **b.** 8^4 **45.** $10^5 = 100,000$ **48.** $10^{-1} = 1/10$
51. $10^5 = 100,000$ **55.** $10^7 = 10,000,000$ **57. a.** 10.8   **58.** A

SECTION 1.2

1. 7010   **2.** 0.0000701   **13.** 0.63850×10^5 **15.** 0.32×10^{-3} **21.** -0.82×10^0 **25.** $10^{25}, 6$ **28.** 4, 6
30. 2, 7   **34.** $0.1 \times 10^{-5} - 0.999 \times 10^5$ **36.** 0.3752×10^{-19} **37.** 0.62×10^4 **41. a.** -0.001 **b.** -0.016%
c. 0.009   **d.** 0.144%   **45.** 6.23; 12.4; 0.625%

SECTION 1.3

1. 9.5   **7.** 512   **13.** 9   **20.** -81 **24.** -2 **25.** (4.) 14   (11.) 15   (15.) 12   **27.** (3.) $X \div Z + Y, 5/11$
(7.) $Z * W * Z, 512$ (13.) $V \div Z \times W, 1$ (23.) $X - Y - Z, -2$ **28.** $Z * X ** 2/(Y + 3)$ **30.** (6.) 9.73
31. (3.) 11.6; 0.727%

SECTION 1.4

1.

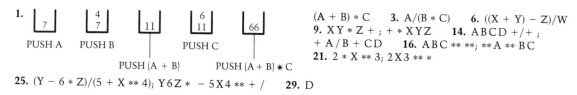

PUSH A   PUSH B   PUSH C

PUSH (A + B)   PUSH (A + B) * C

$(A + B) * C$ **3.** $A/(B * C)$ **6.** $((X + Y) - Z)/W$
9. $XY * Z + ; + * XYZ$ **14.** $ABCD +/+ ;$
$+ A/B + CD$ **16.** $ABC ** **; ** A ** BC$
21. $2 * X ** 3; 2X3 ** *$

25. $(Y - 6 * Z)/(5 + X ** 4); Y6Z * - 5X4 ** + /$ **29.** D

CHAPTER 2

SECTION 2.1

2. F   **4.** F   **11.** T   **15.** F   **19.** F   **24.** T   **29.** T   **39.** F   **43.** {3, 4, 5, 6, 7}   **44.** {4, 5, 6, . . .}
49. {0, 3, 6, 9, . . .}   **54.** $\{x \mid x \geq 4, x \in I\}$ **57.** $\{y \mid y = 5x, x \in \mathbb{N}\}$ **61.** $\{\phi, \{0\}, \{1\}, \mathbb{B}\}$ **64.** F;
$A = C = \{1, 2\}, B = \{1, 2, 3, 4\}$ **67.** T   **70.** F; $B = \{x, y\} = C, A = z$ **73. a.** T   **c.** F

SECTION 2.2

1. D **3.** ϕ **6.** {3, 5, 7, 9}   **11.** {3, 6, 7, 8, 9}   **17.** {3, 6, 7, 8, 9}   **23.** {1, 2, 4, 5, 10}   **29.** U
35. {7, 8, 3}   **41.** $\mathbb{R}$ **47.** {5, 7, 9, 11}   **51.** $\{x \mid -2 \le x \le 11, x \in \mathbb{I}\}$ **54.** {..., −32,770, −32,769,
32,768, 32,769, ...}   **62.** ϕ **64.** A **68.** A' **72.** C **75.** $A = \{1, 2\}, B = \{2, 3, 4\}, C = \{2, 9\}$
79. $C - (A \cup B)$ **81.** $A \cap B'$ double shaded.   **85.** $(A \cup B)'$ shaded.
88. a. WEEKDAYS
d. [ ]

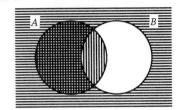
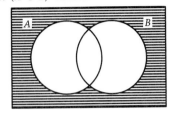

SECTION 2.3

2. a. 32   **3. a.** Or **b.** $x \in A$ **c.** A **d.** $A \cup \phi$ **e.** $x \in A$ **f.** $x \in A \cup \phi$ **g.** $x \in A$ or $x \in \phi$
h. Union   **6.** $x \in A \Rightarrow x \in A$ or $x \in B$ (def. "or")$\Rightarrow x \in A \cup B$ (def. $\cup$)
8. $x \in A \cup U \Rightarrow x \in U$ def. U
  $x \in U$ $\Rightarrow x \in A$ or $x \in U$ def. "or"
       $\Rightarrow x \in A \cup U$ def. $\cup$
16. a. $B' \cup A''$ **b.** Double complement **c.** Commutative **d.** $A \cup A$ **e.** Idempotent
19. $(A' \cup B')' = A'' \cap B''$ DeMorgan
      $= A \cap B$ complement of complement
21. $(A \cup \phi) \cap (A \cup B) = A \cup (\phi \cap B)$ distribution
            $= A \cup \phi$ $\cap$ with ϕ
            $= A$ identity
30. Assume there are two different identities for union, ϕ and Z. Thus, (1) $X \cup \phi = X$ and (2) $X \cup Z = Z$ for any
X. Let $X = \phi$ in (2): $\phi \cup Z = \phi$. Let $X = Z$ in (1): $Z \cup \phi = Z$. Thus, $\phi = \phi \cup Z = Z \cup \phi = Z$, by the
commutative property and equations (1) and (2). We have $\phi = Z$, which is a contradiction.

SECTION 2.4

1. $p \wedge q$ **6.** $\sim p \vee (p \wedge q)$
8. I go to dinner and eat spaghetti and/or salad. I go to dinner and eat spaghetti, and/or I go to dinner and eat
salad. Distributive.   **10.** I go to dinner or I do not go to dinner. No. Complement.   **14.** I do not like pizza or
chicken.   **16.** I like pizza, liver, and chicken.   **18.** $x > y$ or $x > z$ **20.** $x = y$ or $x < y$ **22.** NOT (RATE
GREATER THAN 2 OR COST LESS THAN 6)   **a.** F

25.

p	$p \wedge p$	p	$p \vee p$
T	T	T	T
F	F	F	F

26. a. Commutative **b.** $(\sim q \vee p)$ **c.** $(\sim q \vee q)$ **d.** Complement **e.** $(\sim q \vee p)$ **f.** Commutative

30.

p	$\sim p$	$\sim\sim p$
F	T	F
T	F	T

31.

p	q	r	$q \wedge r$	$p \wedge (q \wedge r)$	$p \wedge q$	$(p \wedge q) \wedge r$
F	F	F	F	F	F	F
F	F	T	F	F	F	F
F	T	F	F	F	F	F
F	T	T	T	F	F	F
T	F	F	F	F	F	F
T	F	T	F	F	F	F
T	T	F	F	F	T	F
T	T	T	T	T	T	T

SECTION 2.5

7.

p_1	p_2	q	$p_1 \lor p_2$	$p_1 \lor p_2 \Rightarrow q$	$p_1 \Rightarrow q$	$p_2 \Rightarrow q$	$(p_1 \Rightarrow q) \land (p_2 \Rightarrow q)$
F	F	F	F	**T**	T	T	**T**
F	F	T	F	**T**	T	T	**T**
F	T	F	T	**F**	T	F	**F**
F	T	T	T	**T**	T	T	**T**
T	F	F	T	**F**	F	T	**F**
T	F	T	T	**T**	T	T	**T**
T	T	F	T	**F**	F	F	**F**
T	T	T	T	**T**	T	T	**T**

8. See Table 2.8, Section 2.6

9.

p	q	$\sim q$	$\sim p$	$p \Rightarrow q$	$\sim q \Rightarrow \sim p$
F	F	T	T	**T**	**T**
F	T	F	T	**T**	**T**
T	F	T	F	**F**	**F**
T	T	F	F	**T**	**T**

13. a.

p	q	$p \Rightarrow q$	$p \land (p \Rightarrow q)$	$p \land (p \Rightarrow q) \Rightarrow q$
F	F	T	F	T
F	T	T	F	T
T	F	F	F	T
T	T	T	T	T

14.

p	q	$p \Rightarrow q$	$q \Rightarrow p$	$(p \Rightarrow q) \Leftrightarrow (q \Rightarrow p)$	Neither
F	F	T	T	T	
F	T	T	F	F	
T	F	F	T	F	
T	T	T	T	T	

19. m: John is married.
o: John is older than 25.
r: John pays a lower insurance rate.
a: John can afford more insurance.
$((m \lor o \Rightarrow r) \land (r \Rightarrow a) \land (\sim a)) \Rightarrow (\sim m \land \sim o)$

m	o	r	a	$m \lor o$	$m \lor o \Rightarrow r$	$r \Rightarrow a$	$\sim a$	$\land$	$\sim m$	$\sim o$	$\sim m \land \sim o$	$\Rightarrow$
F	F	F	F	F	T	T	T	T	T	T	T	T
F	F	F	T	F	T	T	F	F	T	T	T	T
F	F	T	F	F	T	F	T	F	T	T	T	T
F	F	T	T	F	T	T	F	F	T	T	T	T
F	T	F	F	T	F	T	T	F	T	F	F	T
F	T	F	T	T	F	T	F	F	T	F	F	T
F	T	T	F	T	T	F	T	F	T	F	F	T
F	T	T	T	T	T	T	F	F	T	F	F	T
T	F	F	F	T	F	T	T	F	F	T	F	T
T	F	F	T	T	F	T	F	F	F	T	F	T
T	F	T	F	T	T	F	T	F	F	T	F	T
T	F	T	T	T	T	T	F	F	F	T	F	T
T	T	F	F	T	F	T	T	F	F	F	F	T
T	T	F	T	T	F	T	F	F	F	F	F	T
T	T	T	F	T	T	F	T	F	F	F	F	T
T	T	T	T	T	T	T	F	F	F	F	F	T

SECTION 2.6

2. Antecedent: Eating a whole pizza; consequent: my being full   **3.** Antecedent: $n \geq 4$; consequent: $n^2 \leq 2^n$
6. By the definition of even, $x = 2n$ for some $n \in \mathbb{I}$. Squaring both sides, we have $x^2 = 4n^2 = 2(2n^2)$, which is even by the definition of even.   **8.** Case 1: x even $\Leftrightarrow x^2$ even; Case 2: x odd $\Leftrightarrow x^2$ odd.
10. Let $x, y \in \mathbb{N}$. If $x < 41$ and $y < 41$, then $x + y < 83$.   **14.** Counterexample: Let $x = 11$.

PROOF: $x < 41$ and $y < 41 \Rightarrow x + y < 41 + 41 < 83$.

22. $((m \vee o \Rightarrow r) \wedge (r \Rightarrow a) \wedge (\sim a))$
    $\equiv ((\sim r \Rightarrow \sim(m \vee o)) \wedge (\sim a \Rightarrow \sim r) \wedge (\sim a))$ contrapositive
    $\equiv ((\sim r \Rightarrow (\sim m \wedge \sim o)) \wedge (\sim a \Rightarrow \sim r) \wedge (\sim a))$ DeMorgan
    $\Rightarrow ((\sim r \Rightarrow (\sim m \wedge \sim o)) \wedge (\sim r)$ commutative and *modus ponens*
    $\Rightarrow \sim m \wedge \sim o$ commutative and *modus ponens*

CHAPTER 3

SECTION 3.1

1. 6   **4.** 18   **9.** 120   **11.** 120   **17.** $2^8 = 256$ **19. b.** 75   **22. a.** 97   **25. a.** 26^6
26. a. $10! = 3,628,800$ **b.** $7! \, 3! = 30,240$ **c.** $2 \cdot 7! \, 3! = 60,480$ **31. a.** 9×10^4 **b.** $10 \cdot 9 \cdot 8 \cdot 7 \cdot 6$ **32. b.** 256
38. a. 12×60 **39. a.** 6   **40.** $11^{14} \times 2$

SECTION 3.2

3. 720   **4.** 1   **8.** 6545   **10. a.** $P(35, 4) = 1,256,640$ **12. a.** 14!   **c.** $5! \cdot 7! \cdot 2!$ **15. b.** 10!
17. a. 2^4 **b.** $C(4, 2) = 6$ **c.** $C(4, 2) + C(4, 3) = 10$ **e.** 1   **19. a.** $C(52, 13)$ **d.** $C(13, 6) \cdot C(13, 7)$
g. $C(26, 13)$ **h.** $C(52, 13) - C(26, 13)$ **20.** $4 \times 5 \times 3 = 60$ **23. a.** $C(46, 6)$ **b.** $C(25, 3) \cdot C(21, 3)$
g. $C(21, 2)C(25, 4) + C(21, 3)C(25, 3)$ **25.** $C(50, 2) = 1225$ **28.** $2^8 \times 2^8 \times 2^8 = 16,777,216$
30. $x^3 + 3x^2y + 3xy^2 + y^3$ **32.** Let $x = y = 1$ in the Binomial Theorem.

SECTION 3.3

1. 11   **5.** 26   **9.** $432 \times 16 \times 100$ **12. a.** 1   **b.** 16   **13. b.** 20   **14.** 62,668   **16.** B

SECTION 3.4

1. 105.26 MHz   **3.** 127.66 nsec   **7.** 100 Mflops   **8. a.** 19 bpi   **b.** 1458 bytes/sec   **9.** 0.075 in
12. a. 5.317 Mbytes   **13. a.** 24.041 Mbytes   **17.** 300 Kbytes/sec   **19. a.** 130 records/track   **b.** 87 tracks
c. 5 cylinders   **21.** 16,000,000 baud

CHAPTER 4

SECTION 4.1

1. $\{(a, x), (a, y), (a, z), (b, x), (b, y), (b, z), (c, x), (c, y), (c, z)\}$ **3. a.** 12
5. b. $\{(2, 3), (2, 4), (3, 3), (3, 4), (5, 3), (5, 4), (7, 3), (7, 4)\}$ **c.** $\{(2, 4), (3, 9)\}$ **d.** $\{(1, 5), (2, 4), (3, 3)\}$ **k.** $\{(3, 9)\}$
7. ϕ **8. a.** T   **9. c.** F; 1 divides 2, but 2 does not divide 1 with a remainder of 0   **13. a.** 29^{20}
  d. TYPE

```
        FINANCIAL TYPE
          = RECORD
            SSN:      PACKED ARRAY [1..9] OF CHAR;
            AID:      50..2000;
            INCOME:   0..50000
            END;
      VAR
          FINANCIAL RECORD: FINANCIAL TYPE;
          FINANCIAL FILE:   FILE OF FINANCIAL TYPE;
```

SECTION 4.2

1. r: $x \leq x$ for all $x \in \mathbb{R}$; T
i: $x \leq x$ is not true for any $x \in \mathbb{R}$; F; $3 \leq 3$
s: $x \leq y \Rightarrow y \leq x$ for all $x \in \mathbb{R}$; F; $3 \leq 5$ but not $5 \leq 3$
a: $x \leq y$ and $y \leq x \Rightarrow x = y$ for all $x, y \in \mathbb{R}$; T
t: $x \leq y$ and $y \leq z \Rightarrow x \leq z$ for all $x, y, z \in \mathbb{R}$; T

3. r: x is even for all $x \in \mathbb{N}$; F; 5 is not even
i: x is odd for all $x \in \mathbb{N}$; F; 6 is not odd
s: x is even $\Rightarrow y$ is even for all $x, y \in \mathbb{N}$; F; $6 \; \sigma \; 5$, but $5 \; \not\sigma \; 6$
a: x and y are even $\Rightarrow x = y$; F; $4 \; \sigma \; 6$ and $6 \; \sigma \; 4$, but $4 \neq 6$
t: x is even and y is even $\Rightarrow x$ is even; T

6. r: $x \cdot x = 1$ for all $x \in U$; F; $2 \times 2 = 4 \neq 1$
i: $x \cdot x \neq 1$ for any $x \in U$; T
s: $x \cdot y = 1 \Rightarrow y \cdot x = 1$ for all $x \in U$; T
a: $x \cdot y = 1 = y \cdot x \Rightarrow x = y$: F; $2 \times (1/2) = (1/2) \times 2$, but $1/2 \neq 2$
t: $x \cdot y = 1$ and $y \cdot z = 1 \Rightarrow x \cdot z = 1$; F; $(1/2) \times 2 = 1$ and $2 \times (1/2) = 1$, but $(1/2)(1/2) = 1/4 \neq 1$

7. r: $(a, b) \; \beta \; (a, b)$ for all $(a, b) \in \mathbb{R} \times \mathbb{R}$ since $a \leq a$; T
i: $(a, b) \; \beta \; (a, b)$ is not true for any $(a, b) \in \mathbb{R} \times \mathbb{R}$; F; $(2, 3) \; \beta \; (2, 3)$ since $2 \leq 2$
s: $(a, b) \; \beta \; (c, d)$ with $a \leq c \Rightarrow (c, d) \; \beta \; (a, b)$ with $c \leq a$; F; $(2, 7) \; \beta \; (5, 4)$ since $2 \leq 5$, but $(5, 4) \; \beta \; (2, 7)$ is not true since $5 > 2$
a: $(a, b) \; \beta \; (c, d)$ with $a \leq c$ and $(c, d) \; \beta \; (a, b)$ with $c \leq a \Rightarrow (a, b) = (c, d)$; F; $(2, 3) \; \beta \; (2, 7)$ and $(2, 7) \; \beta \; (2, 3)$ since $2 \leq 2$, but $(2, 3) \neq (2, 7)$
t: $(a, b) \; \beta \; (c, d)$ with $a \leq c$ and $(c, d) \; \beta \; (e, f)$ with $c \leq e \Rightarrow (a, b) \; \beta \; (e, f)$ with $a \leq e$; T

12. r: x is a prime factor of x for all $x \in U$; F; 4 is not prime
i: x is never a prime factor of x; F; 3 is a prime factor of 3
s: x is a prime factor of $y \Rightarrow y$ is a prime factor of x; F; 3 is a prime factor of 12, but 12 is not a prime factor of 3
a: x is a prime factor of y and y is a prime factor of $x \Rightarrow x = y$; T
t: x is a prime factor of y and y is a prime factor of $z \Rightarrow x$ is a prime factor of z; T, x must equal y

16. a.   **c.**

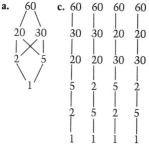

18. a. 7 4 6   **b.** 7!
5 3   **c.** 5!
1 2   **d.** 4

SECTION 4.3

2. a. $-65, 505, 670; 66, -444, 91; 7, 37, 2, 132, -203; 13, 88, 9083, -72; 9, 4, -141, 444$ **b.** 5

3. [2]   **7.** [5]   **10.** F; $0 \notin \mathbb{N}' \cup \mathbb{I}^+$. Therefore, $\mathbb{N}' \cup \mathbb{I}^+ \neq U = \mathbb{I}$.

12. b. r: Condition A has the same value (TRUE or FALSE) as A.
s: If condition A has the same value as condition B, then B has the same value as A.
t: For conditions A, B, and C, if A has the same value as B and B has the same value as C, then A has the same value as C.

16. a. $(2, 4), (3, 6), (4, 8)$ **b.** An infinite number.
c. r: $(x, y) \; \alpha \; (x, y)$ since $x \cdot y = y \cdot x$.
s: If $(x, y) \; \alpha \; (u, v)$, then $x \cdot v = y \cdot u \Rightarrow u \cdot y = v \cdot x \Rightarrow (u, v) \; \alpha \; (x, y)$.
t: If $(x, y) \; \alpha \; (u, v)$ and $(u, v) \; \alpha \; (w, z)$, then $x \cdot v = y \cdot u$ and $u \cdot z = v \cdot w \Rightarrow u = v \cdot w/z \Rightarrow x \cdot v = y \cdot (v \cdot w/z) \Rightarrow x = y \cdot (w/z) \Rightarrow x \cdot z = y \cdot w \Rightarrow (x, y) \; \alpha \; (w, z)$.

17. a. $(5, 2, 7), (2, 5, 7), (2, 7, 5)$ **c.** $C(9, 3)$ **e.** $r!$

CHAPTER 5

SECTION 5.1

1. a. 4, 4, 9, c^2, $(c + 1)^2 = c^2 + 2c + 1$; {4, −4}, ϕ, {b, −b}   **b.** Yes, {$x \mid -5 \le x \le 5, x \in \mathbb{I}$}, $\mathbb{I}$, {0, 1, 4, 9, 25}
c. No, no   **6. a.** 1, 2, 20;   **b.** No, 1 maps to each element of U since 1 divides every integer.
10. a. {15, −15}, {n, −n}; $|a|$, $|2a| = 2|a|$;   **b.** Yes, $\mathbb{I}$, $\mathbb{N}$, $\mathbb{N}$ **c.** No, yes   **14. a.** 32, 2^m, $2^{m+3} = 8 \times 2^m$
b. Yes, $\mathbb{N}$, $\mathbb{N}$, {$2^n \mid n \in \mathbb{N}$}   **c.** Yes, no   **17. a.** 16, $6 + b$; {(0, 5), (1, 4), (2, 3), (3, 2), (4, 1), (5, 0)}
b. Yes, $\mathbb{N}^2$, $\mathbb{N}$, $\mathbb{N}$ **c.** No, yes   **23. a.** 1, 0   **b.** Yes, {$v \mid 0 \le v \le 4$}, $\mathbb{B}$, $\mathbb{B}$ **c.** No, yes
26. a. (6, −41), $(x^2 - 1, y - 1)$ **b.** Yes, $\mathbb{R}^2$, $\mathbb{R}^2$, $\mathbb{R}^2$ **c.** Yes, yes   **29.** 2^n **31.** 50^{35} **34.** If $n \le 2$, 2; else, none.
36. $P(50, 35)$ **39.** $2^n - 2$ **41.** None   **47. a.** $A(n_a, n_r) = n_a - n_r$ **48. a.** 2^8 **b.** $C(8, 3)$ **c.** $2^8 - C(8, 3)$
49. a. 2^n **b.** 2^{2^n}

SECTION 5.2

3. No answer in $\mathbb{R}$ **8.** 8   **13.** 5   **17.** a **21.** $|x|$ **22.** x **26.** $\sqrt{x^2 + y^2}$ **28.** xy
30. a. 1, 3, 4, 0   **b.** 8.57, 5   **35. a.** $\mathbb{R}_c$ **b.** {$x \mid x \ge 0, x \in \mathbb{R}_c$}   **c.** No
38. No. Let L be the collection 3, 3, 5, 5; $\text{freq}_L(3) = \text{freq}_L(5) = 2$ **43. b.** 27
44. a. {$x \mid x \ge 0, x \in \mathbb{R}$} $\times \mathbb{R}^+$ **b.** {$x \mid x \ge 0, x \in \mathbb{R}$}
   c. FUNCTION PERCENT (M, N: REAL): REAL;
```
        BEGIN
             IF (M < 0) OR (N <= 0) THEN
                 PERCENT := -1
             ELSE
                 PERCENT := M * 100/N
        END;
```
48. a. \$1,320,000,000   **51.** 77.1, 312 (truncated)   **54. a.** 24   **d.** 68.4

SECTION 5.3

3. −17   **4.** −18   **5.** 17   **6.** 18   **11.** −17   **12.** −18   **13.** 18   **14.** 18   **18.** −3456
22. a. $\mathbb{R}_c^2$, $\mathbb{R}_c$
23. a. FLOOR(X):
   (1) [Positive] If X > 0 then return TRUNC(X).
   (2) [Whole number] If X = TRUNC(X), then return X.
   (3) [Negative, not whole] Return TRUNC(X) − 1.
25. a. (1) $32.456522 \times 10^3 = 32456.522$;
   (2) Truncate: 32456.
   (3) Divide by 10^3 to get 32.456.
26. a. (1) Add $0.5 \times 10^{-3} = 0.0005$ to get 32.457022;
   (2) Truncate to 3 decimal places to get 32.457.
28. a. No, MAX(2, 3) = MAX(−7, 3)   **33.** A

SECTION 5.4

1. a. $f \circ g$ not defined for $x = 7$ **b.** $g \circ f$ = {(2, 6), (2, 0)}   **5. a.** (1) f^{-1} = {(7, 2), (4, 0)}, g^{-1} = {(2, 4), (6, 7), (8, 9)}
6. $f \circ g(x) = |9x + 1|$, $g \circ f(x) = 9|x| + 1$ **10.** $f \circ g(x) = 9x^2 + 4$, $g \circ f(x) = 3x^2 + 6x + 14$ **11.** Yes
12. No   **19. a.** F(X) = X − TRUNC(X)   **21.** (CDR (CDR X)) = (CDDR X)   **24.** $\sqrt{|3x + 7|}$

SECTION 5.5

1. 6   **3.** 1   **10.** 16044, 74077, 63634   **12. a.** 647, 69, 479, 461, 247
c. 0.6318, 0.0673, 0.4677, 0.4501, 0.2412   **13. b.** Only three numbers are generated.
21. Let $n = n(L)$
   (1) Sort list L to form list L'.
   (2) If $\text{mod}_2(n) = 1$, then return $L'((n + 1)/2)$; else, return the average of $L'(n/2)$ and $L'(n/2 + 1)$.
23. $\mathbb{I}$, {TRUE, FALSE}   **25.** 293   **26. a.** 1000   **b.** 91   **c.** 89   **28. a.** 10   **30.** E   **31. a.** 5   **b.** 11
32. 10493 is a choice.

CHAPTER 6

SECTION 6.1

1. $(21, -28, 56, 0)$ **4.** $(-6, 0, 29, 2)$ **12.** 125   **27.** $a = 8, b = 4, c = 1/2$

28. c. SALARY $:= 0$
      FOR $:= 1$ TO 5 DO
              SALARY $:=$ HOURS[I] * RATE[I] + SALARY;

29. $\mathbf{u} + \mathbf{v} = (u_1, u_2, u_3) + (v_1, v_2, v_3)$ def. $\mathbf{u}, \mathbf{v}$

             $= (u_1 + v_1, u_2 + v_2, u_3 + v_3)$ def. $+$

             $= (v_1 + u_1, v_2 + u_2, v_3 + u_3)$ commutativity of $+$ in $\mathbb{R}$

             $= (v_1, v_2, v_3) + (u_1, u_2, u_3)$ def. $+$

             $= \mathbf{v} + \mathbf{u}$ def. $\mathbf{u}, \mathbf{v}$

39.

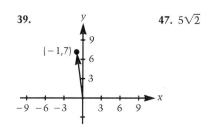

47. $5\sqrt{2}$

SECTION 6.2

1. $10\frac{1}{4}$ **5.** 22   **9.** -4 **12.** 45   **14.** 24,183,900   **17.** 935   **20.** 370,300   **23.** $2^7 = 128$
25. $6! = 720$ **27.** $5^4 \cdot 4! = 15,000$ **34.** a^n **35.** $n!$ **36.** $(\Pi_{i=1}^n u_i)(\Pi_{i=1}^n v_i)$ **40. b.** 118   **44. a.** 6300
48. $\Sigma_4^{36} 2^i$ **54.** No   **55.** Yes, 17   **61.** $\Sigma_{i=1}^6 i$ **64.** $\Sigma_{j=1}^n j^2$ **67.** $\Sigma_{k=1}^n k^3$

SECTION 6.3

1. 3, 5, 9, 17, 33   **4.** $x_1 = 1, x_n = x_{n-1} + 3$ for $n > 1$ **7.** $x_1 = 3, x_n = x_{n-1} + 3$ for $n > 1$
10. $s_1 = 1, s_n = s_{n-1} + 1$ for $n > 1$ **16.** $p(1) = 8, p(n) = 8p(n-1)$ for $n > 1$
18. b. $A(0) = P, A(I) = A(I-1) * (1 + R)$ for $I > 0$ **21.** $x_n = 5^{n+1}$ **24.** $x_n = 5 + n$
28. $x_n = 2 \cdot 3^n + 7((3^n - 1)/2)$ **32.** $x_n = (a^{n+1} - 1)/(a - 1)$ **34.** $2^n \cdot n!$
36. a. Set of all nonempty strings of 0's   **b.** $S \to 0S \to 00S \to 000$
39. a. Set of all palindromes with a 1 in the middle.   **b.** $S \to 0S0 \to 00S00 \to 001S100 \to 0011100$
41. $S \to 00S, S \to 00$ **44. a.** $c_1 = 1, c_n = c_{n-1}$ for $n > 1$ **b.** $S \to S1, S \to 1$
50. $t(0) = 1, t(n) = t(n-1) \cdot 2$ **51.** $GCD(12, 18) \to GCD(12, 6) \to GCD(6, 6) \to 6$
57. a. INTEGER   **b.** REAL   **c.** N $- 1$ **d.** A   **e.** A[N]

SECTION 6.4

1. 1. Prove $P(1)$: $1 = \displaystyle\sum_{i=1}^1 i = 1(1 + 1)/2 = 1.$

     2. Assume $P(k)$: $\displaystyle\sum_{i=1}^k i = \frac{k(k + 1)}{2}.$

        Prove $P(k + 1)$: $\displaystyle\sum_{i=1}^{k+1} i = \frac{(k + 1)(k + 2)}{2}.$

PROOF

$$\sum_{i=1}^k i = \frac{k(k + 1)}{2} \qquad \text{induction hypothesis}$$

$$\sum_{i=1}^k i + (k + 1) = \frac{k(k + 1)}{2} + (k + 1) \qquad \text{add } k + 1$$

$$\sum_{i=1}^{k+1} i = \frac{k(k + 1)}{2} + \frac{2(k + 1)}{2}$$

$$= \frac{k(k + 1) + 2(k + 1)}{2}$$

$$= \frac{(k + 1)(k + 2)}{2} \qquad \text{factor out } k + 2$$

3. 1. Prove $P(1)$: $2^{2(1)} - 1 = 4 - 1 = 3 \equiv 0 \pmod 3$.
   2. Assume $P(k)$: $2^{2k} - 1 \equiv 0 \pmod 3$.
      Prove $P(k + 1)$: $2^{2(k+1)} - 1 \equiv 0 \pmod 3$.

   PROOF

   $$2^{2k} - 1 = 3s \text{ for some integer } s \qquad\qquad \text{induction hypothesis}$$
   $$\Rightarrow 2^{2k} = 3s + 1 \qquad\qquad\qquad\qquad\qquad \text{add 1}$$

   Thus,
   $$2^{2(k+1)} - 1 = 2^{2k+2} - 1 = 2^{2k}2^2 - 1$$
   $$= (3s + 1)2^2 - 1 = 12s + 4 - 1 = 12s + 3$$
   $$= 3(4s + 1) \equiv 0 \pmod 3.$$

10. 1. Prove $P(0)$: $\displaystyle\sum_{i=0}^{0} r^i = (1 - r^1)/(1 - r)$
   $$r^0 = 1.$$
   2. Assume $P(k)$: $\displaystyle\sum_{i=0}^{k} r^i = \frac{(1 - r^{k+1})}{1 - r}.$
      Prove $P(k + 1)$: $\displaystyle\sum_{i=0}^{k+1} r^i = \frac{(1 - r^{k+2})}{1 - r}.$

   PROOF

   $$\sum_{i=0}^{k} r^i = \frac{(1 - r^{k+1})}{1 - r} \qquad \text{induction hypothesis}$$
   $$\sum_{i=0}^{k} r^i + r^{k+1} = \frac{(1 - r^{k+1})}{1 - r} + r^{k+1} \qquad \text{add } r^{k+1}$$
   $$\sum_{i=0}^{k+1} r^i = \frac{(1 - r^{k+1})}{1 - r} + \frac{r^{k+1}(1 - r)}{(1 - r)}$$
   $$= \frac{1 - r^{k+1} + r^{k+1}(1 - r)}{(1 - r)}$$
   $$= \frac{1 - r^{k+1} + r^{k+1} - r^{k+2}}{(1 - r)}$$
   $$= \frac{1 - r^{k+2}}{(1 - r)}.$$

22. a. 1. If $Y = N = 0$, then the loop is not executed and ANS $\leftarrow Z = Z \cdot X^0$.
   2. Suppose if Y reaches the top of the loop with a value K, then ANS $= Z \cdot X^K$. Suppose Y reaches the top of the loop with a value of $K + 1$. After going through the loop once, the value of Z is $Z' = Z \cdot X$ and the value of Y is $Y' = K + 1 - 1 = K$. Now, at the top of the loop again with $Y = K$, we use the induction hypothesis to say that after the loop's execution,
   $$\text{ANS} \leftarrow Z' \cdot X^K = Z \cdot X \cdot X^K = Z \cdot X^{K+1}.$$

27. 1. There exists more than one prime, for instance, 2 and 3.
   2. Suppose there exists more than $k - 1$ primes, say at least $p_1, p_2, \ldots, p_k$. Prove there exists more than k primes.

   PROOF

   By the theorem $p_1 \cdot p_2 \cdots p_k + 1$ is a prime. Moreover, $p_1 \cdot p_2 \cdots p_k + 1$ is greater than each of the p_i, $i = 1, 2, \ldots, k$. Therefore, there exists more than $k - 1 + 1 = k$ primes.

CHAPTER 7

SECTION 7.1

1. a. 2×3 **b.** 0, 3, no such element, -2 **c.** 6 **d.** For $i = 1, 7$; for $i = 2, -4$ **e.** 3
2. $\begin{bmatrix} 18 & 9 & -6 \\ 0 & -24 & 12 \end{bmatrix}$ **5.** $\begin{bmatrix} 5 & 5 & 1 \\ -7 & -6 & 5 \end{bmatrix}$ **9.** $\begin{bmatrix} 0 & 0 & 0 \\ 0 & 0 & 0 \end{bmatrix}$

15. a. I. For $i = 1$ to M do
    A. SUM $\leftarrow 0$
    B. For $j = 1$ to N do
       a. SUM $\leftarrow$ SUM $+ C_{ij}$
    C. Print SUM

21. a. FOR I := 1 TO M DO
    FOR J := 1 TO N DO
      READ (D[I, J]);

24. a. 100   **25. b.** MEM[345] **e.** MEM[347] **h.** A[1, 2]   **26. b.** MEM(347) **e.** MEM(344)
27. d. $n(n + 1)/2$ **28. a.** 16 **b.** 2^{16} **c.** $2^{16} - 2$ **31.** $(3, 4, 2, 5, 1) \cdot (21, 12, 16, 3, 40)$
32. 1. Prove $P(2)$: If A_1 and A_2 are $m \times n$ matrices, then $A_1 + A_2 = A_2 + A_1$. True, by the commutative property.
    2. Assume $P(k)$: If $A_1, A_2, \ldots, A_k$ are $m \times n$ matrices, then

$$A_1 + A_2 + \cdots + A_k = A_k + \cdots + A_2 + A_1.$$

    Prove $P(k + 1)$: If $A_1, A_2, \ldots, A_k, A_{k+1}$ are $m \times n$ matrices,

$$A_1 + A_2 + \cdots + A_k + A_{k+1} = A_{k+1} + A_k + \cdots + A_2 + A_1.$$

PROOF
$$
\begin{aligned}
A_1 + A_2 + \cdots + A_k + A_{k+1} &= (A_1 + A_2 + \cdots + A_k) + A_{k+1} & \text{associative} \\
&= A_{k+1} + (A_1 + A_2 + \cdots + A_k) & P(2) \\
&= A_{k+1} + (A_k + \cdots + A_2 + A_1) & P(k) \\
&= A_{k+1} + A_k + \cdots + A_2 + A_1 & \text{associative}
\end{aligned}
$$

33. $\begin{bmatrix} 1 & 1 & 1 \\ 1 & 0 & 0 \\ 1 & 0 & 0 \end{bmatrix}$ **35. b.** $\begin{bmatrix} 2 & 10 \\ 14 & 6 \end{bmatrix}$ **e.** $D_4 = \begin{bmatrix} 0 & 8 & 2 & 10 \\ 12 & 4 & 14 & 6 \\ 3 & 11 & 1 & 9 \\ 15 & 7 & 13 & 5 \end{bmatrix}$

SECTION 7.2

1. $\begin{bmatrix} 6 & 0 & 8 \\ 18 & 2 & 16 \end{bmatrix}$ **13.** B   **20.** $\begin{bmatrix} 3 & 4 \\ 1 & 2 \end{bmatrix}$ **22.** Switch rows.
29. Multiply the first row by 3 and zero out the second row.
36. Multiply the second column by 4 and place in the first column; zero out the second column.
40. $\begin{bmatrix} 5 & 0 \\ 0 & 7 \end{bmatrix}$ **46.** $\begin{bmatrix} x + 2y \\ 3x + 4y \end{bmatrix}$ **50.** 1, 4   **53.** 6, 7
56. a. $\begin{bmatrix} 1 & 0 & -3 \\ 0 & 1 & 2 \\ 0 & 0 & 1 \end{bmatrix}$ **d.** $(5/2, 4), (-7/2, 5/2), (-5, -3/2), (-7/6, 1), (-2, 7/5)$
60. c. $(8, 5, 7, 3) \cdot (-3, 8, 1, -9)$ **d.** $8(-3) + 5(8) + 7(1) + 3(-9)$
65. $\begin{bmatrix} 0.866 & 0.5 & 0 \\ -0.5 & 0.866 & 0 \\ 0 & 0 & 1 \end{bmatrix}$ **b.** $\begin{bmatrix} 1 & 0 & 1 \\ 0 & 1 & -4 \\ 0 & 0 & 1 \end{bmatrix}$ **c.** $\begin{bmatrix} 1 & 0 & -1 \\ 0 & 1 & 4 \\ 0 & 0 & 1 \end{bmatrix}$
66. d. $\begin{bmatrix} 1 & 0 & -7 \\ 0 & 1 & -3 \\ 0 & 0 & 1 \end{bmatrix}$ **e.** $\begin{bmatrix} 0.259 & -0.966 & 0 \\ 0.966 & 0.259 & 0 \\ 0 & 0 & 1 \end{bmatrix}$ **67. a.** $\begin{bmatrix} 1.5 & 2.598 & 0 \\ -2.598 & 1.5 & 0 \\ 0 & 0 & 1 \end{bmatrix}$ **68. a.** $\begin{bmatrix} 0.707 & -0.707 & 1.707 \\ 0.707 & 0.707 & -0.121 \\ 0 & 0 & 1 \end{bmatrix}$

SECTION 7.3

1. $(2, 0)$ **3.** $(31/5, 6/5, 9/5)$ **8. a.** No solution
b.

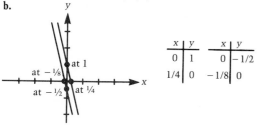

    c. Lines are parallel

x	y		x	y
0	1		0	$-1/2$
1/4	0		$-1/8$	0

13. 45.4 hr on-line, 7.57 min of CPU, 0.151×1000 pages per semester; 3.24 hr on-line, 0.541 min of CPU, 10.8 pages per week
14. 203.6K, 205.6K, 102.8K

SECTION 7.4

1. $\begin{bmatrix} 3 & -2 \\ -1 & 1 \end{bmatrix}$ **3.** Not possible   **6.** $\frac{1}{52}\begin{bmatrix} 6 & -31 & 23 \\ 10 & 9 & -5 \\ -6 & 5 & 3 \end{bmatrix}$ **8.** Not possible

9. a. Using $X = A^{-1}B$, $x_1 = -47$, $x_2 = 37$ **12. a.** YV3CUCLDDFOJNGE

CHAPTER 8

SECTION 8.1

1. d. 3, 2, 4   **2. b.** i, iii, iv, v, vi   **d.** None   **5. b.**

8. b. $\sum\limits_{i=1}^{n-1} i$ **10. a.** $n(n-1)$ **11. a.** 1, 3

12. a. 10   **b.** C(10, 0)

SECTION 8.2

3. $\begin{bmatrix} 0 & 0 & 0 & 0 & 1 \\ 0 & 0 & 1 & 1 & 0 \\ 0 & 1 & 0 & 1 & 0 \\ 0 & 1 & 1 & 0 & 0 \\ 1 & 0 & 0 & 0 & 0 \end{bmatrix}$ **6. a.** $\begin{bmatrix} 0 & 1 & 0 \\ 1 & 1 & 2 \\ 0 & 2 & 0 \end{bmatrix}$ **b.** $\begin{bmatrix} 1 & 1 & 2 \\ 1 & 6 & 2 \\ 2 & 2 & 4 \end{bmatrix}$ **8.** $\begin{bmatrix} 0 & 1 & 0 \\ 1 & 1 & 1 \\ 0 & 1 & 0 \end{bmatrix}$ **10. a.** $\begin{bmatrix} 0 & 1 & 0 & 1 \\ 0 & 0 & 1 & 0 \\ 0 & 0 & 0 & 1 \\ 1 & 0 & 0 & 0 \end{bmatrix}$ **b.** 4

SECTION 8.3

4.

7. a.

8.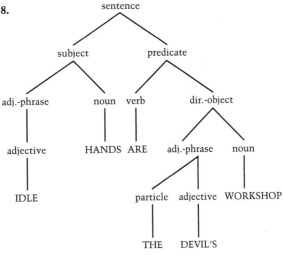

10.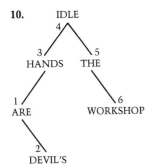

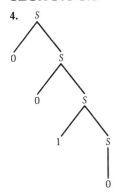

13. RECURSIVE ALGORITHM, POSTORDER(R), FOR POSTORDER TRAVERSAL OF TREE WITH ROOT R:
POSTORDER(R):
   If the tree is empty, then return.
   If there is a node to the left of R, LEFT(R), then
     POSTORDER(LEFT(R)).
   If there is a node to the right of R, RIGHT(R), then POSTORDER(RIGHT(R)).
   Visit the node R.

17. C   **24. d.** 2^3 **26. d.** 15

SECTION 8.4

1. Not possible   **4.** g, e, d, b, a, c, f **9.** Not possible; 4 nodes have odd degree
15. Through vertices 1, 2, 3, 3, 6, 4, 1, 5
17.   **23.**

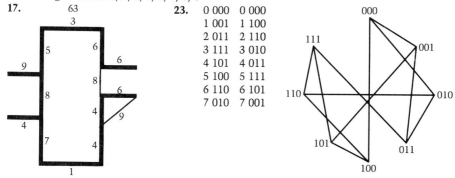

0 000	0 000
1 001	1 100
2 011	2 110
3 111	3 010
4 101	4 011
5 100	5 111
6 110	6 101
7 010	7 001

25. a. 0110   **27. a.** One sequence is 0011 **b.** 00, 01, 11, 10   **32.** No, more than two points have degree 1

SECTION 8.5

1. a. 0011   **d.** Divides by 2—Shift to the right one bit

3.   **6.**   **9.** Yes

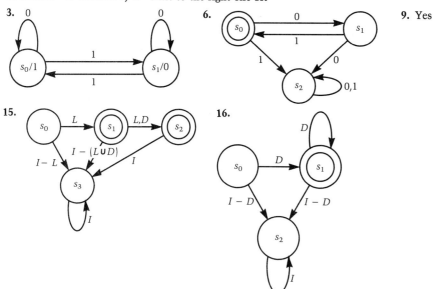

20. $I - \{E\}$

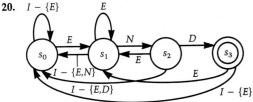

23.

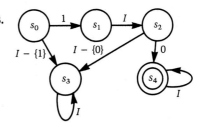

25. $I - \{D\}$

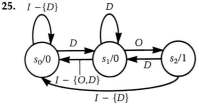

CHAPTER 9

SECTION 9.1

1. (1b.) $xy = yx$

$00 = 0 = 00$
$01 = 0 = 10$
$10 = 0 = 01$
$11 = 1 = 11$

2. (4a.) X OR F = X

F OR F = F
T OR F = T

3. b. 1, 30

4. a. $D = \{1, 3, 7, 21\}$ **b.** 1, 21 **c.** $x' = 21/x$ **8. c.** 01
9. b. $\underbrace{00 \ldots 0}_{n} \underbrace{11 \ldots 1}_{n \text{ factors}}$ **11. c.**

x	$(f_3 + f_4)(x)$
00	1
01	1
10	1
11	0

13. a. $(1, 1, 0, 1, 1, 0, 0)$ **f.** Best choice $= (1, 1, 1, 0, 0, 0, 1)$

SECTION 9.2

1. a. Commutative **b.** $yx + yy'$ **c.** Complement **d.** yx **e.** Identity **f.** Commutative

5.
$$
\begin{aligned}
x + xy &= x \cdot 1 + xy && \text{identity} \\
&= x(1 + y) && \text{distributive} \\
&= x \cdot 1 && \text{add 1} \\
&= x && \text{identity}
\end{aligned}
$$
Dual: $x \cdot (x + y) = x$

7. Suppose 0 and z are two different additive identities. Thus, $x + 0 = x$ and $x + z = x$ for any x. Substitute $x = z$ in the first equation and $x = 0$ in the second, giving $z + 0 = z$ and $0 + z = 0$. Thus, $z = z + 0 = 0 + z = 0$. Since $z = 0$, we cannot have two different identities, but only one.

12. a.
$$
\begin{aligned}
xy + x'y &= (x + x')y && \text{distributive} \\
&= 1 \cdot y && \text{complement} \\
&= y && \text{identity}
\end{aligned}
$$
b. Dual: $(x + y)(x' + y) = y$

16. a.
$$
\begin{aligned}
x + y'z + u'(x + y'z) &= 1(x + y'z) + u'(x + y'z) && \text{identity} \\
&= (1 + u')(x + y'z) && \text{distributive} \\
&= 1(x + y'z) && \text{addition of 1} \\
&= (x + y'z) && \text{identity}
\end{aligned}
$$
b. Dual: $x(y' + z)(u' + x(y' + z)) = x(y' + z)$

19. $x + x'y = (x + x')(x + y)$ distributive
$\qquad\quad = 1(x + y)$ complement
$\qquad\quad = x + y$ identity
　Dual: $x(x' + y) = xy$

22. $(x + y)'(x' + y')' = (x' \cdot y')(x'' \cdot y'')$ DeMorgan
$\qquad\qquad\qquad = (x' \cdot y')(x \cdot y)$ double complement
$\qquad\qquad\qquad = (x' \cdot x)(y' \cdot y)$ associative and commutative
$\qquad\qquad\qquad = 0 \cdot 0$ complement
$\qquad\qquad\qquad = 0$ multiplication by 0

25. $xy'z' + xyz + x'yz + xyz'$
$\qquad = xyz + x'yz + xy'z' + xyz'$ commutative
$\qquad = (x + x')yz + xz'(y' + y)$ distributive
$\qquad = 1 \cdot yz + xz' \cdot 1$ complement
$\qquad = yz + xz'$ identity

SECTION 9.3

1. a. $u = (xy)'$

b.

x	y	xy	(xy)'
0	0	0	1
0	1	0	1
1	0	0	1
1	1	1	0

c. u is 1 if x or y or both are 0.

5. $u = x' + y'$　　**8.** $u = xy$

3. a. $u = xy' + x'y$

b.

x	y	x'	y'	xy'	x'y	xy' + x'y
0	0	1	1	0	0	0
0	1	1	0	0	1	1
1	0	0	1	1	0	1
1	1	0	0	0	0	0

c. u is 1 if x or y is 1 but not both.

12.

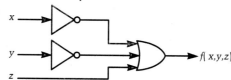

SECTION 9.4

1. a. 111111

2.

x	y	f(x, y)
0	0	1
0	1	0
1	0	1
1	1	0

$f(x, y) = x'y' + xy' = y'$

$y \rightarrow \triangleright\!\circ \rightarrow y'$

4.

x	y	z	f(x, y, z)
0	0	0	1
0	0	1	0
0	1	0	1
0	1	1	0
1	0	0	0
1	0	1	1
1	1	0	0
1	1	1	1

$f(x, y, z) = x'y'z' + x'yz' + xy'z + xyz$
$\qquad\qquad = x'z' + xz$

$x \rightarrow$

$z \rightarrow$

$\rightarrow f(x,y,z)$

6. b. $u = cx'y + cxy' + cxy$ **7. a.**

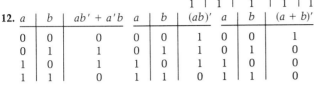

x	y	c_{i-1}	c_i	s
0	0	0	0	0
0	0	1	0	1
0	1	0	0	1
0	1	1	1	0
1	0	0	0	1
1	0	1	1	0
1	1	0	1	0
1	1	1	1	1

10. a. $a = w$
$b = w'x + wx'$
$c = x'y + xy'$
$d = y'z + yz'$

12.

a	b	$ab' + a'b$	a	b	$(ab)'$	a	b	$(a + b)'$
0	0	0	0	0	1	0	0	1
0	1	1	0	1	1	0	1	0
1	0	1	1	0	1	1	0	0
1	1	0	1	1	0	1	1	0

13.

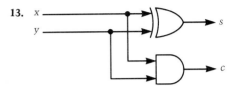

CHAPTER 10

SECTION 10.1

1. 7   **6.** 49   **12.** $2^9 = 512$ **13.** $2^{15} - 1 = 32{,}767$ **14.** 0.75   **17.** 5.625   **21.** 100   **27.** 11001
32. 101000   **38.** 11010, 11000   **40.** 1001, 111   **42.** 100000, 11110   **46.** 0, $2^4 - 1 = 15$
48. 0, $2^{32} - 1 = 4{,}294{,}967{,}295$ **52.** 512; 0.5K   **54.** 4,194,304; 4096K   **57.** 13   **60.** 16   **63. a.** 4
66. a. Moves to the left   **b.** Doubled

SECTION 10.2

2. 226   **6.** 4095   **9.** 10.25   **13.** (2) 1110 0010   (5) 1010 0000 1111 1111   (9) 1010.01   **15.** 18
19. 3E   **22.** B9, BA, BB, BC   **26.** FE, FF, 100, 101   **28.** 999A   **31.** 60C0   **34.** 9998   **36.** EF
40. a. FFF; 4095   **42.** 4A.8   **45.** 1F.FE   **47.** 4C 49 53 50   **53.** 32   **57. a.** 2   **59.** 59   **63.** 15
67. 7, 10, 11, 12   **74. a.** 6777   **75. a.** 777; 511   **76. a.** chmod 710

SECTION 10.3

1. 79   **2.** 0.48   **6.** $D07.B\overline{3}$ **9.** 10111   **10.** 0.1101   **14.** 3F, 111111   **17.** 1C.04, 11100.000001
19. $0.D1 \times 2^0$; 0.00359375, 0.43826%   **21.** $0.D3 \times 2^{-6}$; 0.00002158; 0.1673%   **24.** $0.9D49 \times 16^{-2}$
26. 0.144×16^2 **28. a.** 0.9999   **29.** 71   **32.** 1-10 10-0 100 → 1A4   **35.** 11-01 1-110 → 336
38. a. 81   **39. a.** 34

SECTION 10.4

1. 5B   **2.** A5   **5.** C0   **6.** Doesn't fit   **10.** FFA5   **12.** DF0D   **14.** 8889   **18.** −44   **19.** 99
25. −29,684   **26.** 28,672   **27.** −24,576   **32. a.** From -2^{32} to $2^{32} - 1$ **34.** 3F

SECTION 10.5

1. 100 0001 1100   **2.** 616   **3.** 4CF   **7.** 0001 0000; $-100_{10} + 116_{10} = 16_{10}$
9. 0111 1100; $-32_{10} + -100_{10} \neq 124_{10}$, overflow   **11.** B9; $-118_{10} + 47_{10} = -71_{10}$
13. F7; $-38_{10} - -29_{10} = -9_{10}$ **18.** 7792; $23{,}266_{10} - -7{,}344_{10} = 30{,}610_{10}$ **19. a.** FD   **20.** A061, forward
22. 27   **24. a.** 65,536   **b.** 30_{16}, 48_{10} **25.** Doubled: 0011 1010, Halved: 0000 1110   **28.** 3   **30.** 0.000011

CHAPTER 11

SECTION 11.1

2.

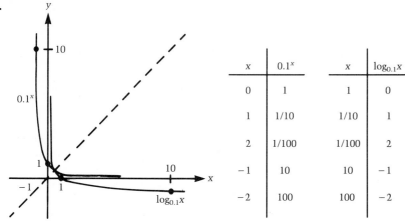

x	0.1^x
0	1
1	1/10
2	1/100
-1	10
-2	100

x	$\log_{0.1} x$
1	0
1/10	1
1/100	2
10	-1
100	-2

5. 7   **12.** 4/7   **16.** 49   **21.** 3   **30.** 4.1127   **33.** $\log_2 512 = 9$ **36.** $4^{3/2} = 8$ **39.** 10^{21} **40. b.** 50
43. a. \$656.87 (rounded)   **c.** 12.078 years   **45.** 4,500,857   **47.** 197.74%   **48. a.** 11.58 years
50. FUNCTION AMOUNT(P, R, T: REAL): REAL;
  BEGIN
    AMOUNT := P * EXP(R * T)
  END;
52. a. -0.00012 **53. a.** 3   **54. b.** 9
58. Let $\log_b u = s$ and $\log_b v = t$. Then by property 1 $u = b^s$ and $v = b^t$. Thus, $uv = b^s b^t = b^{s+t}$. By property 1 we have $\log_b uv = s + t$. So from the definition of s and t, $\log_b uv = \log_b u + \log_b v$.

SECTION 11.2

1. Yes, no   **5.** No, yes   **9.** Yes, no   **13.** No, yes   **17.** $O(n^4)$ **20.** $O(n^2)$ **23.** $O(n^3)$ **26.** $O(N)$
29. a. 2   **32. a.** $O(m)$ **34. c.** $O(n^2)$ **35.** $O(n)$ **39. a.** 1.193   **40. a.** 10^3
41. a. $\{2, 5, 1\}\{6\}\{8, 7, 9\} \rightarrow \{1\}\{2\}\{5\}\{6\}\{7\}\{8\}\{9\}$ **42.** B   **43.** B   **44.** D   **45.** E

INDEX

ABC Computer, 253, 360
ABS, 150, 153
Absolute error, 15–18
Absolute value function, 74, 153
Absorption property, 63, 339
Accepting state, 315
Acyclic, 276
Ada, 64
Additive inverse
    matrix, 237
Address, 364–65, 393
Adjacency matrix, 282
Adjacent, 272–73
Advanced Placement, 27–28, 161,
    169, 181, 295–96, 416–17
AI, 314
al-Khowârizmî, 99
Algebra of propositions, 53–63
Algebra of sets, 46–53
ALGOL, 209
Algorithm, 11, 13, 99
Alphabetize, 290–92, 295
Anagram, 137–39
Analysis of algorithms, 408–17
Analytic engine, 64
Ancestor, 288–89
AND, 54
AND gate, 341
Antecedent, 71
Antisymmetric, 126
AP sample questions, 27–28, 161,
    169, 181, 295–96, 416–17
APL, 18, 22, 262
Arc, 272
Argument
    function, 146
    valid, 67–68, 70, 76–77
Arithmetic mean, 157
Arithmetic
    binary and hexadecimal, 387–93
Array, 120, 186–88, 232–34
    one-dimensional, 186–88, 232

Pascal, 186, 188, 233–34
    two-dimensional, 120, 232–34
Articulation point, 279
Artificial intelligence, 167–68, 314
ASCII, 85, 88, 148, 173, 179, 368,
    370, 372, Appendix A
Assembler language, 372, 390–91,
    393
Assignment
    Pascal, 16
Associative property, 47, 60, 236,
    245, 327
Atanasoff, John Vincent, 253, 360
Atanasoff-Berry Computer, 253, 360
Atom, 168
Attributes, 123
Augmented matrix, 254
Automata theory, 210

B, 3
Babbage, Charles, 64
Backus, John, 147
Backus-Normal Form, 209–10
Bandwidth, 107
Base, 6
Base 2 number system, 360–93
Base 8 number system, 370–73, 380
Base 16 number system, 366–93
BASIC, 103–05, 178, 360
Baud, 107, 304
Baudot, 107, 304
Belongs to, 32
Berkeley, Edmund, 67
Best case, 408
Big oh notation, 408–9
Binary digit, 5
Binary number system, 360–93
    count in, 362
    decrement in, 363
    increment in, 362
Binary relation, 119

Binary search, 411–12, 414
Binary set, 3
Binary tree, 290, 296, 407
Binomial coefficient, 98
Binomial Theorem, 95–96, 98
Bipartite graph, 277
Bit, 5, 364
Bits per inch, 108
Block, 109
BNF, 209–10
Boole, George, 326
Boolean algebra, 326–55
    two-element, 283, 327
Boolean expression, 54–56, 61, 70,
    329, 338, 344
Boolean operators, 22
Boolean type, 180
Boolean variable, 343
bpi, 108
Braille, Louis, 90
Bug, 217
Built-in-function, 164–70
Byte, 6, 108, 364
Bytes per inch, 108

Canonical sum-of-products,
    349–50
CAR, 168–69, 171
Cartesian product, 118
Cases
    proof by, 73
CDC 7600, 8
CDR, 168–69, 171
Ceiling function, 167
Central processing unit, 105
Chain, 128–29
Change of base formula, 404
Charles II, King, 370
Check bit, 333
Child, 288–89
Chip, 341

CHR, 173
Circuit
    combinational, 98, 333, 341–55
    Euler, 300
    graph, 274
    Hamiltonian, 298
    logic, 98, 333, 341–55
Clock cycle, 11
Clock rate, 103, 106
COBOL, 121
Code word, 333
Code, Gray, 298
Codomain, 144
Column major order, 238
Column vector, 189
Combination, 93–99
Combinational circuit, 344–55
Combinational function, 98, 150,
    333, 344–55
Combinatorics, 84
Commutative property, 47, 60,
    235, 327
Compiler, 121
Complement, 41, 47, 60, 312, 327,
    336, 381
    double, 50, 338
Complete graph, 274–75
Complexity, 408–17
Composite function, 170–75
Composition, 170–75
Compound proposition, 57
Compound statement, 57
Compounding, 400, 403
Computability theory, 314
Computer graphics, 42, 98, 154
    190–91, 232–33, 246–48,
    250–53, 293
Computer implementation of
    Pascal sets, 329–31, 334
Computer integer, 2
Computer real numbers, 4
Computer speed, 7–12
Concatenation, 209
Conclusion, 71
Congruence, 137
Conjunction, 57, 60
Connected graph, 274–75
Connection matrix, 283
CONS, 168–69
Consequent, 71
Continuous compounding, 400, 403
Contradiction, 66
    proof by, 72
Contrapositive, 73
Converse, 70
Conversion from decimal, 373–80
Coordinate, 121, 154
COUNT, 157
Count with binary numbers, 362
Count with hexadecimal numbers,
    367

Counterexample, 37, 45, 74
Counting principle, 84–91
CPU, 106
Cray, Seymour, 106, 189
CRAY-1, 106
CRAY-2, 188–89
Cross product, 118, 122
Cryptography, 264–67
Cryptosystem, 176, 181–82
Cycle time, 105–6, 112
Cycle
    graph, 274
Cylinder, 110

Data base, 123, 147, 153, 165
    relational, 147, 153, 165
Data structure, 186
Data transfer rate, 107–10, 112–13
Data type, 2
de Bruijn sequence, 303–5, 309
Dead state, 315
Deck of cards, 93
Decoder, 10, 354, 365
Decoding, 176
Decrement binary number, 363
Decrement hexadecimal number,
    367
Degree, 272–73
Degree of polynomial, 200
Delay, 103–5
DeMorgan's Laws, 50, 60, 337
Density, 109–10, 112
Descartes, René, 119
Descendant, 288–89
Diagonal, 240, 246
Difference, 40
Difference equation, 204–16
Difference
    vector, 190–91
Digraph, 276, 279–88
Dimension, 186
Direct method of proof, 71
Directed graph, 276
Directory, 134
Disjoint, 39, 40, 87
Disjoint-OR counting principle, 87
Disjunction, 57, 60
Disk pack, 8, 110–11
Disk storage, 110–11, 113
Distributive property, 47, 60, 327
Dither matrix, 241
Dithering, 241
Divide, 128
Divisible, 225
Domain, 144, 147
Dot product, 191
Double complement, 50, 60, 338
Doubly-linked list, 380–81, 284
Dual, 49, 334
Dump
    memory, 368

e, 401
EBCDIC, 85, 88, 148, 173
Eckert, Presper, 107
Edge, 272
Eisenhower vs. Stevenson, 342
Empty set, 33
Encode, 176
Encoder, 365
ENIAC, 107
EPSAL, 5
Epsilon, 188
Equal, 59, 122, 345
Equal combinational functions,
    345
Equal sets, 35
Equal vectors, 188
EQUIVALENCE, 136, 139
Equivalence, 65
Equivalence class, 134
Equivalence relation, 133, 139
Equivalent, 59, 344
Equivalent combinational
    functions, 59, 345
Equivalent systems, 255
Error, 15–18
Euclid, 215
Euler circuit, 300–1
Euler path, 300–5, 309
Euler, Leonhard, 289, 403
EVEN function, 175, 180
Exclusive or, 41, 69
Exp, 400
Exponent, 6
Exponential function, 398–406
Expression
    Boolean, 343

Factorial, 86, 206–7
Factorial function, 206–7
Fibonacci sequence, 214
Field, 120
File, 120–21
Final state, 310, 315
Finite geometric series, 199
Finite-state machines, 310–18
Floating point number, 13
Floating-point operations per
    second, 106–7, 112
Floor function, 167
Flops, 106–7, 112
Floyd's algorithm, 287
Fluorinert, 189
Formal grammar, 210–15, 225,
    289–90, 294
FORTH, 25
FORTRAN, 146, 153, 238
Fractional part, 13
Freq, 155
Frequency function, 155
Frequency of computer, 103,
    105–6, 112

Full binary tree, 296
Full m-ary tree, 297
Full-adder, 353–54
Function 144–82
  ABS, 153
  absolute value, 74, 153
  built-in, 164–70
  CAR, 168–69, 171
  CDR, 168–69, 171
  ceiling, 167
  CHR, 173
  combinational, 98, 150–52, 333,
    344–55
  composite, 170–75
  CONS, 168–69
  EVEN, 175, 180
  exponential, 398
  factorial, 206–7
  floor, 167
  frequency, 155
  GCD, 215
  greatest integer, 167
  hashing, 178–80
  INT, 165, 178
  inverse, 173
  logarithmic, 401–7
  MAX, 164–65, 327, 415
  mean, 157
  median, 158
  MIN, 163, 328
  mod, 175–83
  next state, 311
  one-to-one, 147
  onto, 149
  ORD, 173
  output, 311
  Pascal, 145–46, 153, 204, 206
  percent, 155
  PRED, 130, 173
  predicate, 151
  projection, 153–54
  recursive, 205
  ROUND, 166
  SQR, 145
  SQRT, 152
  SUCC, 130
  SUM, 157, 415
  SUM (MACSYMA), 197, 203
  switching, 98, 150–52, 333,
    344–55
  transition, 311
  TRUNC, 165
Fundamental Counting Principle,
  84–91
Fundamental Theorem of
  Arithmetic, 138

g, 7
Gate, 340–48
Gauss, Carl Friedrich, 196, 255
Gaussian elimination, 255, 260

GCD function, 215
Generating rule, 205
Geometric interpretation, 190
Giga, 7
Grammar, 210–15, 225, 289–90,
  294
Graph, 127–30, 272–318
  bipartite, 276
  complete, 274–75
  connected, 274–75
  directed, 276
  loop in, 274
  network, 277
  simple, 274–75
  state, 311
  tree, 54, 276, 287–97, 306–7, 407
Graph of partially ordered set,
  127–30
Graph theory, 272–318
Graphics, 42, 98, 154, 190–91,
  232–33, 246–48, 250–53
Gray code, 298–99, 309, 354–55
Greatest common divisor, 215
Greatest integer function, 167

$\mathbb{H}$, 3
Half-adder, 349–50, 355
Half-life, 407
Hamilton, William R., 298
Hamiltonian circuit, 298
Hamiltonian path, 298–99, 309
Hamming code, 333–34
Hamming, Richard, 333
Hard disk, 110–11
Hashing function, 178–80
Height, 288–89
Hexadecimal number system, 3, 6,
  366–93
  count in, 367
  decrement in, 367
  increment in, 367
Hexadecimal set, 3
Hierarchy diagram, 48, 127
Hopper, Grace, 121
Hypothesis, 71

$\mathbb{I}$, 2
$\mathbb{I}^+$, 3
IBG, 109–10, 112
$\mathbb{I}_c$, 2
Idempotent property, 50, 60, 335
Identifier, 21
Identities, Boolean algebra, 327
  matrix, 236, 246
  proposition, 60
  set, 47
Identity matrix, 246
Image, 144
Implication, 64
In-degree, 276
Incident, 272–73

Inclusion-exclusion principle, 87
Increment binary number, 362
Increment hexadecimal number,
  367
Index, 186, 195
Induction, 216–27
Induction hypothesis, 218
Infix notation, 24
Initial condition, 205
Initial point, 276
Initial state, 311
Initialize, 16
Inner product, 191
Inorder traversal, 291–92
Input alphabet, 311
Inquiry, 398
Insertion sort, 415–16
INT, 165, 178
Integer, 2
  largest unsigned binary, 363
  largest unsigned hexadecimal,
    367
Integer arithmetic, 21, 23
Integrated circuit chip, 341
Inter-block gap, 109–10, 112
Inter-record gap, 109–10
Internal node, 296
Intersection, 39, 50
Intersection with empty set, 50
Inverse, 148, 172
Inverse function, 173
Inverse
  multiplicative, 261
Inverter, 341
$\mathbb{IR}$, 4
IRG, 109–10, 112
Irrational number, 4
Irreflexive, 126
Isolated point, 272–73
Ith element, 121
Iverson, Kenneth, 262

K, 6, 7, 11
Key, 149
Kilo, 6, 7
Knuth, Donald, 12
Königsberg bridge problem, 310
Kruskal's Algorithm, 306, 415
Language, 210
Largest unsigned binary integer,
  363
Largest unsigned hexadecimal
  integer, 367
Law of syllogism, 224
Leaf, 288–89
Lehmer, D. H., 177
Length of path, 274
Length of vector, 192
Level, 288–89
Linear congruential method,
  177–80

Linear equation, 253
Linear ordering, 128–33
Linearly ordered set, 128–33
Linked list, 280–81, 284
LISP, 167–69, 171, 175
List, 168
    doubly-linked, 281
    linked, 280
Logarithmic function, 401–7
Logic, 54–70
Logic circuit, 348–55
Logic gate, 340–48
Logical connectives, 57
Logical expression, 54–56, 61, 70,
    329, 338
Logically equivalent, 59
Loop
    graph, 16, 274
    Pascal, 16, 101
    program, 99–105
Lovelace, Ada Augusta, 60
Lower bound, 195, 199
Łukasiewicz, Jan, 24

m, 7
M, 7
MACSYMA, 197, 203
Magnitude, 14
Mainframe, 7
Mandelbrot set, 31
Mantissa, 13, 378
Mapping, 144
Mark I, 121, 217
Mathematical induction, 216–27
Matrices, 232–67, 281–86
Matrix, 232–67, 281–86
    additive inverse, 237
    adjacency, 282
    augmented, 254
    connection, 283
    dither, 241
    identity, 246
    multiplicative inverse, 260–67
    network, 284
    reachability, 286
    square, 246
    symmetric, 239
    threshold, 240
    zero, 236
Matrix multiplication, 242–53,
    260–62, 265–67
Matrix product, 242–53, 260–62,
    265–67
Matrix representation, 280–87
Mauchly, John, 107
MAX, 164–65, 327, 415
MAXINT, 2
MEAN, 415
Mean, 157, 415
Median, 158

Mega, 7
Megahertz, 106
Memory, 6, 11, 134
Memory dump, 368, 372
Memory size, 6, 11
Metric system, 6, 7
Mflops, 106
MHz, 106
Micro, 7
Microprocessor, 341
Milli, 7
MIN, 163, 328
Minimal spanning tree, 306–7,
    309–10
Mod function, 175–82
Modulo, 137, 175–82
modus ponens, 69
Moore, Charles, 25
Most significant digit, 14
Mozart's dice waltz, 90
Multiplication
    matrix, 242–53, 260–62, 265–67
Multiplicative identity
    matrix, 246
Multiplicative inverse, 261
Mutually exclusive, 87

ℕ, 2
n, 7
$n(S)$, 32
N-ary relation, 122
NAND gate, 355
Nano, 7
Napier, John, 403
Natural logarithm, 401
Natural number, 2
Negation, 57
Nesting of loops, 100
Network, 277
Network matrix, 284
Next-state function, 311
Node, 272
    internal, 296
NOR gate, 355
Norm, 192
Normalize, 13
Normalized exponential notation,
    13–15, 17
NOT, 55
Null set, 33
Number of combinations, 93–99
Number of elements, 32
Number of permutations, 91–92,
    96–97
Number systems, 360–93
Number
    signed, 381–87
    unsigned, 363

Octal number system, 370–73, 380

Odd parity, 150, 316, 351
One's complement, 381
One-dimensional array, 186, 232
One-to-one, 147
Onto, 149
Operating system, 134
Operator precedence, 18–23,
    63–64, 68
    Pascal, 63
OR, 55
OR gate, 341
ORD, 173
Order, 409
Order of priority of operations,
    18–23, 63–64, 68
Ordered *n*-tuple, 121
Out-degree, 276
Output alphabet, 311
Output function, 311
Overflow, 207, 389–90

ℙ, 5
p, 7
Page, 293
Parameter, 146
Parent, 288–89
Parity, 108, 150, 316, 351
Parity bit, 108, 351
Parse, 289
Parse tree, 289–90, 294
Partial order, 125–33
Partially ordered set, 125–33
    graph of, 127–30
Partition, 133–35
Pascal array, 186, 188, 233–34
Pascal assignment, 16
Pascal function, 145–46, 153, 204,
    206
Pascal loop, 16, 101
Pascal operator precedence, 63
Pascal procedure, 416–17
Pascal program, 101
Pascal set, 32–33, 37, 39, 41, 46,
    329–31
Pascal's triangle, 98–99
Pascal, Blaise, 33
Path, 274, 297–305
    Euler, 300–5, 309
    Hamiltonian, 298–99, 309
    length, 274
PDP-8, 106
PEEK, 380
Percent, 155
Perfect square, 4
Permutation, 91–92, 96–97
Pi notation, 201–2
Pico, 7
Pigeonhole principle, 167
Pivot number, 256
Pixel, 241

Point, 272
  initial, 276
  isolated, 273
  terminal, 276
Point articulation, 279
Polish notation, 24–28
Polynomial, 200
  degree of, 200
  zero, 200
Polynomial-time algorithm, 409
POP, 24
Postfix notation, 24–27, 292
Postorder traversal, 292
Power set, 46–47
Precision, 14
PRED, 130, 173
Predicate function, 151
Prefix notation, 24
Preimage, 144
Preorder traversal, 295
Prime, 5, 137–39, 179, Appendix B
Procedure
  Pascal, 416–17
Product
  dot, 191
  inner, 191
  scalar, 191
Product of scalar and matrix, 234
Product of scalar and vector,
  190–91
Production, 210
Program
  Pascal, 101
Program counter, 393
Program loop, 99–105
Program verification, 216, 223–24,
  226
Projection, 153–54
Proof techniques, 70–77, 216–27
Proposition, 57
Propositional calculus, 54–70
Psuedo-random number, 176
Public-key cryptosystem, 176,
  181–82
PUSH, 24

$\mathbb{Q}$, 3
$\mathbb{Q}^+$, 150
Queue, 86
QUIK sort, 415

$\mathbb{R}$, 4
Random number, 176–81, 204
Range, 144
Rational number, 3
$\mathbb{R}_c$, 4
Reachability matrix, 286
Read-only memory, 393
Real number, 4
Recognize, 314–18

Record, 109
Recurrence relation, 204–16
Recursion, 204–16
Recursive function, 204–16
Reflexive, 126
Register, 390
Related, 119
Relational data base, 123–24, 147,
  153, 165
Relational operators, 22
Relative error, 15–18
Relative file organization, 148
Representation, 280–87
Richter scale, 407
RND, 178
ROM, 393
Root, 287
Rooted tree, 287
Rotation, 246, 251–53
Round, 15
ROUND, 166
Row major order, 239
Row vector, 189
Rules, 210

S-expression, 168
Scalar, 190
Scalar product, 191
Scale, 246–48, 250, 252–53
Search
  binary, 411–12, 414
  sequential, 104, 408, 411
Secondary storage, 8
Seed, 177
Sequential search, 104, 408, 411
Set, 2, 32–53
  linearly ordered, 128–33
  Mandelbrot, 31
  Pascal, 32–33, 37, 39, 41, 46,
    329–31
Set difference, 40
Set of computer integers, 2
Set of computer real numbers, 4
Set of integers, 2
Set of irrational numbers, 4
Set of natural numbers, 2
Set of positive integers, 3
Set of positive rational numbers,
  150
Set of prime numbers, 2, 5
Set of rational numbers, 3
Set of real numbers, 4
Set-builder notation, 3
Sets
  computer implementation of,
    329–31, 334
Shannon, Claude E., 340, 413
Shift, 365–66, 391–92
Sigma notation, 195–204
Signed-magnitude, 381

Significant digit, 14
Simple graph, 274–75
Slice, 163
Small-scale integration, 341
Solid modeling, 42, 293
Sort
  insertion, 415–16
  QUIK, 415
  topological, 129, 132–33
Spanning tree, 306–7, 309–10
Speed of tape, 108, 112–13
SPSS, 164
SQR, 145
SQRT, 152
Square matrix, 246
Stack, 24–28, 206–7
Stack overflow, 207
Standard position, 187
Start symbol, 210
State graph, 311
Statement, 57
Storage density, 108–9, 112
Subgraph, 288–89
Subroutine, 8
Subset, 33–34, 93
Subtree, 288–89
SUCC, 130, 173
Suffix notation, 24–27, 292
SUM, 157, 415
  MACSYMA, 197, 203
Sum
  vector, 189
Sum-of-products
  canonical, 349–50
Supercomputer, 7, 189
Swedberg, Emanuel, 370
Switching function, 98, 150–55,
  333, 344–55
Syllogism, 69, 71, 224
Symbolic expression, 168
Symmetric, 126
Symmetric difference, 41, 42
Symmetric matrix, 239
System of linear equations, 253–60

Table, 232
Tape density, 108–10, 112–13
Tape speed, 108–9, 112–13
Tautology, 66
Terminal point, 276
Terminal symbol, 210
Theorem, 70
Theorem-proving program, 50
Threshold matrix, 240
Time-sharing, 92
Top, 24
Topological sort, 129, 133–34
Total order, 128–33
Track of disk, 110
Track of tape, 108

Transition diagram, 311
Transition function, 311
Transitive, 126
Translation, 246–48, 250, 252
Traveling salesperson, 97, 309, 413
Tree, 54, 276, 287–297, 306–7, 407
  binary, 290, 296, 407
  parse, 289–90, 294
  rooted, 287
  spanning, 306–7, 309–10
Tree traversal
  inorder, 291–92
  postorder, 292
  preorder, 295
Trigonometry, 251–53
TRUNC, 165
Truncate, 5, 21, 165–66
Truth table, 55
Turing machine, 314
Turing, Alan, 264, 314
Two's complement, 381–87

Two-dimensional array, 120, 232
Two-element Boolean Algebra,
    283, 327

Unblocked, 109
Union, 38
Union with U, 50
Unique, 53
UNIVAC, 108, 342
Universal set, 33
UNIX, 134, 370–71, 373
Unsigned number, 363
Upper bound, 195, 199

Valid argument, 67–68, 70, 76–77
Variable
  Boolean, 343
Vector, 187, 190
  column, 189
  length of, 192
  norm of, 192

row, 189
Vector difference, 190–91
Vector equality, 188
Vector sum, 189
Venn diagram, 38
Vertex, 272
Vertices, 272
Very small scale integration, 341
Visiting the node, 291
Vocabulary, 210
Voltage, 150
von Neumann, John, 176

Word, 210
Worst case, 408

XOR gate, 355

Zero matrix, 236
Zero out, 236
Zero polynomial, 200

T